AF352265

Laser Diagnostics and Optical Measurement Techniques in Internal Combustion Engines

Other SAE books of interest:

Engine Combustion: Pressure Measurement and Analysis
By David R. Rogers
(Product Code: R-388)

An Introduction to Engine Testing and Development
By Richard D. Atkins
(Product Code: R-344)

Internal Combustion Engine Handbook
By Richard van Basshuysen and Fred Schaeffer
(Product Code: R-345)

For more information or to order a book, contact:
SAE International
400 Commonwealth Drive
Warrendale, PA 15096-0001 USA

Phone: 877-606-7323 (U.S. and Canada only) or
724-776-4970 (outside U.S. and Canada)
Fax: 724-776-0790;

Email: CustomerService@sae.org;
Website: http://books.sae.org

Laser Diagnostics and Optical Measurement Techniques in Internal Combustion Engines

Hua Zhao

Warrendale, Pennsylvania
USA

400 Commonwealth Drive
Warrendale, PA 15096-0001 USA

E-mail: CustomerService@sae.org
Phone: 877-606-7323 (inside USA and Canada)
724-776-4970 (outside USA)
Fax: 724-776-0790

ISBN 978-0-7680-5782-9
SAE Order No. R-406
DOI 10.4271/R-406

Library of Congress Cataloging-in-Publication Data
Zhao, Hua, 1963-
Laser diagnostics and optical measurement techniques in internal combustion engines / Hua Zhao.
    p. cm.
Includes bibliographical references.
ISBN 978-0-7680-5782-9
1. Automobiles—Motors—Exhaust gas—Measurement. 2. Automobiles—Motors—Combustion. 3. Internal combustion engines—Combustion. 4. Optical measurements. 5. Lasers—Industrial applications. I. Title.
TL214.P6Z43 2012
    629.25'04—dc23   2012013193

Information contained in this work has been obtained by SAE International from sources believed to be reliable. However, neither SAE International nor its authors guarantee the accuracy or completeness of any information published herein and neither SAE International nor its authors shall be responsible for any errors, omissions, or damages arising out of use of this information. This work is published with the understanding that SAE International and its authors are supplying information, but are not attempting to render engineering or other professional services. If such services are required, the assistance of an appropriate professional should be sought.

To purchase bulk quantities, please contact:
SAE Customer Service
E-mail: CustomerService@sae.org
Phone: 877-606-7323 *(inside USA and Canada)*
724-776-4970 *(outside USA)*
Fax: 724-776-0790

Contents

Chapter 3. Fundamentals of Laser Spectroscopic Techniques........................ 41

Chapter 4. Principle and Application of LDA and PIV for In-Cylinder Flow Measurements 71

Preface

The increasing concern with CO_2 emissions and energy prices has led to a raft of new CO_2 emission and fuel economy legislation being introduced in major automotive markets. It is therefore imperative for automotive manufacturers and Tier 1 suppliers to develop a new generation of internal combustion (IC) engines with both ultra-low emissions and high fuel efficiency.

Better understanding of the combustion and pollutant formation processes in IC engines is clearly necessary to increase their efficiency and cleanliness and to develop new engines with increased efficiency and lower emissions. As improvements in efficiency and emission abatement have reached points of diminishing returns, there is an increasing need for measurements directly inside the combustion chamber, where the combustion and pollutant-formation processes actually take place. Owing to the wide range of in-cylinder measurement techniques and the cross-disciplinary nature of laser-based optical techniques, it is difficult for postgraduates and research engineers to get an adequate overview on suitable measurement techniques in this field without extensive search and study of numerous publications. The purpose of this book is to provide, between one set of covers, an introduction to the experimental techniques that are suitable for in-cylinder measurements. In addition, it includes sufficient details for the reader to set up and apply these techniques to IC engines and combustion flows.

The spectrum of in-cylinder measurement techniques included in this book primarily covers advanced laser-based optical techniques and, where appropriate, other suitable techniques. The techniques described here are tied to the characterization of the flow and thermodynamic properties before, during, and after the combustion process, in the sequence of the processes taking place inside the cylinder, including flows, fuel injection, mixture formation, combustion, and pollutant formation. This approach ensures that the development and application of measurement techniques are linked to an identified need, as it is the end that justifies the means but not vice versa. This approach is particularly useful for researchers or engineers who are interested in measurement tools to solve a specific research problem. At the same time, by presenting different techniques for the same diagnostic purpose, the merits and limitations of each technique can be understood better.

Based on *Engine Combustion Instrumentation and Diagnostics*, this book focuses on laser-based optical techniques for combustion flows and in-cylinder measurements and includes new chapters on optical engines and equipment, case studies, and updated descriptions of each technique. Chapter 1 is an introduction to optical engines, including optical engine design, optical access, and optical materials. Chapter 2 begins with an overview of laser-based optical techniques, then presents a brief description of lasers, major optical components, and photodetectors. As a prerequisite to the development of spectroscopic techniques, Chapter 3 includes an overview of the salient features of spectroscopy and presents detailed explanations of the fundamentals of laser Rayleigh scattering (LRS), spontaneous Raman scattering (SRS), and laser-induced fluorescence (LIF). Techniques for in-cylinder flow measurements are described in detail in Chapter 4, including laser Doppler anemometry (LDA) and particle imaging velocimetry (PIV). Chapter 5 is dedicated to the principle and application of planar laser-induced fluorescence (PLIF), the most widely used technique for in-cylinder fuel and combustion species measurement. The other species measurement techniques, including LRS and SRS, are considered in Chapter 6. As direct fuel injection is increasingly applied to gasoline as well diesel engines, spray characterization and atomization measurements are required in the development of optimized fuel sprays and will be discussed in Chapter 7. Chapter 8 presents measurement and visualization techniques for detecting the occurrence of combustion through emission/absorption, chemiluminescence imaging, and multiple probes. Schlieren and shadowgraph techniques are included in Chapter 8 because of their prowess in detecting nonluminous flow and combustion processes.

In-cylinder gas temperature provides critical information about autoignition combustion, whether it is desired, as in CAI/HCCI (controlled autoignition/ homogeneous charge compression ignition) engines, or a dreaded, as in the case of knocking/super-knock in highly downsized gasoline engines. Chapter 9 presents the techniques for in-cylinder gas temperature measurements. The chapter begins with classical emission/radiation methods, then presents recent developments in laser-based thermometry, which have remained mostly in the hands of experienced researchers in a few well-equipped laboratories. Chapter 10 concludes with a description of measurement techniques for soot, which is an important issue for direct-injection gasoline engines as well as diesel engines.

I would like to thank all my PhD students and postdoctoral research assistants who have made so many contributions through their research to the case studies in this book. I hope that others will benefit from this book in their search for more efficient and cleaner combustion engines.

Hua Zhao
Brunel University, London
January 2012

chapter 1
Optical Engines

1.1 Introduction

Single-cylinder optical engines are often used to study in-cylinder flow, combustion, and pollutant formation as part of combustion system development, as well as providing validation data for simulations. When a single-cylinder engine is built, it is not mechanically balanced, as a multicylinder engine is. This can be compensated for with balance shafts or weights. In addition, it is a cost-effective way to make use of the cylinder head from a multicylinder engine rather than a specially manufactured single-cylinder head. A multicylinder inline engine may also be modified for in-cylinder measurements by incorporating an extension block into one of the cylinders, dispensing with the costly bespoke single-cylinder engine. In practice, a four-cylinder direct-injection (DI) diesel engine was modified for in-cylinder flow, mixing, and combustion measurements by Hentschel et al. (1989) so that realistic gas dynamics could be reproduced.

1.2 Optical Access

Optical access to the combustion chamber is required for optically based techniques. One optical access is sufficient for direct observation of combustion. For laser-based optical techniques, two or more optical accesses are required. Research engines with different optical accesses have been designed to allow the investigation of in-cylinder flow and combustion phenomena. Their design features are usually matched to the requirements of the optical techniques and the objectives of the investigation. For example, optical engines can be constructed to an atypical production configuration for fundamental studies (Gatowski et al., 1984). However, it is always desirable and usually necessary to preserve the realistic engine geometries and boundary conditions so that information gained from measurements in these engines can be used to validate design software and improve the engine's performance and emissions.

A production engine is normally constructed with multiple cylinders of identical design. Each cylinder consists of a cylinder head, a cylinder block, and a crankcase. There are, therefore, three ways through which to gain optical access to the combustion chamber: the top (cylinder head), bottom (piston), or side (cylinder liner). More than one access point is often required for optical techniques.

1.2.1 Optical Access Through the Cylinder Head

The construction of a cylinder head determines the size and implementation of optical access through the top. Typical two-stroke engines achieve intake and exhaust flows through ports installed in the crankcase and cylinder block. The lack of overhead valves in a two-stroke engine, thus, permits replacement of the cylinder head by an optical window, allowing complete view of the whole combustion chamber from top (Felton et al., 1988; Ghandhi et al., 1994). Alternatively, side valves can be used to operate in the four-stroke mode (i.e., as many early four-stroke spark-ignition engines were constructed). A four-stroke engine has been used for autoignition and spark-ignition combustion studies with a transparent cylinder head (Zhao et al., 2001). The engine has a disk-shaped combustion chamber with a flat-fused silica window on the top, which permits a complete view of the combustion chamber, as shown in Fig. 1.1. There are also two side windows that allow the transmission of a laser beam or a laser sheet. The ingenious parts of the engine design are the specially arranged intake and exhaust valves. The two intake valves are positioned so that the cylinder wall shrouds their outer edges. Hence, two concentric flows may be formed in the combustion chamber. The two inlet streams can be of identical air/fuel mixtures that can give a relatively uniform mixture distribution in the cylinder. Or by having the different air/fuel ratios in the two ports, a stratified charge may be formed. A greater degree of charge stratification can be achieved if one of the intake ports is used to admit only air or EGR (exhaust gas recirculation).

Fig. 1.1 Brunel single-cylinder optical engine with a transparent cylinder head.

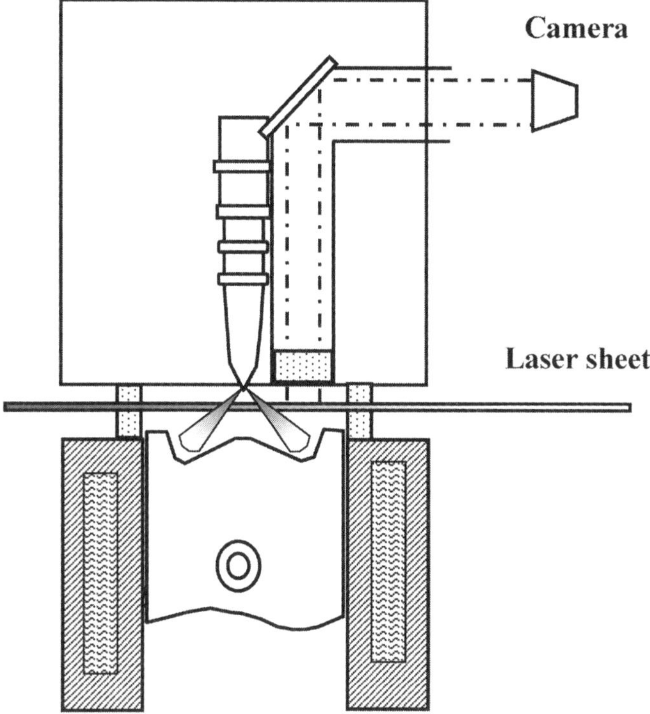

Fig. 1.2 A DI diesel engine with one of the exhaust valves replaced with an optical window.

Automotive four-stroke engines, however, are constructed with overhead valves, which seriously limit the space for optical access from the top. For relatively large DI diesel engines, one of the two exhaust valves can be replaced by a quartz window, as shown in Fig. 1.2. Since the intake valves are unchanged, the in-cylinder flowfield is preserved. By positioning a multihole injector in an orientation, so that the axis of one of the sprays passed directly below the window in the cylinder head, the atomization and combustion of one of the sprays can be studied (Dec et al., 1991).

A swirl chamber in an IDI (indirect-injected) engine can be used to study the interaction of diesel spray with air and diesel combustion. Fig. 1.3 shows the cylinder head geometry of an IDI diesel engine in the author's laboratory. The engine is based on a single-cylinder Lister-Petter DI diesel engine. Its original cylinder head consists of two parts. The top half is cast iron and contains the valve gear, which is retained in the modified engine. The lower half is replaced with an aluminum block, in which is machined a swirl chamber and valve seats. The swirl chamber is connected to the main combustion chamber via an interchangeable throat of variable diameters. The swirl chamber is fitted with optical windows on both ends, through which diesel spray atomization,

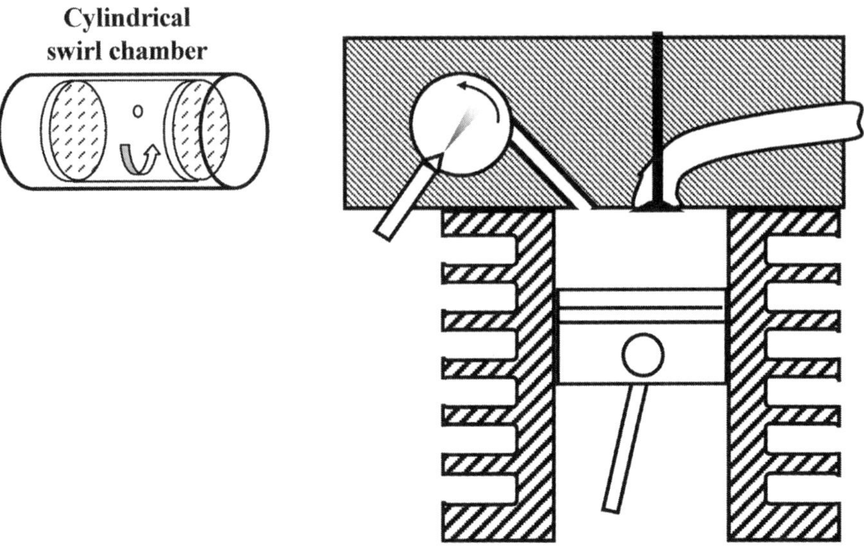

Fig. 1.3 An IDI diesel engine with a swirl chamber and optical windows.

combustion, and soot formation can be studied by means of shadowgraph, light extinction, and the two-color method. The swirl ratio can be altered by either the engine speed or the throat diameter. The third window in the orthogonal direction provides the optical access for the laser beam when right-angle detection is required. As the windows are shielded from the fuel sprays by the rotating airflow, continuous engine operation can be carried out for several minutes, limited only by the rise in the cylinder head temperature. It takes a couple of hours of accumulated engine running before the windows need to be taken off and cleaned.

1.2.2 Optical Access Through the Piston

The use of overhead valves in modern engines has led to the most widely used and well-established optical access through the piston top (Fig. 1.4). An elongated piston, with a quartz window in its crown and a slot on one or both sides, reciprocates about a slanted mirror at 45°, which is fixed to the engine block. The light received through the window by the mirror is then reflected horizontally to an imaging device. This design is often referred as the Bowditch extension, named after the engineer at General Motors who first proposed and developed the setup (Bowditch, 1961). The benefit of this configuration is in the use of a realistic cylinder head without the need for modifications and optical access to a large portion of the combustion chamber.

For four-stroke spark-ignition engines, the top piston surface is normally flat. Direct gasoline engines use a complicated piston shape to create the desired air/fuel mixture and efficient combustion. Figure 1.5 shows one example of

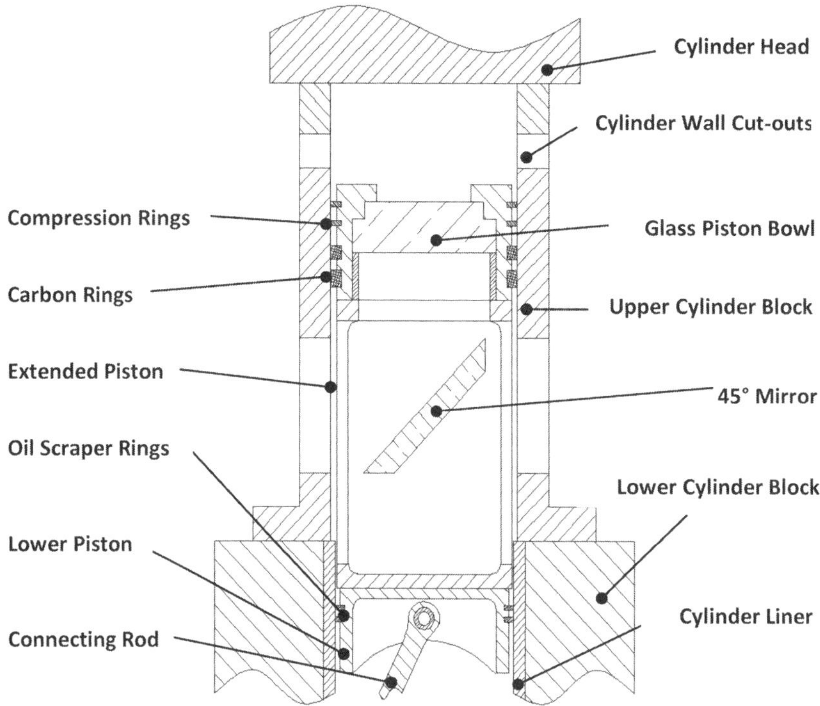

Fig. 1.4 Typical construction of an optical engine with an extended piston.

Fig. 1.5 Optical piston crown in a DI gasoline engine.

an optical piston crown design to study the air/fuel mixing in a stratified direction gasoline engine (Richter, 2006). In the case of modern diesel engines, the piston top is often characterized by a reentrant-type piston bowl geometry. As a compromise, it is common to maintain the lip region in the piston top but install a flat window at the bottom of the combustion bowl. To obtain realistic mixing and combustion conditions, a complete glass piston top with real piston bowl geometry has been used (Kim et al., 2008), but optical distortion has to be corrected in this case.

1.2.3 Optical Access Through the Sides

To apply spatially resolved laser diagnostics to internal combustion engines, additional windows are mounted between the cylinder head and cylinder block for the admission and exit of a laser beam or sheet as shown in Fig. 1.1 and Fig. 1.4. Alternatively, a quartz ring can be installed to provide a 360° view from the side. Because any contact between the piston and the glass ring may result in window destruction, the piston trajectory should be well centered to avoid any piston slap onto the glass liner. It may be desirable to keep the piston rings from contacting the optical window. Lubrication-free sealing between the optical liner and piston can be achieved by using piston rings made from polyamide-imide (Torlon®) or Victrex® PEEK directly or supported by an inner steel expansion ring.

When a larger side view into the cylinder is required, the complete cylinder liner may be replaced with a glass liner (Fig. 1.6). The design of the liner assembly must accommodate the extreme hardness of the quartz or sapphire liner by appropriate mounting structures and thermal properties that can lead to large and destructive thermal stresses. Thermal stresses can be decreased by skip-firing at elevated temperatures. A detailed description on the design and installation of a sapphire liner was given by Bates (1988).

The curved walls of the optical liner, however, produce distorted images of in-cylinder flow and combustion, particularly near the wall, which need to be corrected for quantitative measurements.

To apply the Schlieren technique to an engine with realistic cylinder geometry, Richman and Reynolds (1984) used a hologram to correct for the Schlieren image distortion caused by the curved surface of a quartz cylindrical liner. However, the correction technique involves a complex optical setup, and the improvement in the resulting Schlieren image was limited. A more successful approach is to machine the quartz liner so that its outer surface is shaped as a corrective lens (Fujikawa et al., 1988).

Fig. 1.6 A single-cylinder DI gasoline engine with a glass liner and side windows in the cylinder head (Richter et al., 1999).

Image distortion because of the curved surface of a cylindrical liner can also introduce errors to particle imaging velocimetry (PIV) measurements, when particle images are recorded through the curved surface.

In many applications, research engines are equipped with a combination of optical access described here. Side windows are normally used in laser-based techniques, with a third window on the side, top, or piston for the observation of scattered light.

1.3 Endoscopic Access

One of the major limitations of engines with large optical access is their low speed. Although such engines can be designed and operated at more than 4000 rpm, they are typically used at engine speeds ranging from 1200 to 1600 rpm and at moderate load. In addition, they are primarily operated as single-cylinder engines and hence cannot be operated with realistic engine operating conditions. As a consequence, increased interest and effort have been directed at the development and application of laser-based techniques through endoscopic access to engines. Figure 1.7 shows a schematic of endoscopic detection of a laser-induced and combustion emissions setup. In this case only small optical access to the engine is required for the endoscope so that high engine speed and high load operation are possible. It is easy to use and does not involve time-consuming alignment. However, the larger *f*-number (ratio of the focus length to aperture) and lower ability to collect light from an endoscopic probe restrict it to brighter emissions. In addition, attempts to increase the field of view result in image distortion. Therefore, the rule of thumb is to select endoscopes as big as possible and as small as necessary.

For imaging in the UV (ultraviolet) region, such as chemiluminescence and UV LIF (UV laser-induced fluorescence), the limited lens material suitable for UV restricts the ability to achieve achromatic correction with conventional refractive optics. This leads to significant step-down in the aperture in order to keep an acceptable image resolution. With recent advances in the production of diffractive optical elements (DOE) by micro- and nano-fabrication technologies, Gessenhardt et al., (2009) developed and demonstrated hybrid endoscopic wide-angle imaging optics for UV LIF measurements with optical access of less than 9 mm. The hybrid endoscopic optics comprise two parts, a front endoscope probe fixed in the cylinder head behind a sealed sapphire window and a separate 50-mm DOE relay lens between the endoscope probe and an ICCD (intensified charge-coupled device) camera. Thus the camera and engine were not mechanically connected. It was shown that this hybrid endoscope collected 1.3 times more light than a standard F-number = 4.5 UV lens at the same image magnification. The 266-nm laser sheet from an Nd:YAG laser was produced by a beam-shaping endoscope, which comprises a diffractive and aspheric shape lens optimized for the conversion of the Gaussian laser beam into a near-top-hat profile and a commercial plano-concave cylindrical lens for producing the light sheet, as well as pair of cylindrical lenses selected to reduce the laser sheet thickness through a telescope setup. Successful OH chemiluminescence imaging and PLIF (planar laser-induced fluorescence) fuel distribution measurements were achieved in a four-cylinder production engine.

Fig. 1.7 Detection of laser-induced scattering and emission through endoscopy in the spark plug hole (Richter et al., 1998).

1.4 Optical Materials

There are a number of transparent materials that can be used as optical windows. They may be classified as plastics, optical glass, and crystals. Plastics include materials made of polymethylmethacrylate (e.g., Plexiglas®) and polycarbonate (Lexan®). They are strong enough to be used for motoring operations, but they cannot be used in firing operations because of their high pressure and temperature.

For applications in the near-ultraviolet (UV) and visible range, optical glasses have been used as optical windows. BK7 is a low index of refraction crown glass. It exhibits excellent transmission from 350 to 2000 nm. At the UV end of the spectrum, the material is not usable at wavelengths lower than 300 nm. The others are fused silica and fused quartz (Pyrex®) (both are SiO_2). Both materials exhibit transmission characteristics that bear some similarity to the optical glass with which they are closely related. However, their simpler chemical composition leads them to have improved transmission in the UV region of the spectrum. The fundamental difference between these two materials is the method of manufacture. Most fused quartz is produced by melting and refusing silica sand and natural quartz. By comparison, fused silica is produced by flame hydrolysis of a silica halide. The resultant material is much purer and free from the sites that could cause absorption in the UV or problems with internal absorption in the use of high-power lasers. Consequently, the transmission range of fused silica is extended to 170 nm, while that of fused quartz is limited to 250 nm. Fused silica has an extremely low coefficient of expansion and can be used at high temperatures (~1000°C continuous).

Sapphire (Al_2O_3) is single-crystal aluminum oxide. It has excellent mechanical and optical properties and is extremely hard. Its strength is retained at elevated temperatures but sensitive to thermal shock. In addition, it is resistant to most forms of chemical attack. It has a transmission band that extends from the UV at around 180 nm, throughout the visible, and to the IR (infrared) region at 5500 nm. Because of its higher refractive index, Sapphire suffers more reflection losses. In addition, its transmission deteriorates more rapidly in the UV region, where UV grade fused silica exhibits much higher transmission down to 170 nm. The high cost of this versatile material, however, tends to limit its use to special applications.

1.5 Operation of Optical Engines

Having selected the appropriate optical materials, another consideration is the clean operation of optical windows. In practice, skip-firing, lubrication-free operation, and cleaner fuel are some of the measures. The lubrication-free operation includes the use of unlubricated bronze Teflon® or graphite-plastic piston rings, minimizing the oil loss through the valve stems into the combustion chamber and minimizing oil pumping from the crankcase into the mirror housing in the extended piston. At higher engine speeds, the extraction of oil mist from this area may be necessary. The presence of lubrication oil in the cylinder, especially when deposited on the viewing window, can cause significant interference with the PLIF measurements because of strong fluorescence emission from the oil droplets. In the author's laboratory, two groups of nonfluorescing lubricants (PIB: poly-iso-butylene; PAO: poly-alpha-olefin) were tested. The use of PIB oil was found to reduce this interference by 2 orders of magnitude.

In addition, special fuels are sometimes used in place of normal fuels. In spark-ignition engines, iso-octane/heptane mixtures are often used to minimize the window fouling from combustion deposit, which permits continuous engine operation for several minutes or longer without cleaning the windows. In diesel engines, the optical windows, especially the one in the piston crown, are obscured in a matter of seconds as a result of soot deposit. In this case, the fuel injection event should be synchronized with the laser and photodetector so that measurements can be done before the window is covered with soot. Oxygenated fuels help to suppress soot formation, which will prolong engine operation.

Occasionally, fluorescence can be generated from an optical window when a UV laser is used and there is impurity in the optical material. As the UV laser power is increased, window breakdown can occur. In this case the laser intensity in the window has to be reduced by an expanded laser beam and/or extended laser pulse by some form of pulse stretcher (Zhao and Zhang, 2008).

Finally, bear in mind that there are some differences between a normal engine and one modified with optical access:

- Differences in surface temperature.
- Different heat transfer, which effects combustion.
- Larger clearances between piston and liner.
- Lower ring packs and larger crevice volumes.
- Lower compression ratio compared with original.
- Heavy piston extension and thus lower engine speed (typically 1000–3000 rpm).
- Good optical access comes with limited load capacity.

A good understanding of the differences between optical and normal engines is required to develop appropriate strategies to compensate for such differences, as demonstrated by Kashdan and Thirouard (2009).

References

Bates, S. (1988). "A Transparent Engine for Flow and Combustion Visualization Studies." SAE Paper No. 880520, SAE International, Warrendale, PA.

Bowditch, F. W. (1961). "A New Tool for Combustion Research—A Quartz Piston Engine." *SAE Trans.* Vol. 69, pp. 17–23.

Dec, J., Zur Loye, A. O., and Siebers, D. S. (1991). "Soot Distribution in a DI Diesel Engine Using 2-D Laser Induced Incandescence Imaging." SAE Paper No. 910224, SAE International, Warrendale, PA.

Felton, P. G., Mantzaras, J., Bomse, D. S., and Woodin, R. L. (1988). "Initial Two-Dimensional Laser Induced Fluorescence Measurements of OH Radicals in an Internal Combustion Engine." SAE Paper No. 881633, SAE International, Warrendale, PA.

Fujikawa, T., Ozasa, T., and Kozuka, K. (1988). "Development of Transparent Cylinder Engines for Schlieren Observation," SAE Paper No. 881632, SAE International, Warrendale, PA.

Gatowski, J. A., Heywood, J. B., and Deleplace, C. (1984). "Flame Photographs in a Spark-Ignition Engine." *Combustion and Flame*, Vol. 56, pp. 71–81.

Gessenhardt, C., Zimmermann, F., Schulz, C., Reichle, R., Pruss, C., and Osten, W. (2009). "Hybrid Endoscopes for Laser-Based Imaging Diagnostics in IC Engines." SAE Paper No. 2009–01–0655, SAE International, Warrendale, PA.

Ghandhi, J. B., Felton, P. G., Gajdeczko, B. F., and Bracco, F. V. (1994). "Investigation of the Fuel Distribution in a Two-Stroke Engine with an Air-Assisted Injector." SAE Paper No. 940394, SAE International, Warrendale, PA.

Hentschel, W., Hessel, A., and Shindler, K. (1989). "Experimental Investigation of Spray Formation and Combustion in a Real Diesel Engine." Autotech Seminar, C399/19, IMechE. London.

Kashdan, J., and Thirouard, B. (2009). "A Comparison of Combustion and Emissions Behaviour in Optical and Metal Single Cylinder Diesel Engines." SAE Paper No. 2009–01–1963, SAE International, Warrendale, PA.

Kim, D., Ekoto, I., Colban, W., and Miles, P. (2008). "In-cylinder CO and UHC Imaging in a Light-Duty Diesel Engine during PPCI Low-Temperature Combustion." SAE Paper No. 2008–01–1602, SAE International, Warrendale, PA.

Richman, R. M., and Reynolds, W. C. (1984). "The Development of a Transparent Cylinder Engine for Piston Engine Fluid Mechanics Research." SAE Paper No. 840379, SAE International, Warrendale, PA.

Richter, M., Axelsson, B., and Alden, M. (1998). "Engine Diagnostics Using Laser Induced Fluorescence Signals Collected Through an Endoscopic Detection System." SAE No. Paper 982465, SAE International, Warrendale, PA.

Richter, M., Axelsson B., Alden, M., and Josefsson, G. (1999). "Investigation of the Fuel Distribution and the In-cylinder Flow Field in a Stratified Charge Engine Using Laser Techniques and Comparison with CFD-Modelling." SAE Technical Paper No. 1999–01–3540, SAE International, Warrendale, PA.

Richter, M. (2006). Eco-Engine Summer School notes, private communication.

Zhao, H., Peng, Z., Williams, J., and Ladommatos, N. (2001). "Understanding the Effect of Recycled Burned Gases on the Controlled Autoignition (CAI) Combustion Process." SAE Paper No. 2001–01–3607, SAE International, Warrendale, PA.

Zhao, H., and Zhang, S. (2008). "Quantitative measurements of in-cylinder gas composition in a controlled auto-ignition combustion engine." *Meas. Sci. Technol.* Vol. 19, pp. 1–10.

chapter 2

Introduction to Laser-Based Optical Instruments

2.1 Overview of Laser-Based Optical Instruments

The laser-based optical instrument for flow and combustion measurements comprises principally the laser, transmitting and receiving optics, and the photodetector. Figure 2.1 shows a typical laser diagnostic setup for measurement techniques with 90° detection. The laser-based optical instrument is characterized by a high spatial and temporal resolution and the ability to carry out in situ measurements. There are two types of lasers based on their mode of operation: pulsed and continuous wave (CW). As discussed later, most lasers used for reacting flow measurements have a pulse duration of ~10 ns, which is shorter than most physical and chemical time scales. Such a high temporal resolution means that the measurement can be performed in such a short time that change in the physical state of the molecules and chemical conversion can be considered negligible.

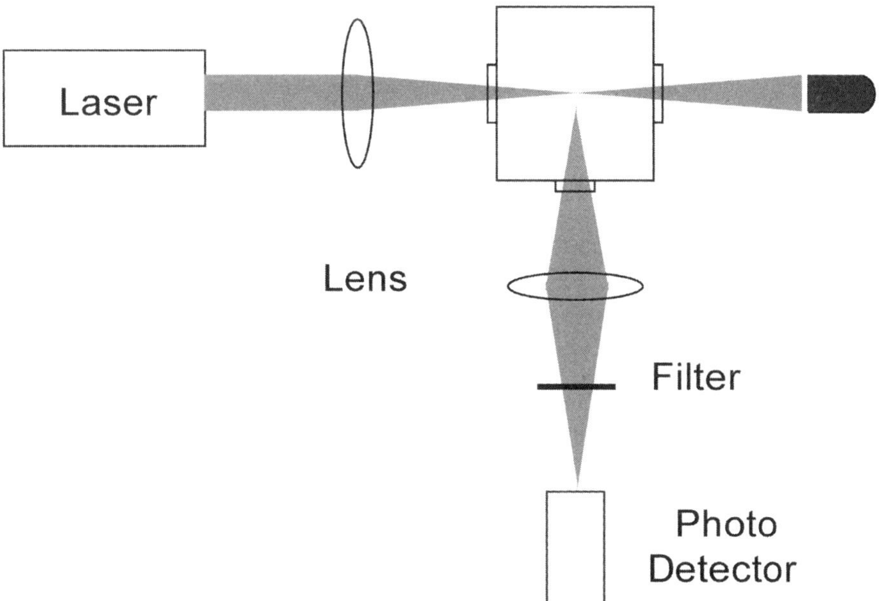

Fig. 2.1 Schematic of the laser-based optical instrument with 90° detection.

The spatial resolution is related to the small flow structure and local species concentration, which can be resolved in one measurement. It can be limited by the laser beam dimensions, optical setup, and the photodetector. The spatial resolution is also dependent on the measurement technique. For a single-point measurement setup shown in Fig. 2.2, the beam waist of a focused laser beam normally defines the spatial resolution. For a typical laser beam of Gaussian intensity distribution, the beam waist d can be estimated by

$$d \approx \frac{2\lambda f}{D} \qquad (2.1)$$

Where λ is the wavelength, D is the beam diameter before the focusing lens, and f is the focal length of the lens. For example, when a green laser ($\lambda \sim 503.2$ nm) of beam diameter of 5 mm is focused by a positive lens of 500 mm, the resulting beam waist will be about 100 μm.

The single-point measurement can be extended to multiple measurements along a section of the focused laser beam by using an array detector (Fig. 2.3). In this case, the number of measurement points and spatial resolution of each measurement point will depend on the number of photosensing elements (pixels) as well as the focused laser beam dimension.

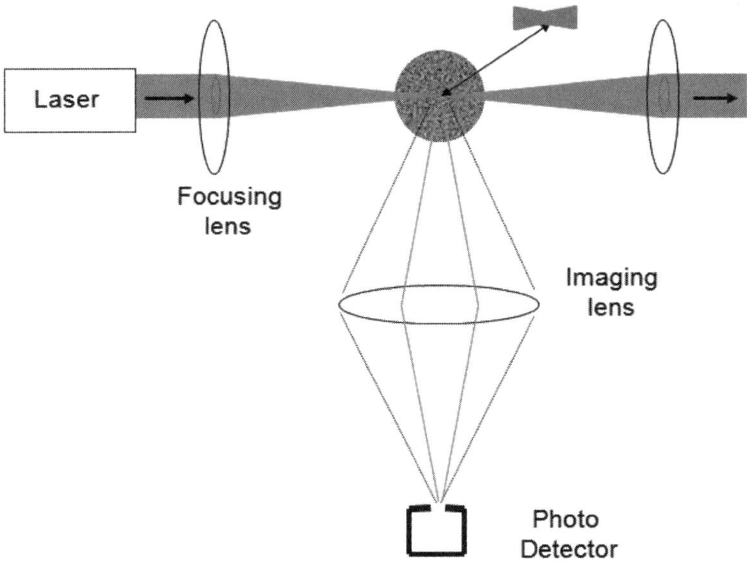

Fig. 2.2 Laser-based single-point optical instrument.

In most applications it is desirable to measure the flowfield, species, and temperature distribution across a large area through the use of a laser sheet (Fig. 2.4). In this case, a combination of spherical and cylindrical lenses are used to form a laser sheet in which the measurement is performed by a two-dimensional detector array (camera). The spatial resolution is then determined by the laser sheet thickness in the y direction and the imaging system in the other two dimensions.

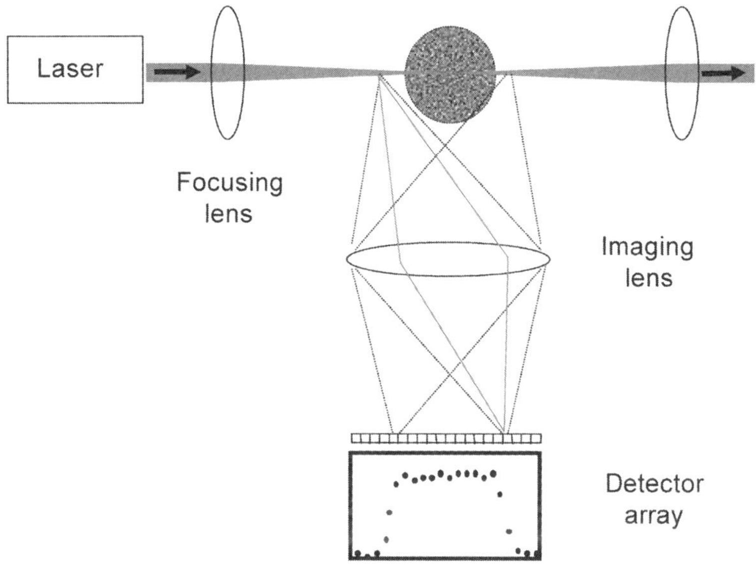

Fig. 2.3 Laser-based multipoint optical setup.

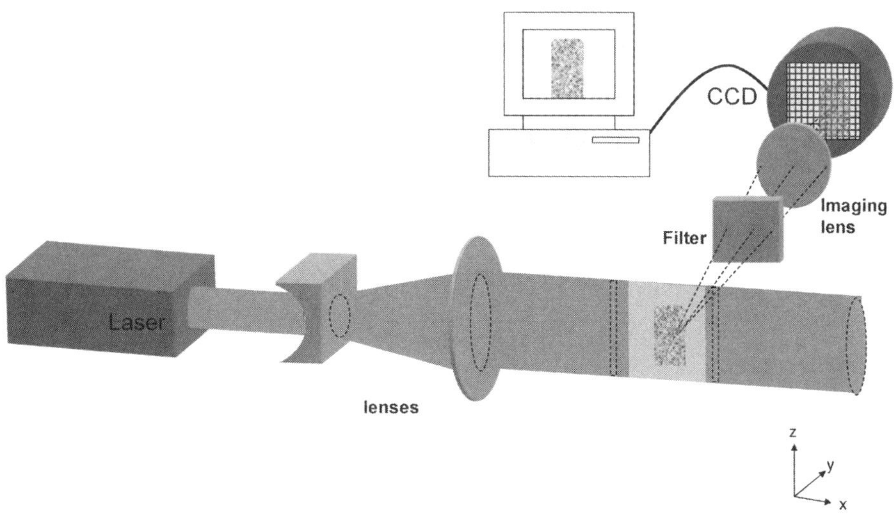

Fig. 2.4 Laser-based two-dimensional optical setup.

Finally, the line-of-sight detection is performed with the laser absorption and extinction technique which measures the spatially averaged value along the laser beam. The direct detection of flame radiation and chemiluminescence is another common form of the line-of-sight detection.

2.2 Lasers

From a practitioner's point of view, a laser can be considered as a source of light with very narrow spectral bandwidth and directionality. In contrast with other light sources, the laser beam from a single spectral line output laser contains a narrow range of wavelengths and can be considered monochromatic for almost all practical applications. In the case of multiple spectral line lasers, each spectral line in the laser output can be treated as an independent monochromatic laser beam. Most laser beams have a divergence angle of about 1 milliradian, that is they spread to about 1 meter in diameter after traveling a kilometer. This means that the laser output power is concentrated onto a very small area, and its intensity would be 1 million times as bright as that of the candle light if you were looking at a candle at a distance of 1 meter. The beam from high-power lasers can be easily focused to high enough intensities to cut and drill many materials. One of the most unique properties of the laser beam is its coherence, the property that the light waves it contains are in phase with one another or photons within the beam are of the same type. For conventional light sources, the coherence length is effectively zero, whereas it is a fraction of a meter or more for most lasers, which are essential for measurement techniques.

The word *laser* is an acronym for light amplification by stimulated emission of radiation, which effectively defines the operating principle of a laser, as shown in Fig. 2.5. A laser can be considered in its simplest form to consist of a gain medium and a set of mirrors. When photons have been generated in the active

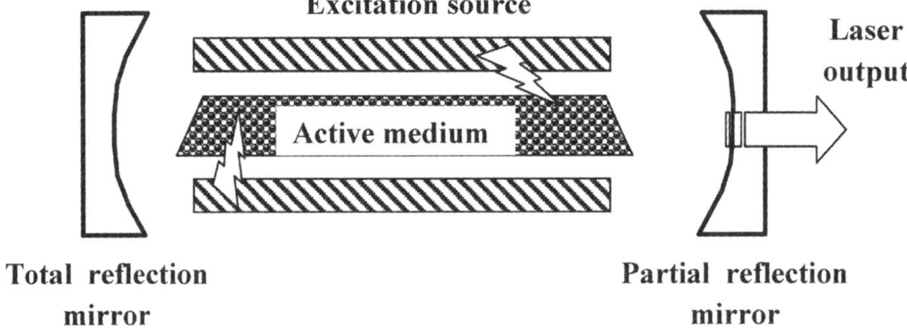

Fig. 2.5 Basic principle of a laser.

medium by other means through a process known as stimulated emission, remarkably the emitted photons are in phase and have the same polarization state and propagate in the same direction. Their numbers are amplified in the amplifying medium through the optical feedback provided by reflective surfaces (e.g., mirrors in many lasers) or other physical processes (e.g., cascading in quantum cascade lasers). Since the first successful demonstration of a ruby laser by Maiman in July 1960, a variety of lasers has been invented, ranging from the tiny semiconductor laser for telecommunications to the powerful laser capable of shooting down missiles. They can be grouped into solid-state lasers, gas lasers, and liquid lasers, according to the type of amplifying medium used. Alternatively, lasers can be defined as pulsed or CW depending on the mode of operation.

2.2.1 High-Energy Pulsed Lasers

High-energy pulsed lasers are commonly used in light-scattering, particle imaging velocimetry (PIV), and laser-induced fluorescence (LIF) measurements. Their high intensities can significantly increase signal-to-noise ratios when gated detection is employed. Flashlamp-pumped Nd:YAG lasers and excimer lasers are the two most common high-energy pulsed lasers used in laser diagnostics.

Flashlamp-pumped high-energy pulsed Nd:YAG lasers produce output primarily at 1.064 μm and are often used to generate laser beams at three wavelengths in the visible and UV (ultraviolet) region by frequency doubling or tripling in optical crystals. The three wavelengths are 532 nm (second harmonic), 355 nm (third harmonic), and 266 nm (fourth harmonic). The most common type of high-energy pulsed Nd:YAG lasers is the flashlamp-pumped Q-switched oscillator-amplifier system. A Q-switched Nd:YAG laser can produce 1 joule at 532 nm in a single pulse of 10 ns. Such lasers are typically operated at a fixed repetition rate at 10 or 20 Hz.

For Nd:YAG lasers, direct access triggering (DAT), as specified in some manufacturers manuals, or external mode should be selected when the application requires laser output to be synchronized with other events, such as a camera. In the single-shot mode, the laser is triggered externally by a single pulse. Lead time between the input fire command and laser output will be around a couple of hundreds microseconds. Alternatively, the input of two triggering pulses can be used for more accurate control between input commands and laser output. The first trigger signal, called fire command, will cause the flashlamp to discharge. This signal precedes lasing by about 180 μs. The second trigger, called Q-switch command, fires the Pockels cell in the laser cavity and precedes lasing by about 170 ns. When running the system in the external mode, the flashlamp must be operated at the frequency specified by the manufacturer (typically, 10 or 20 Hz). Operating the flashlamp at another frequency will cause poor beam quality because of the thermal lensing effect of the YAG rod. For example, to use a 10-Hz Nd:YAG laser in an engine experiment, the engine should be operated at 1200 rpm so that a single laser pulse is generated at a specified crank angle in every cycle.

The active media in excimer lasers are composed of the combination of a rare-gas atom (such as Ar, Kr, or Xe) and a halogen atom (Fl, Cl, or I). Different wavelengths can be achieved by selecting the appropriate rare gas and halogen as the lasing medium in an excimer laser. The most frequently used excimer lasers are XeCl, 308 nm; KrF, 248 nm; and ArF, 193 nm. They are available with pulse energies of a few hundred millijoules and repetition rates up to several hundred hertz. Excimer lasers generally have a broad spectral output. Certain specially designed excimer lasers can be tuned over a limited wavelength range (1 nm). The tunable excimer lasers have been used in UV Raman and LIF applications. Excimer lasers tend to be large, heavy, and expensive, and they require a fair level of maintenance. However, they are well suited for two-dimensional LIF-imaging applications because of their high pulse energy in UV wavelengths. The excimer laser can be triggered directly by an external device, and its frequency of operation is only limited by the maximum repetition rate.

In LIF experiments, fluorescence emission requires a precise excitation frequency. The laser frequency must closely overlap a resonant transition in the species of interest. In some cases, fixed frequency lasers exhibit such overlaps, such as the XeCl laser (OH) and ArF laser (NO). In many applications, tunable dye lasers must be used. The output wavelength of a dye laser covers the whole visible spectrum and can be altered by proper dye selection. Wavelength in the UV range can also be obtained by frequency doubling and wave mixing. Dye lasers are generally radiation-pumped by a flashlamp or another laser. Most pulsed dye lasers intended for spectroscopic applications are laser-pumped by the harmonic output of an Nd:YAG or excimer laser, because they possess better beam quality and spectral characteristics, for example, narrower line width and better frequency stability, than flashlamp-pumped varieties.

2.2.2 High-Repetition Pulsed Lasers

High-repetition pulsed lasers generate laser pulses at high repetition rates in the tens of kilohertz and are well suited for high-speed flow and spectroscopic imaging applications. The copper vapor laser was particularly favored for high-speed visualization experiments because of their high repetition rate (up to 50 kHz) and short pulse width (~30 ns). The wavelengths of Cu vapor lasers are within the yellow (578 nm) and green (510 nm) spectrum. The beam diameter is typically 40 mm. These lasers are characterized by their high average power output (typically 1–30 W). When synchronized with a high-speed video camera, images can be recorded at 50,000 frames per second. Whereas most lasers are cooled, metal vapor lasers need thermal insulation so that the operating temperature for vaporizing the metal can typically reach 1500°C.

Relatively unknown to the flow and combustion research community, certain flashlamp-pumped CW Nd:YAG lasers can be operated at a high repetition rate on the order of 20 kHz with 200- to 300-ns pulses and produce as much as 20 watts of power. Thus, they provide an economical option to copper vapor lasers for high-speed visualization experiments.

However, because of their high efficiency and compactness, diode-pumped solid-state (DPSS) YAG or YLF lasers are replacing ion lasers, copper vapor lasers, and some low-power flashlamp-pumped lasers in both commercial and scientific applications. In place of the flashlamp in a YAG or YLF laser, a diode laser provides the pumping energy in the DPSS laser. The lifetime of laser diodes is long compared with that of discharge lamps—typically many thousands of hours. The main disadvantage of diode pumping (as compared with lamp pumping) is the significantly higher cost per watt of pump power. This is particularly severe for high power. For this reason, flashlamp-pumped Q-switched YAG lasers are still used in cases where high pulse energy is needed, such as LIF imaging and PIV measurements involving a large measurement area and small scattering particles.

2.2.3 Continuous Wave Lasers

CW lasers are used in flow and combustion diagnostics. The two important CW gas lasers are argon-ion and He-Ne lasers. Argon-ion lasers are gas lasers and can supply over 10 W in the blue-green range (mainly 488 and 514 nm). Emission is produced at several wavelengths through the use of broadband laser mirrors. Individual wavelength can be selected by means of Brewster prisms or an etalon in the laser resonator. Argon-ion lasers are frequently used for multicomponent laser Doppler velocimetry (LDV) systems. He-Ne lasers are the most common and most efficient lasers in the visible range (633 nm). The power of commercial models ranges from less than 1 mW to more than 10 mW. Their application is limited to some LDV systems and light extinction experiments.

The semiconductor diode laser is another important type of CW laser. These lasers are similar in structure to incoherent light-emitting diodes (LEDs) and function as LEDs in cases when the drive currents are low. They are electrically pumped and amplified by an electrical current flowing through a p–n junction of the materials, which are semiconductors doped with impurities to produce the positive (p-type) and negative (n-type) carriers. In such a heterostructure, electrons and holes can recombine, releasing the energy portions as photons. Most laser diodes emit in the near-infrared (NIR) spectral region, but others can emit visible (particularly red or blue) light or mid-IR light. These are mainly found in the laser pointers, CD/DVD devices, and the telecommunications industry.

Diode lasers can be tuned by adjusting their operating temperature or altering the supplying current. Although temperature changes allow tuning over 100 cm^{-1}, the scanning rate is limited to a few hertz. In comparison, adjusting the supply current can provide tuning at rates up to gigahertz, but restricted to a smaller tunable range (about 1 to 2 cm^{-1}). Additional tuning and linewidth narrowing can be achieved by the use of extracavity dispersive optics. Adding an external optical cavity forces the diode laser to operate in a single longitudinal mode by creating a wavelength-dependent loss within the laser cavity. These external-cavity designs yield continuous tuning over the wide gain curve of the diode laser element with a very narrow linewidth. The tuning range depends on the gain element used in the cavity: At 630 nm, the tuning range is 10 nm; while at 1550 nm, the tuning range can be more than 70 nm. In both cases, the linewidth is less than 300 kHz. The inherent efficiencies of diode lasers make the external-cavity diode laser an attractive replacement for conventional dye and solid-state tunable technologies. As a result, these lasers have quickly found use in the applications of atomic spectroscopy, metrology, and combustion and emissions monitoring.

However, since most combustion gases have strong ground-state vibrational absorption lines beyond 2 µm, development of quantum cascade lasers (QCLs) allows the detection of such gases through the stronger absorption lines in the midwave and long-wave IR spectral range ($\lambda = 3.5 - 25$ µm). QCLs were first invented at Bell Labs in 1994 (Normand et al., 2007). They differ from traditional semiconductor diode lasers in that they do not need a *p-n* junction for light emission. A multilayered semiconductor material structure is used to form multiple active regions or quantum wells, in which the photons are generated by injected electrons and amplified as they cascade from one well to another. By using several tens or even 100 quantum wells in a series (a cascade), a higher optical gain and multiple photons per electron are obtained at the expense of a higher required electrical voltage. The operation voltage can easily be of the order of 10 V, whereas few volts are sufficient for ordinary laser diodes. As the transition energies are defined not by fixed material properties but by design parameters (particularly by layer thickness values of quantum wells), QCLs can be designed for operating wavelengths ranging from a few microns to well above 10 µm, or even in the terahertz region. Variable pulse lengths from 10 ns up to 10 µs allow the laser to be operated either in true pulsed or quasi-CW regimes. The distributed feedback (DFB) type quantum cascade laser has very narrow line widths that are suitable for gas detection. Fabry-Perot lasers generate a wide wavelength range and much higher power output than DFB lasers. There are also external-cavity lasers, where a wavelength tuning element such as a diffraction grating is part of the resonator. Whereas continuously operating room-temperature devices are normally limited to moderate output power levels in the milliwatt region, output of multiple watts is achieved with liquid or Peltier cooling.

In addition to the suitable wavelength range, QCLs usually feature a relatively narrow linewidth and good wavelength tunability, making them very suitable for spectroscopic detection of trace gases.

2.3 Laser Beam Delivery and Focusing

In most laser-based optical instruments for flow and combustion diagnostics, it is necessary to modify or shape the laser beam using lenses and other optical elements.

2.3.1 Gaussian Laser Beam

The output of most lasers (of stable laser resonators) has a Gaussian transverse intensity distribution as a function of radius r (Fig. 2.6):

$$I(r) = I_0 e^{-2r^2/w^2} = \frac{2P}{\pi w^2} e^{-2r^2/w^2} \tag{2.2}$$

Where P is the total beam power, and w is the radius where the intensity falls to $1/e^2$ (or 0.135) of its center line value. Diffraction causes light waves to spread transversely as they propagate. The beam spreading is described by the beam radius as a function of the distance z:

$$w(z) = w_0 \left[1 + \left(\frac{\lambda z}{\pi w_0} \right)^2 \right]^{1/2} \tag{2.3}$$

Where w_0 is the radius of the $1/e^2$ irradiance contour at the plane where the wave front is flat. This position is commonly known as the beam waist, and w_0 is called the waist radius. A waist occurs naturally at the midplane of a symmetric confocal cavity in a laser.

At a distance sufficiently away from the beam waist ($z \gg \pi w_0/\lambda$), $w(z)$ can be approximated by

$$w(z) = \left(\frac{\lambda z}{\pi w_0} \right) \tag{2.4}$$

Since z is much larger than $\pi w_0/\lambda$ so that the $1/e^2$ intensity contours asymptotically approach a cone of half angle θ,

$$\theta = \frac{w(z)}{z} = \frac{\lambda}{\pi w_0} \tag{2.5}$$

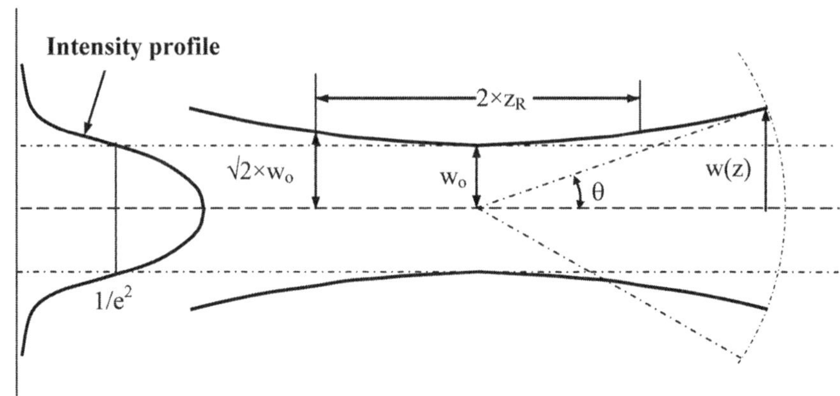

Fig. 2.6 Gaussian beam intensity distribution and geometry.

This value determines the far field divergence of the Gaussian beam. Far field is an optical term for the region at a greater distance from an aperture, as well as that which occurs at the focus of a lens.

A Gaussian beam will remain collimated for some finite distance and diverge with the spreading half angle θ. The Rayleigh range is the distance over which a laser beam can be maintained collimated before diffraction spreading takes over, that is, the distance on either side of the waist before the beam area doubles, and is given by

$$z_R = \frac{\pi w_0}{\lambda} \tag{2.6}$$

2.3.2 Focusing of a Gaussian Laser Beam

The transformations of a laser beam through simple optics can be found by calculating the Rayleigh range and beam waist location following each individual optical element (Self, 1983). These parameters are calculated using a formula analogous to the well-known standard lens formula:

$$\frac{1}{s} + \frac{1}{s'} = \frac{1}{f} \tag{2.7}$$

Where s and s' are the distances from the object and image to the lens, and f is the focal length of the optical lens. For a Gaussian beam, assuming that the waist of the input beam represents the object and the waist of the output beam represents the image, the formula is expressed in terms of the Rayleigh range of the input beam:

$$\frac{1}{s + z_R^2/(s-f)} + \frac{1}{s'} = \frac{1}{f} \tag{2.8}$$

The magnification is given by

$$m = \frac{w_0{'}}{w_0} = \frac{1}{\sqrt{\left\{ \left[1 - \left(\dfrac{s}{f}\right)\right]^2 + \left(\dfrac{z_R}{f}\right)^2 \right\}}} \tag{2.9}$$

The Rayleigh range of the output beam depends on m^2 and is given by

$$z_R{'} = m^2 z_R \tag{2.10}$$

Therefore, the spot size and the focal position of a Gaussian beam can be determined from Eqs. 2.8 and 2.9. As the beam divergence angle θ is normally quoted by the laser manufacturer, the direct relationship between the beam waist and divergence (Eq. 2.5) can be used to substitute for the beam waist w_0 in Eqs. 2.8 and 2.9.

Two cases of particular interest are when $s = 0$ (the input waist coincides with the surface of the lens) and $s = f$ (the input waist is at the front focal point of the optical system). For the case of $s = 0$, the equations for image distance and waist size reduce to

$$s{'} = \frac{f}{1 + \left(\lambda f/\pi w_0^2\right)^2} \tag{2.11}$$

$$w_0{'} = \frac{\lambda f/\pi w_0}{\left[1 + \left(\lambda f/\pi w_0^2\right)^2\right]^{1/2}} \tag{2.12}$$

For the case of $s = f$, the results are

$$s{'} = f \tag{2.13}$$

$$w_0{'} = \lambda f/\pi w_0 \; or \; w_0{'} = f \cdot \theta \tag{2.14}$$

Substituting typical values into these equations yields nearly identical results. Therefore, for most applications, the simpler second set of equations can be used.

Since most laser beams (except excimer lasers) spread at the diffraction-limited angle of $2.44\lambda/D$, or some multiple of the diffraction angle for high-power lasers, the beam waist of a focused laser beam can be estimated by

$$w_0' \approx \frac{\lambda f}{D},$$

where D is the diameter of the laser beam incident on the lens. Therefore, for a given focal length lens, the larger the incident beam diameter is on the lens, the smaller the focal spot will be.

The beam diameter of a laser beam can be easily adjusted by a telescopic lens pair (Fig. 2.7). For a high-power laser beam, an astronomical telescope is rarely used to avoid laser-induced gas breakdown at the focus, which occurs when the gas molecules are ionized by the intense irradiation of the laser beam, but it can be used to clean up a low-power laser beam by placing an aperture at the focal point. The Galilean arrangement is generally used for laser beam magnification. In addition, both telescopes can be employed to adjust the

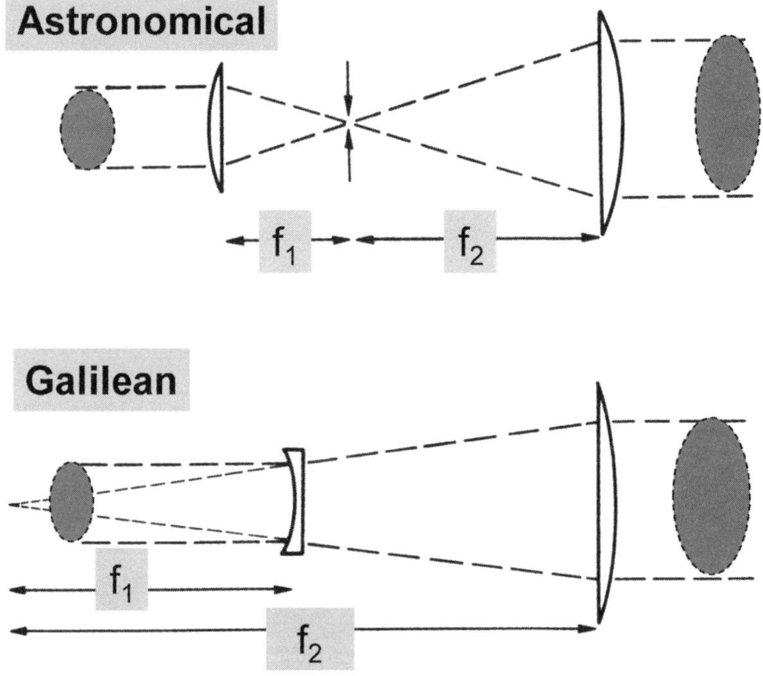

Fig. 2.7 Telescopic beam-expanding optics.

convergence or divergence of a laser beam and hence the focal location of the laser beam produced by another focusing lens.

2.3.3 Laser Sheet Optics

Two-dimensional imaging techniques require the illumination of a measurement section by a thin laser light sheet. The essential element for the generation of a light sheet is a cylindrical lens. When using a laser with a sufficiently small beam diameter and divergence, such as the Argon-ion laser, one cylindrical lens may be sufficient to generate a light sheet of appropriate shape. For other lasers, a combination of different lenses is usually required to generate a thin light sheet of high intensity. At least one additional lens has to be used for focusing the laser beam to an appropriate thickness, shown in Fig. 2.8(a). Frequently, another cylindrical lens is added to generate a light sheet of constant height, Fig. 2.8(b).

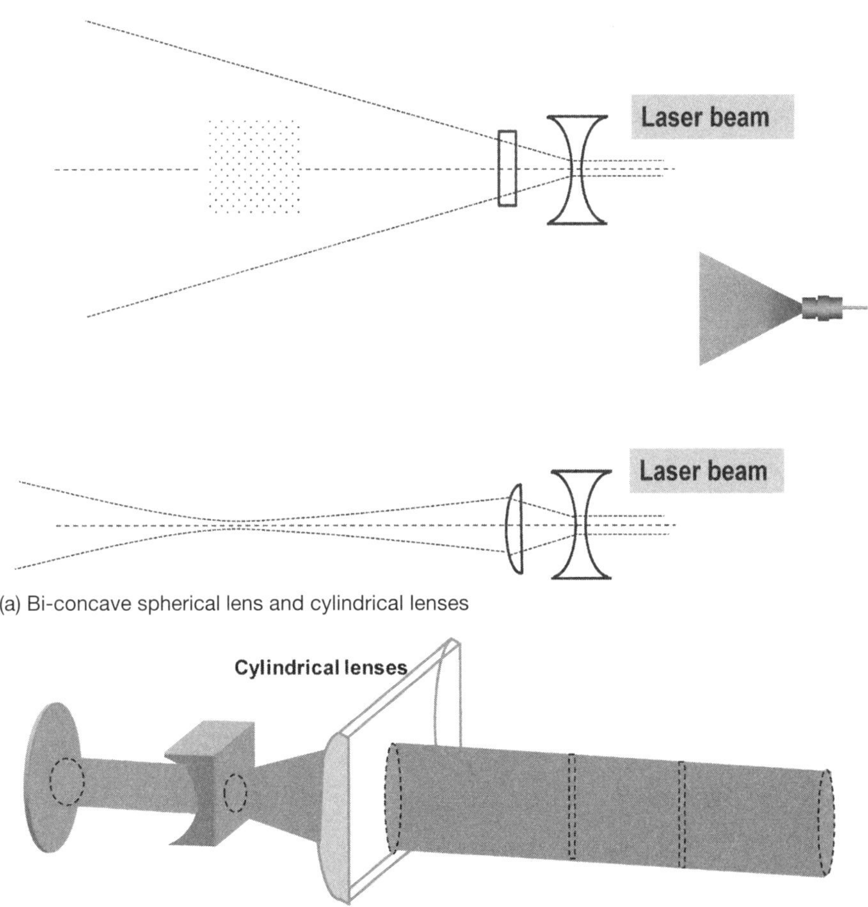

(a) Bi-concave spherical lens and cylindrical lenses

(b) Long focus spherical lens and two cylindrical lenses

Fig. 2.8 Examples of laser sheet optics.

The chosen optical system should be aimed at producing a thin sheet of approximately uniform illumination. In the configurations shown in Fig. 2.8(b), a long-focal-length spherical lens loosely focuses the beam, ensuring that the beam thickness will vary little over the imaged section. The position and thickness of the focused beam waist in the test section is determined by the laser beam profile as well as the sheet-forming optics. For example, the typical beam profile of an Nd:YAG laser is characterized by the Gaussian distribution, and the divergence of the beam is very small (<0.5 mrad). Therefore, the laser beam can be focused to a very thin sheet (<200 μm). Given the nature of electrical discharge in an excimer laser, its beam profile is normally characterized in one direction as "top hat" and in the other direction as nearly Gaussian. The divergence in these two orthogonal directions is relatively larger (<3 and 1 mrad, respectively). Because of the larger divergence, the thickness of the laser sheet is restricted to more than 1 mm. To obtain a smaller focused beam diameter (thinner sheet), unstable resonator optics are sometimes used in excimer lasers. However, this is accompanied by a decrease in the laser output and further reductions in the focused section. The total loss can be more than 50%.

2.4 Optics and Photodetection Systems

2.4.1 Optics

This section describes basic optical components, including optical lenses (and their selection), optical filters, mirrors, and polarizing optics (Fig. 2.9).

2.4.1.1 Optical Lenses

The singlet lens is the simplest optical element and comprises one piece of glass only. Glass exhibits dispersion (i.e., change of refractive index with wavelength), and thus the singlet lens suffers from poor color performance. When focused with high-dispersion glass, the red focus will be found beyond the blue focus because of the longer focal length of the red light. This is of no concern if a lens is used to deliver a single-colored laser beam. So to improve the color performance, an achromatic doublet is required. The doublet lens is manufactured from two different glass types and is specifically corrected at two wavelengths, for which the focal position is identical. As a result, the change in focal position with color can be reduced. In addition, the doublet lens is also characterized with much-reduced spherical aberration and coma than a singlet of the same *f*-number, and hence better spatial resolution. In addition to the singlet and doublet, there are complex lenses for specific functions, such as the telephoto lens, wide-angle lens, and lenses for microscopic applications.

The lens material should be chosen according to the wavelength. BK7 and fused silica are the most common lens materials used in the visible and UV wavelength regions. BK7 is a low-loss glass suitable for most applications in the visible to the NIR region. For the transmission of high-power or UV lasers, fused silica must be used. Specially manufactured UV-grade fused silica is necessary for the transmission of UV light below a wavelength of 200 nm.

At the glass surface, light is not only bent (as a result of the sudden change in the refractive index) but also lost by reflection. Such loss is about 8% for a singlet and more than 20% for an air-spaced triplet lens. Reflection losses also reduce the contrast of an image. Antireflective (AR) coatings are applied to lenses to counter this loss and thus improve transmission. Broadband multilayer AR coatings are typically applied to optimize transmission over a wavelength range. In addition, the narrow-band AR coating or V-coat can be optimized for a specific wavelength and angle of incidence, which is particularly useful with lasers. Many standard commercial coatings can be specified for the common laser types.

In selecting a lens system for imaging purposes, the major trade-off is between its resolution and light-gathering ability. One of the established methods to assess resolution is imaging of black and white lines. The quantity of discernible lines you can image defines the resolution. The spatial frequency response or modulation (contrast) transfer function is a parameter used to define the resolution of an optical component or imaging system. Modulation (contrast) is defined as

$$\text{Modulation (contrast)} = (I_{max} - I_{min})/(I_{max} + I_{min}),$$

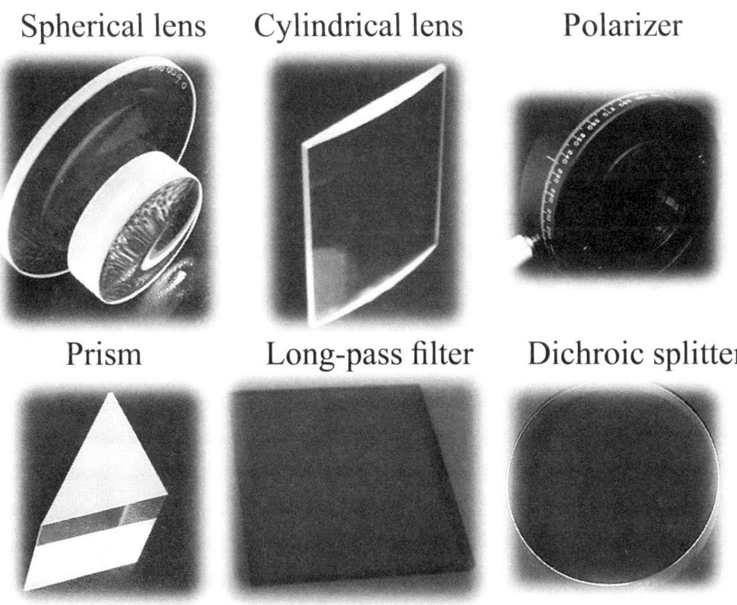

Fig. 2.9 Common optical components.

where I_{max} and I_{min} are maximal and minimal intensity.

Imagine a series of images with increasing numbers of alternating dark and bright lines. In the first few images there are some dark and bright lines that are clearly discernible. As the number of lines increases, the lines will start to merge and will eventually merge fully, reaching zero contrast. The frequency of lines at which lines start merging, for example, at 50% modulation, is defined as the spatial frequency response. For a lens with finite aperture D and focusing error δ (the absolute value of the difference between the "in-focus" object distance and the actual object distance), it is given by (Paul et al., 1990)

$$f_{L} = 1.22 \frac{f_{\#}(1+M)}{M^2 \delta} \tag{2.15}$$

where M is the magnification (image size/object size) of the system, and $f\#$ is the f-number of the lens given by $f_{\#} = {f}/{D}$. Thus, the larger the lens, the smaller the $f\#$ becomes.

According to Eq. 2.15, the spatial response (resolution) of a lens is enhanced by a large f-number and low magnification. The reverse is true for optimizing photon collection efficiency and increasing the signal-to-noise ratio, because of the large collection solid angle arising from low f-numbers and large magnifications.

2.4.1.2 Mirrors and Filters

For laser beam steering, metal and dielectric coatings are applied to optical flat or curved substrates. Gold, silver, and aluminum are primarily used for reflecting broadband light source with relative low energies at a wide range of angles. Aluminum coating is particularly suited for UV wavelength range. As metal coating is sensitive to moisture and scratches, a protective dielectric layer can be applied for the visible and NIR wavelength range.

Dielectric coating stacks are used in high-energy laser mirrors for a given laser wavelength. Since the reflection of the dielectric coating stacks depends on the angle of incidence, properties of coating and substrate, and the state of polarization, it is important to specify the correct component and polarization state when selecting a laser mirror.

In addition, right-angle prisms can be used as mirrors to direct the high-power laser beam at 90° using total internal reflection at the hypotenuse face. They can also be used as a 45° mirror with the hypotenuse face coated with metallic or dielectric coating.

In laser-based optical systems, interference filters are often used for the rejection of ambient light or in spectroscopic measurements where the laser and stimulated emission are relatively close in wavelength. Fig. 2.10 shows the spectrum of a typical narrow-band interference filter and the important parameters associated with the filter. A typical interference filter has a 50% passband around 7–10 nm and a peak transmission of 50%. The bandwidth is determined by the number of layers of coating: The more there are, the narrower the transmission range is. The blocking rate outside the transmission range can be as high as 10^{-5}. Note that the transmission characteristics of such filters depend on the angle, and peak transmission will decrease noticeably with as the angle of incident light goes beyond 30°.

To isolate regions of a spectrum, short-pass and long-pass filters are often used as emission filters in fluorescence measurements and as stray light filters to eliminate unwanted radiation. Figure 2.11 shows the transmission curves of colored filter glass, where a long-pass filter transmits wavelengths longer than a specific wavelength, the cutoff wavelength for 50% transmission is often specified, and a short-pass filter blocks the transmission of longer wavelengths. A band-pass filtering arrangement can be readily achieved by combining long-pass and short-pass filters. In addition to the colored filter glass, dielectric-coated edge-pass filters can be used for applications in which a precipitous fall in the transmission curve is needed. Unlike colored filter glass, these dielectric long-pass and short-pass filters will shift to shorter wavelengths with an increase in the incident angle, a feature that can be employed to fine-tune the location of the cutoff wavelength.

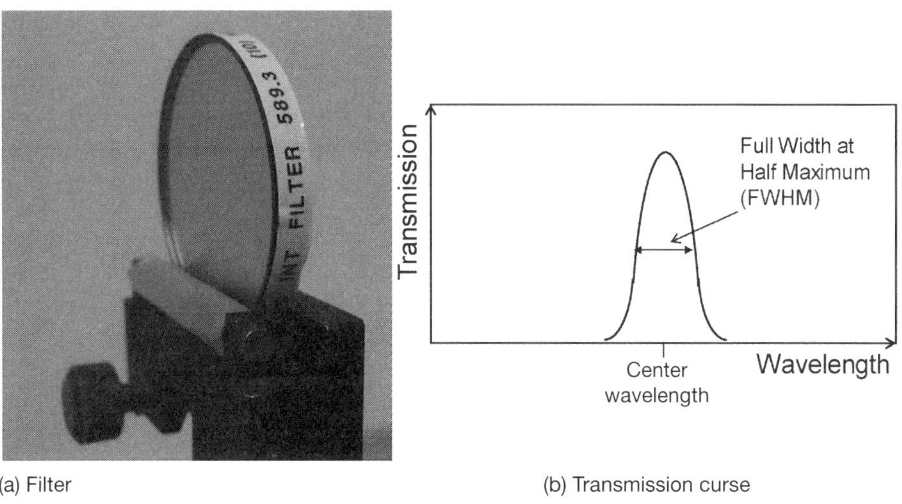

(a) Filter   (b) Transmission curse

Fig. 2.10 Interference filter and its transmission characteristics.

In a multichannel detection or double-image system, neutral-density filters of metal-coated glass can be used to provide a balanced radiation level to photodetectors. They comprise a polished glass substrate with an Inconel® metallic coating and characterized by uniform attenuation over a broad spectral range. The optical density D of these filters is defined by the logarithm of the transmission ratio, and hence the transmission T is related to the optical density by $T = 10^{-D}$. The value of D typically varies from 0.1 ($T = 79\%$) to 4 ($T = 0.01\%$).

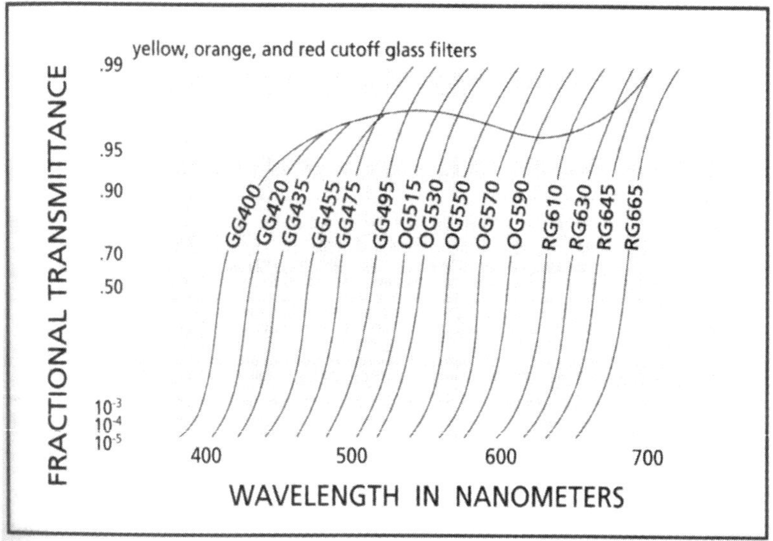

(a) Transmission curve of long-pass colored glass

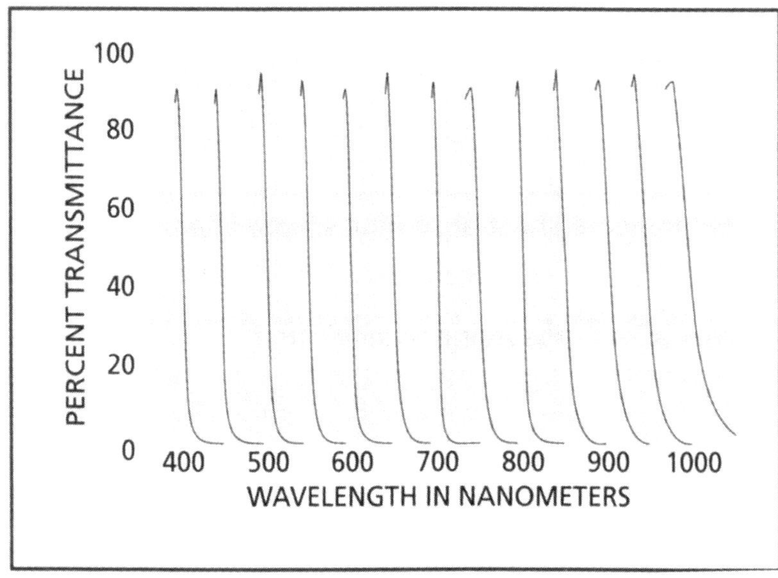

(b) Transmission curve of short-pass colored glass

Fig. 2.11 Long-pass and short-pass filter glass.

2.4.1.3 Beam Splitters/Separators and Beam Samplers

Both beam-splitting cubes and plate beam splitters can be used as nonpolarizing beam splitters of laser beams (Fig. 2.12). Antireflection coating is applied to all four faces of the cube and the rear surface of the plate. The beam-splitter coating is on the diagonal internal surface of the cube and the front surface of the plate. The plate beam splitter is more cost effective but suffers from ghosting (reflection from the second surface) and shifting of the beam, whereas the beam-splitting cube provides virtually no shifting of the beam and is mostly free from the ghosting effect. By leaving the front surface uncoated, an optical plate can be used to pick up as much as 10% of the incident laser beam at 45° for monitoring the laser power and beam quality. The back surface is wedged to eliminate internal fringes and antireflection coated to remove ghosting. Fused silica should be used for the cubes and plates for applications where better thermal stability and less distortion to the beam quality are required.

Special beam splitters, known as dichroic filters/mirrors, are often used to separate or combine laser beams and signals of different colors (Fig. 2.13). They can be supplied with different dielectric coatings according to the wavelengths of the laser beams to be separated or combined.

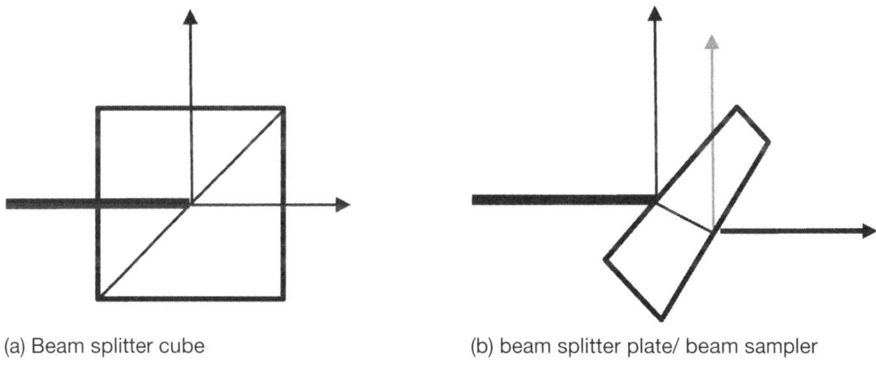

(a) Beam splitter cube   (b) beam splitter plate/ beam sampler

Fig. 2.12 Broadband beam splitter and beam sampler.

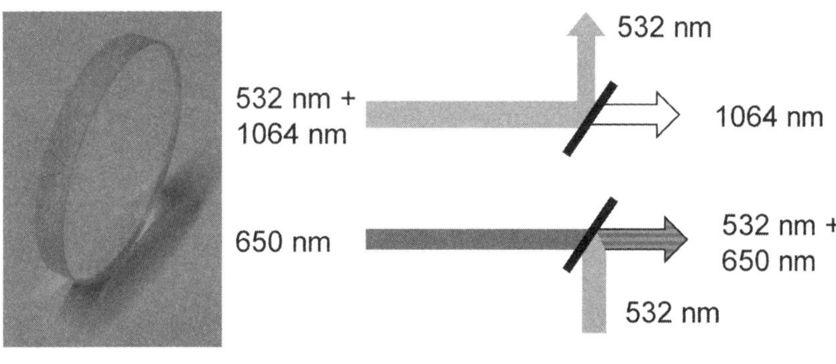

(a) Dichroic filter/mirror   (b) beam separation arrangement

Fig. 2.13 Dichroic filter/mirror for beam separation/overlapping of laser beams or separation of color in imaging.

2.4.1.4 Polarization Optics

Unlike natural light, whose electric field varies rapidly and hence is unpolarized, laser output is polarized. The state of laser polarization may be linear, circular, or elliptical. This state must be considered and altered for many laser-based optical measurement techniques, through the use of a class of optical elements known as retarders (Fig. 2.14). A retardation plate that introduces a relative phase difference of 180° or a retardation of $\lambda/2$ (half a wavelength) between the two orthogonal components of a light wave is known as a half-wave plate. This is used for rotating the input polarization state, for example, the angle of the polarization of linear laser beam or the handedness of circular or elliptical light from right to left, or vice versa. In contrast, the quarter-wave plate is used to convert linearly polarized light into circularly polarized light when the polarization of the incident light is at 45° to the optical axis of the plate. Similarly, an incoming circular polarized beam will emerge linearly polarized.

In addition to the delivery of laser beams, the polarization state associated with a particular laser-induced emission and scattering process, for example, Raman scattering, can be employed to discriminate any unwanted light by incorporating a polarizer in the receiving optics so that only the light of the desired polarization state will be detected.

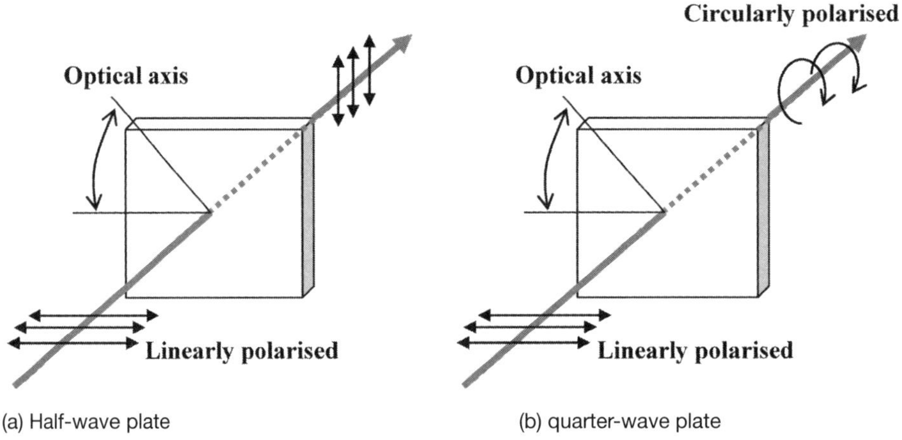

Fig. 2.14 Quarter-wave and half-wave plates.

2.4.2 Photon Detectors and Charge-Coupled Device (CCD) Cameras

This section discusses some of the photon detectors commonly used for laser-based optical instruments. The photodiode is the most compact optical detector. It works by emitting photoelectrons from its surface after absorption of a number of suitable photons. The material for photoelectron emission directly affects the region of the spectrum that can be detected and the quantum efficiency of the detector. These detectors are available in the visible and NIR regions. The PIN photodiode is the most popular one available for general purpose detection because of its faster response and improved linearity than the conventional p- or n-type. The salient features of the PIN photodiode include (1) responsivity, as measured by A/W, amps of the output current for a given radiant power falling on it; (2) the useful spectral range of detection; (3) noise limits or detectivity; and (4) speed of response. For applications that require greater sensitivity than the traditional PIN photodiode, such as monitoring and characterization of a pulsed laser on a nanosecond scale, the avalanche photodiode can provide a very sensitive but more expensive solution.

A photomultiplier tube (PMT) offers the most sensitive single-element photon detection. As shown in Fig. 2.15, the incident light falls on a photocathode material from which electrons are generated and then accelerated toward a series of dynodes, thus giving rise to a great number of secondary electrons. Many types of PMT are available with a range of different spectral sensitivities and rise times. Note that an operating PMT should be protected from ambient light to avoid being damaged.

The CCD (charge-coupled device) is now the most widely used imaging device in scientific research. In principle, it can be considered as an array of photon detectors (Fig. 2.16) or pixels that produce an electric charge by photoelectric effect. The electric charge generated in each pixel is then moved and stored in a neighbor cell before being transferred to the shift register, where the signals are clocked out. The charge is preamplified and digitized before it is transferred to a computer for image processing via an interface card.

The sensitivity of a CCD camera can be increased more than 20,000 times using an image intensifier. In an image intensifier, the photon-electrons produced by the incident photons are accelerated away from the photocathode by an applied voltage, which can be gated down to nanoseconds. These electrons undergo cascading amplification in the same way as the PMT does and finally strike onto a phosphor-coated surface, thereby leading to an output image brighter than the original image. Such an image intensifier, referred to as an MCP (microchannel plate; see Fig. 2.17), comprises millions of small-diameter (10-μm) channels, where electrons are accelerated by a few hundred volts (potential) and are multiplied before striking onto the phosphor plate. The phosphor image produced is then optically coupled to an array detector. The coupling can be achieved via an optical lens or coherent fiber-optic bundles. Although lens coupling can alter the magnification easily and hence different array detectors can be used, it has poor collection efficiencies and poor spatial resolution compared to the fiber-optically coupling.

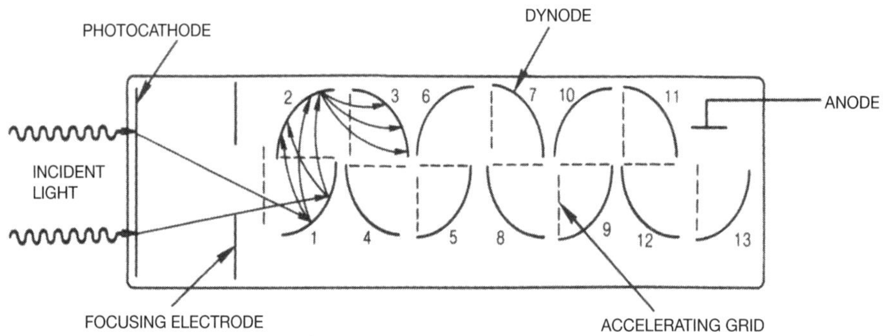

Fig. 2.15 Operation of a photomultiplier tube (PMT).

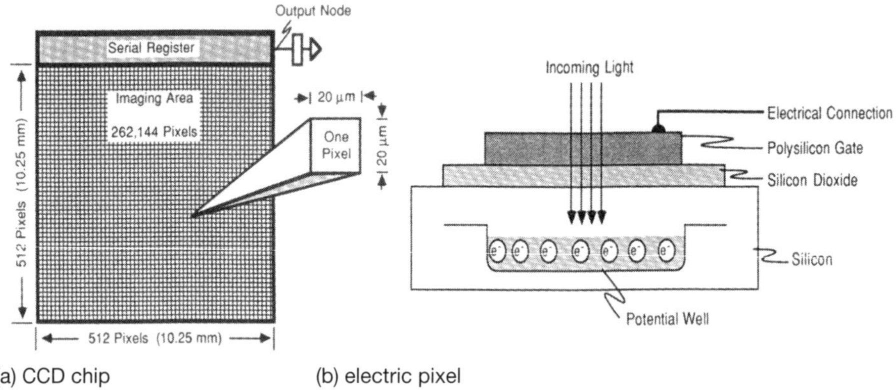

(a) CCD chip   (b) electric pixel

Fig. 2.16 CCD camera chip.

The intensifier unit should be selected carefully to match the ICCD (intensified charge-coupled device) system performance to the requirement of the experiment. First, the quantum efficiency (QE) of the photocathode determines the fraction of incoming photons that lead to the release of electrons. The manufacturers normally provide a photocathode that may be optimized for UV to blue region of the spectrum, or blue to visible region, or NIR applications. Second, the gate speed should be considered. Fast gating or shuttering capability of intensifiers allows capture of transient emission, and it is also used to eliminate the ambient light. Typically, high-speed gating is achieved by switching the high voltage between the photocathode and MCP, and it can be as low as nanoseconds. Furthermore, for high frame rates (>1 kHz), P46 phosphors with fast decay time (<2 µs) should be selected instead of P43, which has a 3-ms decay time.

The modern, gated, intensified CCD systems feature much more compact design with a built-in gate pulse generator and camera controller. For low-light detection, ICCDs are generally cooled thermoelectrically by a built-in Peltier cooler to reduce the noise level and hence increase the dynamic range of the system. However, ICCD cameras are limited to a framing rate of a few hertz. To speed up the readout time and obtain an increased signal-to-noise ratio, *binning* of pixels may be considered, which means combining the charges in a number of pixels, for example, 4 × 4, through a setting in the camera setup.

For some applications, for example, spontaneous Raman scattering and CARS (coherent anti-Stokes Raman scattering) measurements, it is often required to record the signal across a large spectral region to determine the species concentration and/or temperature from the detailed spectral features. This used to be achieved by scanning a monochromator/spectrometer so that one wavelength region at any one time is detected by a photomultiplier attached at its exit. This approach has been superseded by the simultaneous detection of the entire spectrum using an optical multichannel analyzer (OMA; Fig. 2.18).

In a modern OMA system, a linear array of detectors is attached to a spectrograph to record the entire spectrum with much higher spectral resolutions. The linear arrays used often include self-scanned photodiode arrays (PDAs) or CCDs. The major differences between them are the sizes of the photosensing elements, that is, the linear versus two-dimensional values

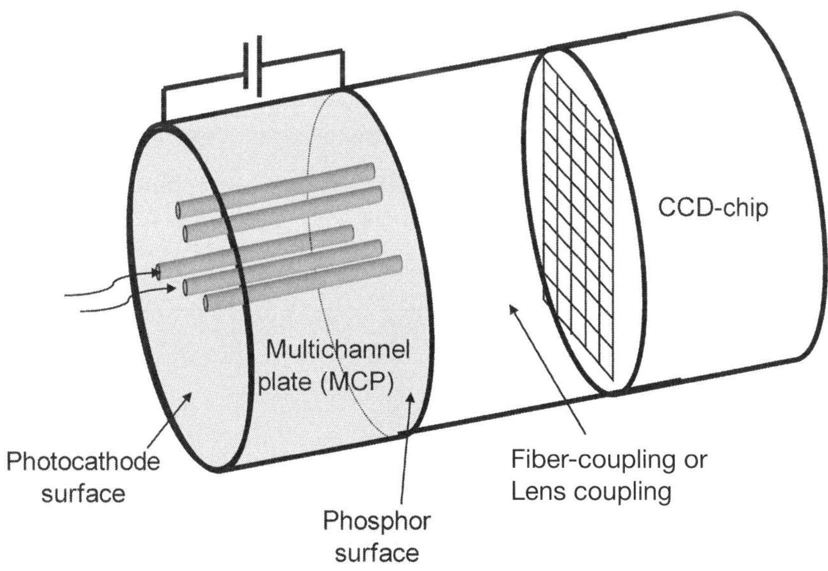

Fig. 2.17 Multichannel plate image intensifier.

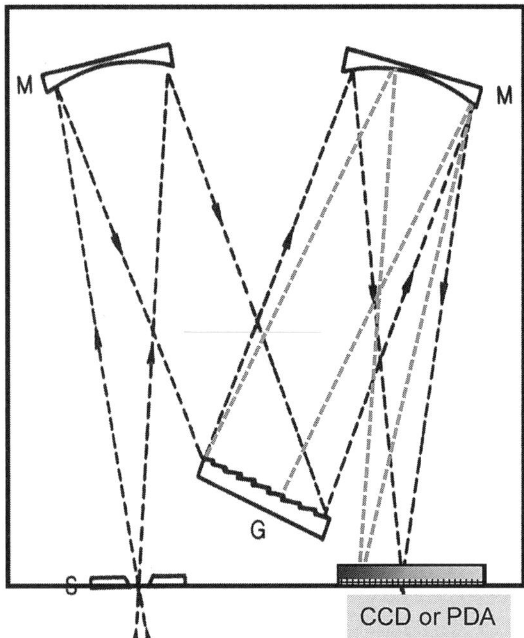

Fig. 2.18 An optical multichannel analyzer (OMA) comprising a spectrometer and PDA/CCD array.

of the arrays. CCDs are very sensitive and generally possess a large dynamic range and very low noise characteristics. When a one-dimensional spatial imaging mode is used, a two-dimensional array can be used with an imaging spectrograph, which will be discussed in the application of spontaneous Raman scattering in Chapter 6.

2.4.3 High-Speed Imaging

Over the last decade, the solid-state high-speed imaging device has completely replaced the conventional film camera as a result of the rapid advances in the CCD and CMOS (complementary metal-oxide semiconductor) image sensors. In a CCD sensor, every pixel's charge is transferred through typically one node to be converted to voltage, buffered, and sent off-chip as an analogue signal. Since all of the pixel can be devoted to light capture, its output uniformity is high. CCDs are normally employed as the performance benchmarks in photographic, scientific, and industrial applications that demand the highest image quality (as measured in quantum efficiency and noise). In a CMOS sensor, each pixel has its own charge-to-voltage conversion, and the sensor often also includes amplifiers, noise-correction, and digitization circuits. As charge-to-voltage conversion takes place on each pixel, uniformity across the sensor is lower. However, speed is one area where CMOS imagers have demonstrated considerable strength because of the relative ease of parallel output structures.

There are broadly two categories of high-speed imaging devices with solid-state image sensors. The first one is the high-speed digital camera with a fast CMOS sensor and high throughput. A state-of-the-art, high-speed digital camera can deliver frame rates of several thousand frames per second (fps) at full resolution of 1 megapixel and several hundred thousand fps at reduced resolution. Depending on the onboard memory, many seconds of high-speed video images can be recorded continuously or intermittently with each burst of video images synchronized with the external event to be recorded.

For ultra-high-speed imaging, multichannel framing cameras can be used. They comprise multiple camera modules and complex internal optics and thus achieve significantly higher framing rates. There are two basic types: the beam splitter and the rotating mirror. In the rotating-mirror CCD system, the light from a single objective lens is distributed through a rotating mirror to multiple CCD sensors. Such a camera can capture several hundred thousand frames per second at full resolution using an electric mirror drive. They are also available with a helium-driven turbine mirror drive for full resolution framing rates up to 25 million per second. In the beam-splitter camera, the image is distributed from a single objective lens to multiple CCDs. Alternatively, ICCD sensors can be used when light gain and precise control of the timing and duration of exposure are desired. In addition, the beam splitter with multiple ICCD cameras will capture images down to nanosecond intervals, equivalent to hundreds of millions frames per second. However, given the cost and complexity of CCD and ICCD sensors, multichannel framing cameras are typically limited to 16 frames.

2.5 Image Processing and Calibration

If qualitative information is expected from images of laser-induced scattering or emission, the experiment and data analysis will be straightforward. To obtain quantitative results, however, a number of additional measures need to be taken.

The first process involves the removal of systematic errors present in the imaging system. According to Seitzman and Hanson (1993), the digitally recorded detector signal for a single pixel of a CCD camera, S_{pix} (number of digitization levels), can be expressed as

$$S_{pix} = \left[N_P \left(\frac{P_c}{P_o} \right) R_{tot} + S_b + S_d \right] \qquad (2.16)$$

where

$N_p = P_s/h\upsilon$, the number of photons leaving the collection volume within the collection solid angle

P_s = the laser-induced scattering or emission signal within the collection volume

P_c = the laser energy fluence within the collection volume

P_o = a nominal laser fluence

R_{tot} = the total responsivity of the detection system

S_b = the background signal associated with laser-induced processes (e.g., Rayleigh and Mie scatterings and window fluorescence) and ambient light (e.g., flame emission)

S_d = the dark current signal

The random noises in a digital imaging system, which include the shot noise in the photodetector and electronic noises in the detection system, are considered negligible in most imaging applications.

Solving Eq. 2.16 for the fluorescence signal N_p produces the following:

$$N_p = \left[S_{pix} - \left(S_b + S_d\right)\right]\left(\frac{1}{R_{tot}}\right)\left(\frac{P_o}{P_c}\right) \tag{2.17}$$

Or

$$P_s = h\upsilon\left[S_{pix} - \left(S_b + S_d\right)\right]\left(\frac{1}{R_{tot}}\right)\left(\frac{P_o}{P_c}\right) \tag{2.18}$$

Equations 2.17 and 2.18 illustrate the appropriate procedure for correcting systematic errors in the recorded image. First, a combined image of the background and dark current signals ($S_b + S_d$) is subtracted from the recorded image (S_{pix}). Then, the new image is multiplied by a correction image accounting for the spatial variations in the responsivity of the detector. Finally, nonuniformities in the laser sheet energy are corrected by another multiplicative factor.

A combined dark and background image ($S_b + S_d$) can be recorded in the absence of the laser-induced emissions. In the case of planar LIF measurement of fuel distribution, this is achieved by removing the fluorescing molecules in the gas mixture when the engine is motored. Since background may vary from image to image, average of the background can be used to minimize the effect. The responsivities are typically measured by uniformly illuminating the detector system with a light source having a spectral distribution similar to the fluorescence, as the spatial variation is often a nonuniform function of wavelength. When two-dimensional measurements are carried out inside an engine, it is more convenient to record the combined effect of responsivity $(1/R_{tot})$ and nonuniformities in the laser sheet (P_c/P_o). The combined effect is obtained from an image recorded from a homogeneous medium (S_{uni}) minus the background image $(S_b + S_d)$. Hence, Eq. 2.18 can be rewritten as

$$P_s = hv\left[S_{pix} - \left(S_b + S_d\right)\right]\Big/\left[S_{uni} - \left(S_b + S_d\right)\right] \qquad (2.19)$$

In the cases of high-power pulsed lasers used for planar imaging applications, such as excimer, Nd:YAG, and dye lasers, shot-to-shot variations in the laser pulse energy and spatial profile can be significant. To correct for the shot-to-short variation in the pulse energy, a fast response photodiode, such as an avalanche photodiode, can be employed to record the laser beam energy from a beam sampler in the optical path. However, it is more difficult to correct for the spatial variation in the laser beam profile when the spatial intensity distribution suffers from temporal fluctuations. One solution is to render the laser beam profile through a specially designed laser beam homogenizer, also known as a beam diffuser or beam shaper, which is made up of diffractive elements or microlens arrays. Pfadler et al. (2006) demonstrated that by the application of the microlens arrays to the inhomogeneous excimer laser profile, an almost uniform beam profile with a remaining nonuniformity of less than 5% could be achieved. The beam homogenization reduced the total measurement error close to the minimum practical achievable uncertainty of the applied ICCD camera.

Although systematic errors such as steering of the laser beam by density gradients, attenuation of the laser sheet across the imaged region, laser-produced perturbations, and partial saturation and radiative trapping (absorption of the fluorescence within the combustion gases) are often avoidable or negligible, their effects must be considered.

2.6 Summary

In this chapter an overview of the laser-based optical instrument was given. The basic characteristics of the most commonly used lasers and Gaussian laser beam delivery were presented. There are hundreds of different optical components; some are generic, and many are specific to a particular application. Recently, many optical manufacturers have prepared and published extensive online technical notes and guides to their electro-optical products, which provide excellent guidance.

References

Normand, E., Howieson, I., and McCulloch, T. (2007). "Quantum-Cascade Lasers Enable Gas-Sensing Technology." *Laser Focus World*, Vol. 43, No. 4, pp. 90–92.

Paul, P.H., Van Cruynighen, I., Hanson, R. K., and Kychakoff, G. (1990). "Applications of Planar Laser Induced Fluorescence Imaging Diagnostics to Reacting Flows.", AIAA/SAE/ASME/ASEE 26th Joint Propulsion conference, Orlando, FL, AIAA-90-1844.

Seitzman, J. M., and Hanson, R. K. (1993). "Planar Fluorescence Imaging in Gases." Chapter 6 in *Instrumentation for Flows with Combustion*, edited by A. Taylor. London: Academic Press.

Self, S. A. (1983). "Focusing of Spherical Gaussian Beams." *Appl. Optics.*, Vol. 22, No. 5, p. 658.

Pfadler, S., Beyrau, F., Löffler, M., and Leipertz, A. (2006). "Application of a beam homogenizer to planar laser diagnostics." *Optics Express*, Vol. 14, No. 22, pp. 10171–10180.

Further Reading

Chang, W. (2007). *Principles of Lasers and Optics.* London: Cambridge University Press.

Hecht, J. (1999). *The Laser Guidebook,* 2nd Ed. New York: McGraw-Hill.

Hecht, E. (2001). *Optics,* 4th Ed. San Francisco: Addison Wesley.

Tkachenko, N. V. (2006). *Spectroscopic Measurement: An Introduction to the Fundamentals.* The Netherlands: Elsevier.

Websites

Edmund Optics: http://www.edmundoptics.com/technical-support/technical-library/ (last accessed January 2012)

Newport Technical Notes: http://www.newport.com/Technical-Notes/980099/1033/content.aspx (last accessed January 2012)

Coherent: http://www.coherent.com/Products/index.cfm?834/Lasers (last accessed January 2012)

Andor Technology: http://www.andor.com/learning/digital_cameras/?docid=326 (last accessed January 2012)

chapter 3
Fundamentals of Laser Spectroscopic Techniques

3.1 Introduction

Laser-induced fluorescence (LIF), emission and absorption, and Raman and Rayleigh scattering are the most frequently used combustion diagnostic techniques. To understand the principles of these techniques, a brief introduction to molecular spectroscopy is presented in this chapter. It provides a descriptive and nonmathematical introduction to the general theory and nomenclature of molecular spectra for the reader who is unacquainted with spectroscopic theory. Then, the fundamental principles of laser-based linear optical techniques will be described and compared, including Raman and Rayleigh scattering and LIF.

3.2 Fundamentals of Molecular Spectroscopy

Laser spectroscopic techniques involve the interaction and exchange of energy between photons and molecules or atoms. In molecular spectroscopy, the nature of the interaction is governed by the internal energy system of the molecule. Therefore, a good understanding of the internal energy system of a molecule is essential for the application of laser spectroscopic techniques.

3.2.1 Internal Energy of a Molecule

The internal energy of a molecule is shared among its electronic energy, vibrational energy, and rotational energy. When the average distribution of these types of internal energy for an ensemble of molecules is equal, the system is said to be in thermal equilibrium. In all three modes of excitation, the energy levels are discrete. The molecules can only contain a precise quanta of energy, which is determined by the energy levels associated with each energy mode, that is, $E_{total} = E_{electronic} + E_{vibrational} + E_{rotational}$.

Figure 3.1 shows the energy levels for the ground ($X^2\Pi$) and first excited electronic states ($A^2\Sigma$) of OH. In each electronic energy state, a number of vibrational levels (v' and v'') exist, in which a number of rotational energy levels (J and K) reside. It is noticed that the energy levels are discrete as dictated by quantum mechanics. Thus, the radiation associated with changes in energy states is not continuously distributed, but is discrete or quantified. The electron configurations and their vibrational and rotational states determine the energy levels possessed by a given configuration. Each configuration possesses a number of energy levels. Change from one energy level to another results from absorption (low to high) or emission (high to low) of radiation. Spectroscopy basically consists of a set of rules governing the allowable configurations, configurational changes, and energies associated with these changes.

Radiative transitions between energy levels occur at an optical frequency determined by

$$E_1 - E_2 = h\nu = hc\bar{\nu} \tag{3.1}$$

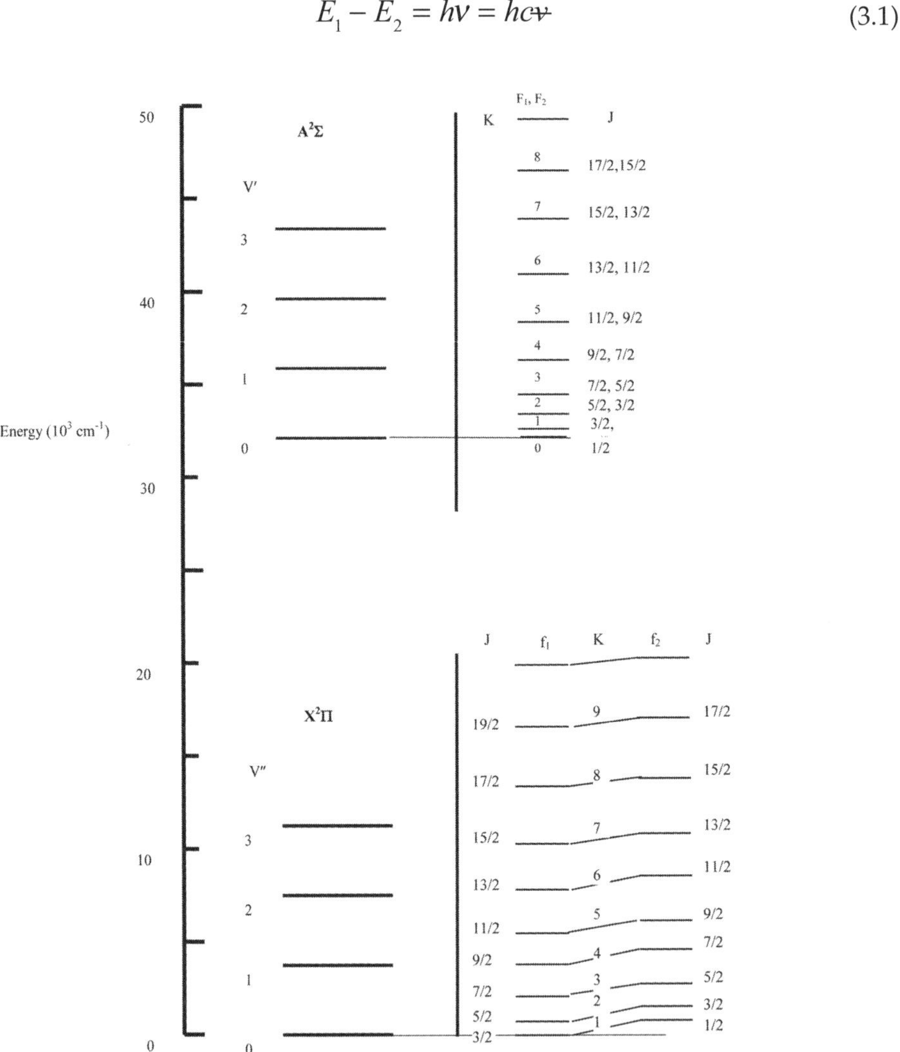

Fig. 3.1 Energy-level diagram for the ground and first electronically excited states of OH radical.

Where h is a universal constant known as the Planck's constant, ν is the frequency of the light wave, and c is the speed of light. $\bar{\nu}$ is termed wavenumber (in cm^{-1}), typically used by spectroscopists, and is a particularly convenient energy measure as used in Fig. 3.1. For a given wavelength of light, the number of wavenumbers, $\bar{\nu}$, is just the reciprocal of the wavelength, that is, the number of wavelengths per unit distance. E_1 and E_2 refer to the energy levels possessed by the molecule involved in the transition.

The electronic states of diatomic molecules, as well as their vibrational and rotational energies, are discussed in the next sections because of their simplicity and relevance to most of combustion diagnostics.

3.2.1.1 Rotational Energy of Diatomic Molecules

Molecules can rotate only at precise frequencies, and successively higher frequencies form a ladder of energies as shown on the right side of Fig. 3.1 and Eq. 3.2:

$$E_r = hcBJ(J+1) \tag{3.2}$$

Where J is the rotational quantum number, which takes the integral values 0, 1, 2, 3, etc., and B is termed the rotational constant. If centrifugal effects are taken into account, more accurate description of rotational energy is given by

$$E_r = hc\left[BJ(J+1) - DJ^2(J+1)^2\right] \tag{3.3}$$

Where D is introduced as the centrifugal distortion constant. Photon energies corresponding to pure rotational energy changes are such that the wavelength is in the microwave region.

3.2.1.2 Vibrational Energy of Diatomic Molecules

Motion of an atom relative to another in a molecule occurs through specific vibrational motions. The associated vibrational energies are modeled as harmonic oscillators and are also quantized:

$$E_v = h\nu(v+1/2) \tag{3.4}$$

The symbol v is the vibrational quantum number, which takes integer values 0, 1, 2, etc.

For more accurate calculation of vibrational energy, vibrating diatomic molecules should be modeled as anharmonic oscillators, which will result in higher order terms appearing in the allowed energy:

$$E_v = hc\left[\omega_e\left(v+\frac{1}{2}\right) - \omega_e x_e\left(v+\frac{1}{2}\right)^2 + \omega_e y_e\left(v+\frac{1}{2}\right)^3\right] \tag{3.5}$$

Where $\omega_e = \upsilon/c$ is the equilibrium energy in cm^{-1} and $\omega_e x_e$, $\omega_e y_e << \omega_e$.

3.2.1.3 Coupling Between Vibrational and Rotational Modes

Normally, a molecule simultaneously rotates and vibrates. The molecular vibration will lead to a change in the rotational moment of inertia, which will cause a change in the rotational energy storage because of vibration. The effect is modeled by the following equations:

$$B_v = B_e - \alpha_e\left(v + 1/2\right) \tag{3.6a}$$

$$D_v = D_e - \beta_e\left(v + 1/2\right) \tag{3.6b}$$

Where the subscript e refers to the equilibrium positions of the nuclei and α_e, β_e are small correction constants. Therefore, the rotational constant B and its centrifugal correction D in Eq. 3.2 will depend on the magnitude of vibration and therefore on v. The constant α_e is called the vibration-rotation interaction constant, and it is important for laser Raman diagnostics.

3.2.1.4 Electronic States of Diatomic Molecules

The motion of electrons is also quantized; however, unlike rotations and vibrations, the energy levels do not form a simple ladder. Figure 3.2 shows the low-lying electronic states of OH.

The symbols for the electronic energy levels indicate the character of the energy state, for instance, spin and symmetry. A brief description on the physical implications of these symbols and electronic structure is warranted here for readers to appreciate the salient features of molecular spectroscopy.

Three quantum numbers n, l, and λ are employed to describe the state of individual electrons. The number n represents the approximate radial distance of an electron from the nuclei, known as the principal quantum number; and l is associated with the value of the angular momentum of the electron in its

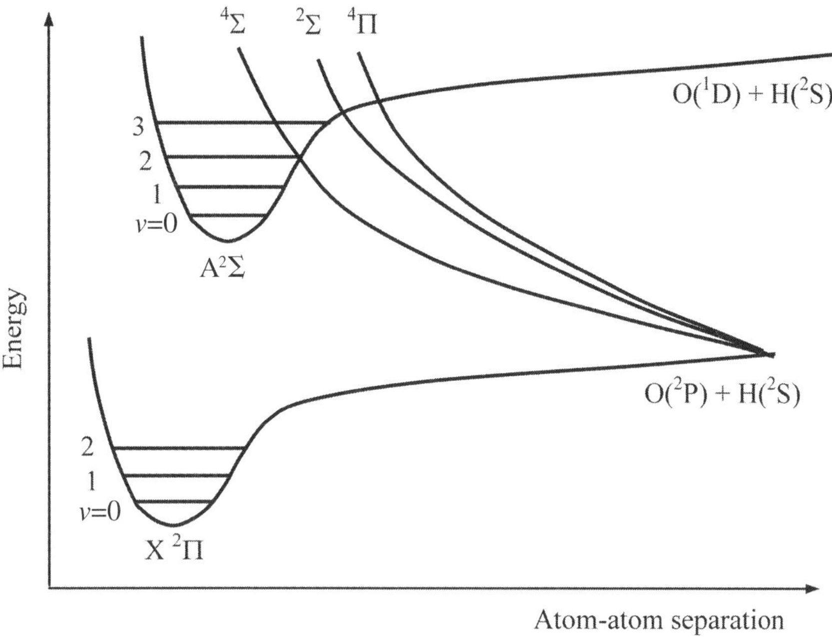

Fig. 3.2 Energy-level diagram for the OH radical with electronic and vibrational levels.

orbit, which is quantized. The number l takes the value 0, 1, 2, 3, etc., up to a maximum value of $n - 1$ and is known as the azimuthal or orbital angular momentum quantum number. According to the value of l, the electrons are named as $s(l = 0)$, $p(l = 1)$, $d(l = 2)$, $f(l = 3)$, and so on, alphabetically. The third quantum number λ is given by $\lambda \equiv |m_l|$, where m_l is related to the quantized component of the orbital angular momentum along the internuclear axis, assuming the integer values $l, l - 1 \ldots - l$. The significance of λ is that the axial component of orbital angular momentum is $h\lambda/2\pi$. Electrons with $\lambda = 0, 1, 2, 3, \ldots$, etc., are termed σ, π, δ, ϕ electrons. Note that in the case $\lambda \neq 0$, there are two values m_l, either positive or negative. Thus, all such states are doubly degenerated. As a result of the degeneration, these levels will be split, a situation termed *lambda-doubling*, when the coupling of the electronic and rotational modes is considered.

In summary, individual electrons are identified by writing $n\ l\ \lambda$, thus $1s\sigma$, $3d\pi$, etc., in the united atom approach (Herzberg, 1971). In addition, a spin quantum number, m_s, is also used to describe the spin momentum of an electron, which takes half integral values $\pm 1/2$, depending on whether the spin is with or opposed to the orbital motion.

When several electrons are involved in a molecule, each electron can be considered separately and described by the quantum numbers n_i, l_i, λ_i. In general, several electronic states emerge from a given electronic configuration.

Each electronic state of a molecule can usually be described by two quantum numbers, Λ and S. Λ is related to the total orbital angular momentum of the electrons along the internuclear axis, given by $\Lambda = |\Sigma m_l|$. It takes the integral values 0, 1, 2, 3, etc. With $\Lambda = 0, 1, 2, 3$, etc., the electronic state is referred to by Greek capital letters Σ, Π, Δ, etc. The other quantum number of importance is the resultant spin of the electrons S, which may be integral or half-integral according to whether there are an even or odd number of outer electrons. The resultant spin of the electrons, $S = |\Sigma m_s|$, determines the multiplicity $M = 2S + 1$ of the state. Thus, any electronic state is represented by the multiplicity, written as a superscript on the left, and its Λ value. Therefore, a $^2\Sigma$ state (read as doublet sigma) has $M = 2$ (i.e., $S = 1/2$) and $\Lambda = 0$.

In denoting a transition between two electronic states of a molecule, it is customary to write the symbol for the state of higher energy first. To distinguish between electronic states of similar character, it is also common to name them by writing a letter, either small or capital, before the symbol. The ground electronic state is denoted by X, except in the case of C_2, in which the true lowest lying ground state was discovered only after a slightly higher energy level was assigned as the X state. A, B, C, and so forth, designate the excited electronic states, which have the same multiplicity. States of multiplicity different from the ground state, whose transitions to the ground state are normally forbidden, are denoted by lowercase letters, a, b, c, and so forth.

Now we can see how the electronic states of OH, as shown in Fig. 3.1, are decided. The configuration of OH in the ground electronic state is $(1s\sigma)^2$ $(2s\sigma)^2$ $(2p\sigma)^2$ $(2p\pi)^3$, where superscripts indicate the number of electrons for a particular combination of $n\,l\,\lambda$. The first two shells $(1s\sigma)^2$ $(2s\sigma)^2$ are closed shells where a shell reaches the maximum number of electrons that it can contain. The closed shells are very stable. It is generally the outer electrons, which are less firmly bound to the nuclei because of the shielding of the nuclear charge by the inner electrons, that become excited, that is, raised to higher lying energy states. The three $2p\pi$ electrons in the OH ground electronic state have values of $n = 2$, $l = 1$ and $\lambda = 1$. The m_l could take values of $+1$ or -1. From the Pauli exclusion principle, no two electrons can possess the same n, l, m_l, m_s. Given the two possible values of m_s, the possible values of m_e are 1, 1, -1 or -1, -1, 1. Therefore, Λ is 1 ($|\Sigma m_l|$) leading to a Π state. Similarly, m_s possibilities are $-1/2$, $+1/2$, $-1/2$ or $+1/2$, $-1/2$, $+1/2$, so the state multiplicity is 2. The ground state is therefore denoted by $^2\Pi$. The first excited state forms as one of the $2p\sigma$ electrons is excited into the $2p\pi$ level forming a closed shell. The single $2p\sigma$ electron then determines the first excited state term as $^2\Sigma$.

3.2.1.5 Coupling Between Rotational and Electronic Modes

Up to this point, the rotational and electronic modes have been treated independently of each other. In general, the coupling between the rotational and electronic motions needs to be considered in order to determine the detailed rotational energy levels. As rotational lines are rarely resolved or used in engines combustion measurements, we shall not discuss the detailed structure of rotational energy levels.

3.2.1.6 Spectral Structure and Selection Rules

Spectral lines arise from transitions among differing electronic, vibrational, and rotational levels. Not all transitions are possible, and transitions are only allowed between electronic states for certain electron spins and certain electronic symmetry. Λ may change by 0 or ±1 only. The multiplicity, that is, M or S, does not change at all for strong band systems. For example, the following are permitted transitions:

$$^{1}\Sigma \rightarrow {}^{1}\Sigma, \qquad {}^{2}\Sigma \rightarrow {}^{2}\Sigma, \qquad {}^{1}\Sigma \rightarrow {}^{1}\Pi, \qquad {}^{1}\Pi \rightarrow {}^{1}\Sigma$$

and the following would not be allowed:

$$^{1}\Sigma \rightarrow {}^{1}\Delta, \qquad {}^{3}\Sigma \rightarrow {}^{3}\Delta, \qquad {}^{1}\Sigma \rightarrow {}^{3}\Sigma$$

For a diatomic molecule, the change of vibrational quantum number is unrestricted during the electronic transition, that is, $\Delta v = 0, \pm1, \pm2, \ldots,$ etc., where the most important transitions for diagnostics are $\Delta v = \pm1$.

Transitions between rotational states have the following selection rules that are independent of Δv:

$$\Delta J = 0, \pm1, \pm2 \text{ only}$$

Transitions in rotational energy levels are grouped together according to the change in J values. Transitions for $\Delta J = 0$ are termed Q-branch transitions and are only observed if $\Delta v = \pm1$. When the change ΔJ of the rotational quantum number is ±1, two line series are formed, referred to as the R branch and P branch. Thus, the P branch lies in the shorter frequency (longer wavelength) side of Q branch, whereas the R branch resides in the higher frequency (shorter wavelength) side of Q branch. Similarly, if the rotational quantum number changes by ±2, the two line series formed are named the O branch and S branch. Therefore, from low to high frequency, the five line series are O branch ($\Delta J = +2$), P branch ($\Delta J = +1$), Q branch ($\Delta J = 0$), R branch ($\Delta J = -1$), and S branch ($\Delta J = -2$).

3.2.2 Population of Energy Levels

The previous discussion outlined the possible energy states a molecule could attain by rotation, vibration, and electronic excitation. Of primary importance in laser diagnostics is how the molecules in any ensemble are distributed over these states. The distribution is prescribed by the Boltzmann equation:

$$N_j = N \cdot \frac{g_j e^{-E_j/kT}}{\sum_j g_j e^{-E_j/kT}} \tag{3.7}$$

Where N_j is the number density of molecules in the jth state of energy, E_j. N is the total number density, and k is the Boltzmann's constant. g_j is the degeneracy of the state; the degeneracy indicates the number of molecules that can occupy any given energy state. The denominator is termed the partition function and is denoted by Q.

The population of individual energy levels within a rotational manifold of J is given by

$$N_J = \frac{N}{Q_{rot}} g_J (2J+1) \exp\left(-B\,J(J+1)hc\Big/kT\right) \tag{3.8}$$

Where g_J is the nuclear spin degeneracy factor, which is important only for homonuclear diatomics, such as N_2, and has a value of 6 for even J levels and 3 for odd J levels for N_2. The rotational partition function Q_{rot} is described as

$$Q_{rot} = \sum_{J=0}^{\infty} (2J+1) \exp\left[-hcBJ(J+1)/kT\right] \tag{3.9}$$

In the case of large kT/hcB, the summation can be replaced by integration, and one obtains

$$Q_{rot} = \frac{kT}{hcB} \tag{3.10}$$

The fractional population of an individual vibrational level is given by

$$N_v = N \exp\left(-hc\,\mathrm{v}\,\nu_{vib}\Big/kT\right)\left[1 - \exp\left(-hc\,\nu_{vib}\Big/kT\right)\right] \tag{3.11}$$

Where the last term is the vibrational partition function, $Q_{v'}$ that is,

$$Q_v = \left[1 - \exp\left({-hc\, \nu_{vib}}\middle/{kT} \right) \right] \qquad (3.12)$$

The partition of population among the rotational levels within a given vibrational state is therefore given by Eq. 3.8 with N replaced by $N_{v'}$ that is,

$$N_J = \frac{N_v}{Q_{rot}} g_J (2J+1) \exp\left({-B_v J(J+1)hc}\middle/{kT} \right) \qquad (3.13)$$

The essential feature of population factors is that they are a function of temperature. It is, in fact, precisely this feature that enables the measurement of temperature by inelastic scattering, such as Raman scattering and LIF. Population factors will be discussed in more detail in other chapters.

3.3 Raman and Rayleigh Scattering

When a laser beam is passed through a transparent gas mixture, a small amount of the radiation energy is scattered. The scattered energy will consist mostly of radiation of the incident wavelength, referred to as *Rayleigh scattering*, and a very small amount of light at certain discrete wavelengths above and below that of the incident beam, called *Raman scattering*.

The occurrence of Rayleigh and Raman scattering may be most easily understood in terms of the quantum theory of radiation. This treats radiation of frequency v as consisting of a stream of particles, the so-called photons, having energy hv. Photons can be imagined to undergo collisions with molecules, and if the collision is perfectly elastic, they will be deflected unchanged, which accounts for the process of Rayleigh scattering. In this case, a detector placed to collect energy at an angle to the incident beam will thus receive photons of energy hv, that is, radiation of frequency v.

However, it may also happen that energy is exchanged between the photon and the molecule during the collision, called the inelastic process or Raman scattering. The molecule may gain or lose some energy in accordance with the quanta laws; thus the energy exchange, ΔE, must be the difference in energy between two of its allowed states. The photon will be scattered with energy $hv\pm \Delta E$, and the equivalent radiation will have frequency $v\pm \Delta E/h$.

3.3.1 Rayleigh Scattering and Its Measurement

The fundamental principle governing Rayleigh scattering is the elastic collisions between gas molecules and incident laser light. The scattered light has the same wavelength as the incident light. The scattered light signal, in general, is directly proportional to laser power, gas density, and a differential cross section depending on the gas that is being measured. If the scattered light from a gas mixture is detected, the above relationship about the scattered light signal can be written as

$$P = \eta_c I_i \Omega V_c N \left(\frac{\partial \sigma}{\partial \Omega} \right) \tag{3.14}$$

where

η_c = the detection efficiency of the collection system

I_i = the incident laser intensity, (W/cm^2)

N = the total number density of the gas molecule, (cm^{-3})

Ω = the solid angle of collection optics (sr)

V_c = the collection volume, (cm^3)

$(\partial\sigma/\partial\Omega)$ = the differential cross section, (cm^2/sr) given by

$$\left(\frac{\partial \sigma}{\partial \Omega} \right) = \frac{3}{8\pi} \sigma \, (\cos^2\phi \cos^2\theta + \sin^2\phi) \tag{3.15}$$

Where φ is the polarization angle of the incident light relative to the scattering plane, defined by the direction of propagation of the incident light and the scattering direction, and θ is the scattering angle, both of which are illustrated in Fig. 3.3.

The term σ is the Rayleigh scattering cross section, given by

$$\sigma = \frac{8\pi}{3} \left[\frac{\pi^2(n^2-1)^2}{N_{STP}^2 \, \lambda^4} \right] \tag{3.16}$$

Where n is the real part of the refractive index of a gas at standard temperature and pressure (STP), N_{STP} is the molecular number density for an ideal gas at STP and has a value of 2.687×10^{19} cm^{-3}, and λ is the wavelength in cm.

In practice, Rayleigh scattering is normally detected at 90°, where the scattering volume is minimum, as defined by the intersection of the laser beam and the

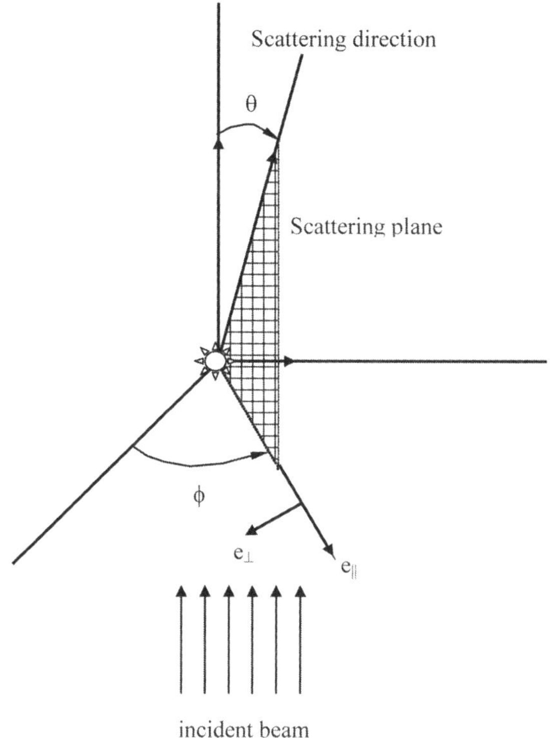

e⊥: vertical polarisation; e∥: horizontal polarisation

Fig. 3.3 Definition of the scattering plane and directions of polarization.

image of the aperture placed in front of the detector. From Eq. 3.15, note that the incident laser beam should be vertically polarized (polarization normal to the scattering plane) for Rayleigh scattering to be detected at a right angle. In this case, the Rayleigh scattered light signal for a mixture of gases is given by

$$P_{\perp} = \eta_c I_i \, \Omega \, V_c \, N \sum_{i=1}^{j} \chi_i \, \sigma_i \tag{3.17}$$

Where $\Sigma \chi_i \sigma_i$ represents the effective Rayleigh cross section for the mixture of gases, j is the number of species, and χ_i is the mole fraction of species i in the mixture and follows $\sum_{i=1}^{j} \chi_i = 1$.

The total molecular number density N is related to the pressure, temperature, and Avogadro's number, $A_o = 6.023 \times 10^{23}$ molecules/mol, by

$$N = \frac{P\,A_o}{R\,T} \tag{3.18}$$

For a given optical detection system, η_c, Ω, and V_c are constants, and the variations in the Rayleigh scattered light signal are a result of the mixture number density (i.e., the variation of temperature or pressure, or both) or the species variations, or even both. Hence, an unambiguous interpretation of the Rayleigh scattered light signal requires that experiments be designed for certain special situations to supply useful information on the interesting variables. This has limited the wider application of laser Rayleigh scattering (LRS) to practical combustion systems, such as internal combustion (IC) engines.

For a binary mixture of fuel vapor and air, Eq. 3.17 becomes

$$P_{fa} = \eta\,I_i\,N[\chi_f\,\sigma_f + \chi_a\,\sigma_a] \tag{3.19}$$

Since $\chi_f + \chi_a = 1$, it can be written as

$$P_{fa} = \eta I_i N[\chi_f\,\sigma_f + (1-\chi_f)\sigma_a] \tag{3.20}$$

Where η is a constant that account for the values of η_c, Ω, and V_c.

A calibration experiment can be performed at a standard condition of pressure p_o and temperature T_o. The Rayleigh scattering signal from a known fuel mole fraction χ_{fo} in the binary mixture is P_{fo}, and the Rayleigh scattering signal from air only is P_{ao}, that is,

$$P_{fo} = \eta\,I_i\,N_o\,[(1-\chi_{fo})\,\sigma_a + \chi_{fo}\sigma_f] \tag{3.21}$$

$$P_{ao} = \eta\,I_i\,N_o\,\sigma_a \tag{3.22}$$

From above equations, the mole fraction of fuel vapor in a binary mixture can be derived as follows:

$$\chi_f = \chi_{fo}\,\frac{P_{fa}\frac{N_o}{N} - P_{ao}}{P_{fo} - P_{ao}} \tag{3.23}$$

Or

$$\chi_f = \chi_{fo} \frac{P_{fa} \frac{p_o}{p} \frac{T}{T_o} - P_{ao}}{P_{fo} - P_{ao}}$$

(3.24)

It is clear from Eq. 3.24 that the mole fraction of fuel χ_f can be obtained from the Rayleigh scattering signal P_{fa} if calibration signals P_{fo} and P_{ao} are measured in advance. It is also necessary to know the temperature and pressure to derive the mole fraction of fuel from the Rayleigh signal. Generally, the pressure can be obtained with the use of a pressure transducer without any difficulty. However, measuring the temperature itself is a daunting task, which is one of the limitations of further applications of the Rayleigh scattering to IC engines.

3.3.2 Raman Scattering Spectrum

When energy is exchanged between the photon of laser beam and molecule during the collision, a Raman scattering process takes place and the photon will be scattered with energy $hv \pm \Delta E$ and the equivalent radiation will have frequency $v \pm \Delta E/h$. Radiation scattered with a frequency lower than that of the incident beam is referred to as Stokes radiation, while that at higher frequency is called anti-Stokes. Since the former is accompanied by an increase in molecular energy, which can always occur subject to certain selection rules, whereas the latter involves a decrease in molecular energy, which can only occur if the molecule is originally in an excited state, Stokes radiation is generally more intense than anti-Stokes. Overall, however, the total radiation scattered at any but the incident frequency is extremely small, that is, Raman scattering is many orders of magnitude weaker than Rayleigh scattering.

Raman spectra depend on the molecular structure of the molecules involved. For simplicity, only diatomic molecules will be considered to illustrate the principle features of Raman spectra. A Raman spectrum of a diatomic molecule may involve changes in either or both of rotational and vibrational energy levels, but not electrical energy levels. A pure rotational Raman spectrum exists when only the rotational energy of a molecule is affected. However, when the molecule changes its state from one vibrational level to another, both vibrational and rotational energies are involved.

3.3.2.1 Pure Rotational Raman Spectrum

In Raman spectroscopy, the precision of measurements does not normally warrant the retention of the term involving D in Eq. 3.3, the centrifugal distortion constant. Thus, the rotational energy is given by Eq. 3.2: $E_r = hcBJ(J + 1)$.

Following the usual practice, we define ΔJ as $(J_{\text{final state}} - J_{\text{initial state}})$. Transitions between different rotational energy levels follow the formal selection rule for a diatomic molecule: $\Delta J = 0$, or ± 2 only. The transition $\Delta J = 0$ is trivial since this represents no change in the molecular energy, and hence Rayleigh scattering if no vibrational energy change is involved. For $\Delta J = +2$ (the Stokes transition), the change of energy level is given by

$$\Delta \nu_r = \frac{\Delta E_r}{hc} = B\left(4J + 6\right)\ \text{cm}^{-1} \tag{3.25}$$

Which happens when molecule gains rotational energy from the photon during collision. J is the rotational quantum number in the lower state and takes values of $0, 1, 2, \ldots$, etc. This forms a series of lines (S branch) to the lower frequency side of the exciting line (Fig. 3.4):

S branch: $\qquad \nu_{r} = \nu_0 - B(4J + 6), J = 0, 1, 2, \ldots$

Similarly, a series of lines (O branch) will appear to the higher frequency side of the exciting line, the anti-Stokes transition, when $\Delta J = -2$ or molecule losses energy to the photon:

O branch: $\qquad \nu_{r} = \nu_0 + B(4J - 2), J = 2, 3, 4, \ldots$

Note that the first Stokes and anti-Stokes lines are separated from the incident light frequency by $6B$ and that successive lines are separated from each other by $4B$, as shown in Fig. 3.4.

The intensity of pure rotational Raman scattering is given by

$$I \propto b_{J\pm 2,J}\ \gamma_0^2 \tag{3.26}$$

Where γ_0 is the anisotropy of the equilibrium polarizability tensor. The $b_{J',J''}$ are the so-called Placzek-Teller coefficients, given by

$$b_{J+2,J} = \frac{3\left(J+1\right)\left(J+2\right)}{2\left(2J+1\right)\left(2J+3\right)} \tag{3.27}$$

$$b_{J-2,J} = \frac{3J\left(J-1\right)}{2\left(2J+1\right)\left(2J-1\right)} \tag{3.28}$$

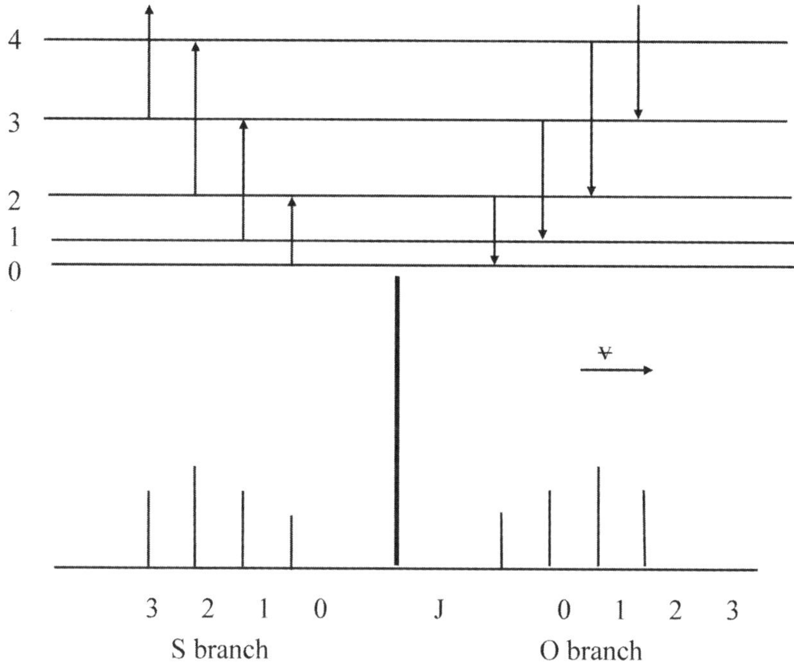

Fig. 3.4 The rotational energy levels of a diatomic molecule and the rotational Raman spectrum arising from transitions between them.

$$b_{J,J} = \frac{J(J+1)}{(2J-1)(2J+3)} \tag{3.29}$$

The last coefficient is included for the vibrational-rational Raman scattering.

3.3.2.2 Vibrational Raman Spectra

The structure of vibrational Raman spectra is easily discussed. For every vibrational mode, we can write an expression of the form Eq. 3.4:

$$E_v = hc\left[\omega_e\left(v+\frac{1}{2}\right) - \omega_e x_e\left(v+\frac{1}{2}\right)^2 + \omega_e y_e\left(v+\frac{1}{2}\right)^3 + \ldots\right](v = 0, 1, 2, \ldots) \tag{3.30}$$

Where ω_e is the equilibrium vibrational frequency expressed in wavenumbers in cm^{-1} and x_e is the anharmonicity constant. Such an expression is very general, whatever the shape of the molecule or the nature of the vibration. For a diatomic molecule, the third and higher terms can be ignored.

Quite general, too, is the selection rule: $\Delta v = 0, \pm 1, \pm 2, \ldots$, etc.

Among the possible transitions between the various vibrational energy levels, the ones from $v_{upper} = 0$ to $v_{lower} = 1$, and vice versa (fundamentals), are the strongest. The probability of $\Delta v = \pm 2, \pm 3, \ldots$, decreases rapidly.

For the fundamentals of vibrational Raman scattering ($\Delta v = \pm 1$), the spectral positions are therefore given as

$$\nu = \frac{E_v}{hc} = \nu_0 \pm \left[\omega_e - 2\omega_e x_e \left(v + 1 \right) \right] \tag{3.31}$$

Where the minus sign represents the Stokes line, and the plus sign refers to the anti-Stokes lines. The vibrational number v in Eq. 3.31 is that of the initial vibrational energy state. The anti-Stokes lines are normally much weaker than the Stokes lines at low temperatures, at which very few molecules exist in the excited state $v = 1$.

When a photon collides with a molecule, both the rotational and vibrational energies of the molecule will be affected. Therefore, the vibrational-rotational Raman scattering will take place, where the change in vibrational energy is accompanied with changes in rotational energy. For a diatomic molecule, Q-branch transitions ($\Delta J = 0$) are much stronger than the O and S branch transitions. Furthermore, since the O and S branches are spectrally quite separated from the Q-branch lines, they can be neglected in the limited spectral regions about the Q-branch. Ignoring centrifugal distortion but accounting for vibration-rotation interaction, the fundamental ($\Delta v = \pm 1$) Q branch ($\Delta J = 0$) vibrational-rotational spectral lines can be detected by Eq. 3.32:

$$\nu = \frac{E_v}{hc} = \nu_0 \pm \left[\omega_e - 2\omega_e x_e \left(v + 1 \right) - \alpha_e J \left(J + 1 \right) \right] \tag{3.32}$$

Note that v and J are the vibrational and rotational quantum numbers of the initial energy states.

For diagnostic purposes, Eq. 3.32 can be simplified to

$$\nu = \nu_0 \pm \nu_{vib} \tag{3.33}$$

The second term in Eq. 3.32 is replaced by the vibrational frequency, υ_{vib}, in cm^{-1}, which is species specific. Table 3.1 lists the vibrational frequencies (cm^{-1}) for

Table 3.1 Characteristic Raman Signal Wavelength and Cross Sections of Some Species at Four Laser Excitation Wavelengths

Species	Vibrational Frequency (cm^{-1})	Spectral Position of Q-Branch Stokes Line (nm)				Vibrational Cross Sections $(\times 10^{-30}\ cm^2/sr)$			
(Excitation wavelength)		248 nm	308 nm	355 nm	532 nm	248 nm	308 nm	355 nm	532 nm
N_2	2331	263.2	331.8	387.0	607.3	13.0	5.2	2.79	0.46
O_2	1556	257.9	323.5	375.8	580.0	16.6	6.7	3.69	0.65
CO_2	1388	256.8	321.7	373.4	574.4	15.1	6.1	3.38	0.60
	1285	256.1	320.7	371.9	571.0	11.1	4.5	2.49	0.45
CO	2145	261.9	329.8	384.3	600.5	13.3	5.3	2.88	0.48
H_2O	3657	272.7	347.1	408.0	660.5	30.8	11.7	6.15	0.89
CH_4	2915	267.3	338.4	396.0	629.6	80.2	31.2	16.6	2.61
	3017	268.0	339.5	397.6	633.7	53.7	20.9	11.1	1.72
C_3H_8	2890	267.1	338.1	395.6	628.6	88.1	34.3	18.3	2.87
	1465	257.3	322.5	374.5	577.0	35.3	14.3	7.9	1.4
	857	253.5	316.3	366.1	557.4	20.8	8.5	4.8	0.89
C_8H_{18}	2939	267.5	338.6	396.3	630.6				
	1452	257.5	322.4	374.2	576.5				
	750	252.7	315.3	364.7	554.1				

molecules of interest to IC engines. The excitation wave-number or frequency υ_0 represents the incident light frequency in cm^{-1}. Equation 3.32 defines the spectral positions of the Q-branched vibrational Raman spectrum of a diatomic molecule. The plus and minus signs represent the anti-Stokes lines (frequency upshifted) and Stokes lines (frequency downshifted), respectively.

In summary, for a diatomic molecule at the ground electronic state (Σ state), the selection rules for Raman scattering are given by $\Delta v = 0, \pm 1$ and $\Delta J = 0, \pm 2$. Figure 3.5 illustrates the rotational-vibrational Raman spectrum of a diatomic molecule. The Raman spectrum consists of a Stokes ($\Delta v = +1$) and an anti-Stokes ($\Delta v = -1$) component, each with three branches: S ($\Delta J = +2$), Q ($\Delta J = 0$), and O ($\Delta J = -2$). Since the rotational lines in the Q branch are stronger and not normally resolved, this branch is usually two orders of magnitude more intense than the S and O branches and is thus the most characteristics feature of the Stokes and anti-Stokes components of the vibrational-rotational Raman spectrum.

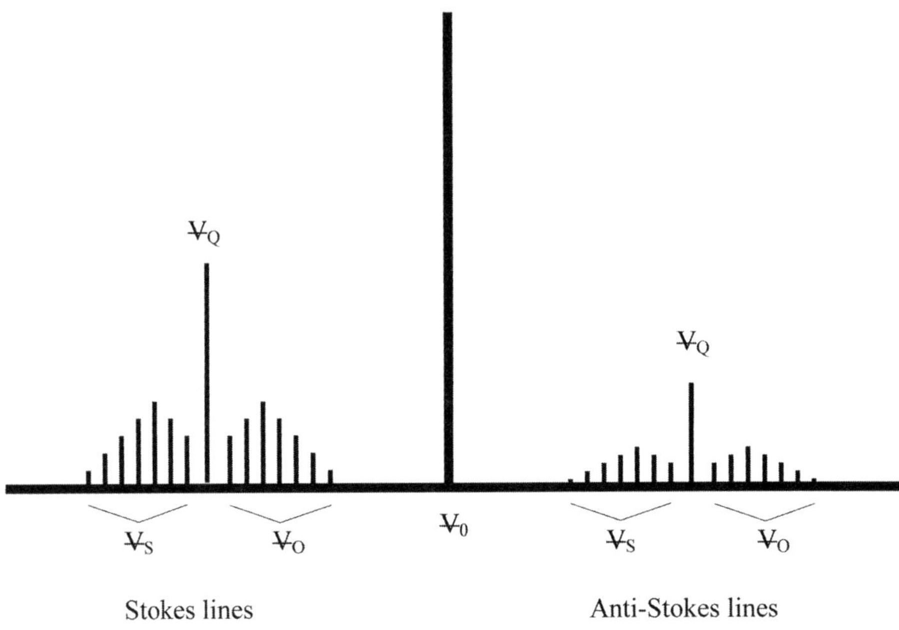

Fig. 3.5 The rotational-vibrational Raman spectrum of a diatomic molecule.

3.3.3 Measurement of Raman Scattering

3.3.3.1 The Intensity of Raman Scattering

The Raman signal P from molecular species i and scattered into a solid angle Ω is usually expressed in terms of a differential scattering cross section $(\partial\sigma/\partial\Omega)$:

$$P = \eta_c I_i \Omega V_c N \left(\frac{\partial\sigma}{\partial\Omega} \right) \tag{3.34}$$

where

η_c = the detection efficiency of the collection system

I_i = the incident laser intensity, (W/cm^2)

N = the number density of the Raman scattering gas molecule, (cm^{-3})

Ω = the solid angle of collection optics (sr)

V_c = the collection volume (cm^3)

The Stokes and anti-Stokes fundamental ($\Delta v = \pm 1$) vibrational Raman scattering cross sections for a diatomic molecule are given by (Long, 1977):

$$\left(\frac{\partial \sigma}{\partial \Omega}\right)_s = \frac{h(\bar{v}_0 - \bar{v}_{vib})^4 \, f(\alpha'^2, \gamma'^2, \rho)}{8mc^4 \bar{v}_{vib}\left[1 - \exp\left(-hc\bar{v}_{vib}\Big/kT\right)\right]} \tag{3.35}$$

$$\left(\frac{\partial \sigma}{\partial \Omega}\right)_{as} = \frac{h(\bar{v}_0 + \bar{v}_{vib})^4 \, f(\alpha'^2, \gamma'^2, \rho)}{8mc^4 \bar{v}_{vib}\left[1 - \exp\left(-hc\bar{v}_{vib}\Big/kT\right)\right]} \tag{3.36}$$

where

h = Planck's constant (6.6256×10^{-34} J s)

c = the speed of light in vacuum (2.998×10^8 m/s)

k = Boltzmann's constant 91.38×10^{-23} J/ K)

T = temperature (K)

$f(\alpha'^2, \gamma'^2, \rho)$ is a function of the mean (α') and anisotropy invariants (γ') of the derived polarizability tensor (Long, 1977), and the polarization state of the incident laser and scattered light ρ. For vertically polarized incident light, the values of $f(\alpha'^2, \gamma'^2, \rho)$ for the vertical, parallel, and total scattered light are given as

$$f\left(\alpha'^2, \gamma'^2, {}^\perp\rho_\perp\right) = \frac{45(\alpha')^2 + 4(\gamma')^2}{45} \tag{3.37}$$

$$f\left(\alpha'^2, \gamma'^2, {}^\perp\rho_{\mathrm{II}}\right) = \frac{(\gamma')^2}{15} \tag{3.38}$$

$$f\left(\alpha'^2, \gamma'^2, {}^\perp\rho_{\perp,\mathrm{II}}\right) = \frac{45(\alpha')^2 + 7(\gamma')^2}{45} \tag{3.39}$$

The exponential terms in Eqs. 3.35 and 3.36 indicate that the higher the gas temperature is, the larger the Raman scattering cross section becomes as the higher energy vibrational states become more populated. For polyatomic molecules, more complex expressions for the scattering cross section arise, but the conclusions drawn from these two equations regarding the temperature dependence of the cross section are still valid. It is also worth noting from the above $f(\alpha'^2, \gamma'^2, \rho)$ values that the Raman scattering intensity is independent of the scattering angle.

Table 3.1 lists room temperature vibrational Raman cross sections for species of interest in IC engines. The cross sections were taken from the tabulation by Inaba and Kobayasi (1972) for an N_2 laser at 337 nm, except those of propane (C_3H_8), which were obtained from Schrotter and Klockner (1979). The values at 248 nm and 308 nm (Excimer lasers), 355 nm and 532 nm (Nd:YAG lasers) are calculated from the original values by scaling according to $(\upsilon_0 - \upsilon_{vib})^4$. The scattering cross sections listed in Table 3.1 are obtained for the linear polarized incident and scattered light perpendicular to the scattering plane as defined in Fig. 3.3. These cross sections are valid and unchanged for any scattering angle between the incident beam and scattering observation axes.

When a photon interacts with a molecule, energy exchange may occur between both the rotational as well as vibrational energy levels and the photon, as discussed in the previous paragraph. This leads to rotational sidebands about the vibrational frequencies in the spectrum of scattered light, as shown in Fig. 3.5. The shape of the sidebands is affected by changes in temperature. Because Raman intensity is normally determined by integrating over only a portion of the sidebands, this change in shape by temperature can influence the signal levels. In addition, the Raman scattering cross section is affected by temperature as shown by Eq. 3.35 or 3.36. These temperature effects on the Raman signal can be modeled by a bandwidth factor $f(T)$, accounting for the fraction of the total Raman scattering actually detected. In general, Eq. 3.34 can be written as

$$P = \eta I_i \, Nf(T) \tag{3.40}$$

Where η is a factor dependent on the Raman cross section, wavelength, geometry, and optical collection efficiency, and is usually determined by an experimental calibration. Equation 5.48 shows that the Raman signal is directly related to the concentration of Raman active species N when Raman scattering is detected at the species Raman frequency defined by Eq. 3.32.

The bandwidth factor $f(T)$ depends on the spectral location, shape and bandwidth of the detection system, and laser line spectral profile. For example, it was shown by Eckbreth (1996) that for N_2 Raman scattering measurement, with a detection spectral bandwidth about 5 nm, the requirement of knowing temperature can be eliminated to obtain quantitative Raman scattering measurements.

In certain cases, the temperature effect may be ignored if the gas temperature is much lower than the characteristic vibrational temperature of the species to be measured, given by $T_v = (hc\omega_e/k)$. The grouping hc/k has a value of 1.44 K×cm^{-1}.

3.3.3.2 Measurement of Major Species Concentration by Raman Scattering

The foregoing analysis of Raman spectra and Raman scattering signal provides the fundamental knowledge and formulation upon which Raman species concentration and temperature measurements are based.

In general, the species concentration is measured from the Raman scattered signal in some spectrally integrated portion of the Raman spectrum as defined in Eq. 3.32. After determining η by calibration from Raman scattering in a known sample by independent measurements, the species concentration N can be calculated from the measured Raman scattering from Eq. 3.40. The application of spontaneous Raman scattering for in-cylinder species measurements will be covered in Chapter 6.

3.3.3.3 Gas Temperature Measurement by Raman Scattering

Temperature measurements by spontaneous Raman scattering can be achieved via two main approaches. The first involves least-squares fitting a library of theoretical spectra to the measured vibrational-rotational spectrum. The theoretical spectrum producing the best fit identifies the temperature at the measurement point. Alternative approaches to temperature measurements employ spectral band ratio strategies, where various portions of the Raman spectrum can be monitored and the temperature determined from the resultant intensity ratio. This can be performed on ratio of the first hot band (Δv: $1 - 2$) to ground state band (Δv: $0 - 1$) in the Stokes spectrum.

A common technique is to ratio the Stokes to anti-Stokes vibrational scattering,

$$\frac{P_{r\,s}}{P_{r\,as}} = \frac{(v_0 - v_{vib})^4}{(v_0 + v_{vib})^4} \exp(-hc\Delta v / kT) \frac{f^{s}(T)}{f^{as}(T)} \tag{3.41}$$

Where $\Delta \tilde{v}$ is the energy difference between the ground and first excited vibrational levels ($\Delta v = 1$) for the fundamental Q branch ($\Delta J = 0$). The Stokes/anti-Stokes method provides the maximum collected signal because of a wider spectral band-pass. However, the ratio becomes moderately temperature sensitive only above 700 K because of the exponential dependence of the excited state population of the anti-Stokes signal with temperature.

In addition to the spontaneous Raman scattering, the nonlinear coherent anti-Stokes Raman scattering (CARS) has proved to be a more robust and accurate gas temperature measurement technique. Detailed discussion on the application of Raman scattering techniques for in-cylinder gas temperature measurements will be given in Chapter 9.

3.4 Laser-Induced Fluorescence

Laser-induced fluorescence (LIF) is the emission of light from an atom or molecule following the excitation by a laser beam. In a LIF experiment, the molecule is initially at a lower electronic state before it is excited to an upper electronic energy level by a laser source. The wavelength of the laser light needs to be chosen so that it coincides with the absorption wavelength.

Typically, the lower state is in the ground electronic level, as the other electronic levels are normally negligibly populated at combustion temperature.

There are five important processes following the excitation of a molecule by a laser source, as shown in Fig. 3.6. First, the molecule can be returned to its original quantum state by laser-induced stimulated emission, denoted as $B_{21}I_v$. Second, absorption of an additional photon can excite the molecule to higher states, even to ionized levels, shown as Q_{ion} in Fig. 3.6. Third, the internal energy of the system can be altered in inelastic collisions with other molecules, producing rotational and vibrational energy transfers, represented by $Q_{rot,\,vib}$ in Fig. 3.7. In many cases, the inelastic collisions with other molecules also result in electronic energy transfer Q_{elec}, which is often referred to as quenching. Fourth, interactions between the individual atoms of the molecule produce internal energy transfer and dissociation of molecule. The dissociation in this case is called predissociation because it is produced by a change from a stable to a repulsive electronic repulsive (unstable) state. Finally, the originally populated state and nearby states indirectly populated through collisions return to lower states through the emission of light, producing the LIF.

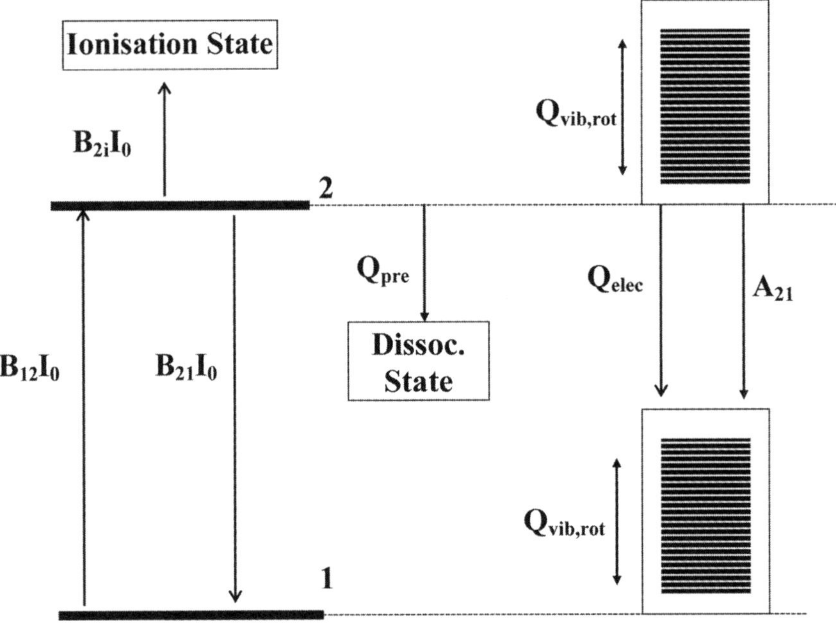

Fig. 3.6 Main energy transfer processes in LIF.

Notes: A_{21} = spontaneous emission from the upper to lower electronic states; $B_{21}I_0$ = stimulated emission; $B_{12}I_0$ = stimulated absorption; Q_{ion} = photo-ionization; Q_{pre} = predissociation; and Q_{elec} = electronic collisional quenching; $Q_{vib,rot}$ = vibrational and rotational collisional quenching.

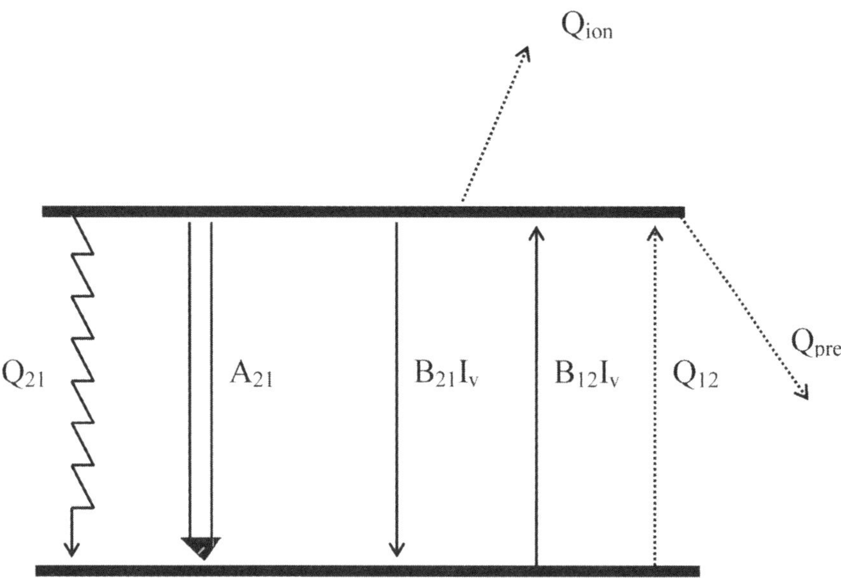

Fig. 3.7 Simple two-energy-level diagram for LIF modeling.

The LIF signal can be related to specific properties, for example, species concentration, of the absorbing species through modeling of these state-to-state transfer processes. Because fluorescence is a function of the upper state's population, this modeling requires solving the state-dependent population dynamics. Most treatments of LIF are based on a semiclassical rate equation analysis developed by Piepmeier (1972). It is adequate for most of LIF measurements that are applicable to engine combustion.

A two-energy-level model is often used as a first approximation in actual data analysis. Such a model is appropriate to LIF detection of some molecular systems with fully equilibrated or fully frozen rotational level manifolds. It also serves to illustrate the essential properties of LIF. Figure 3.7 depicts such a two-energy-level model. For this two-energy-level model, the rate equations for the lower and upper energy level become

$$\frac{dN_1}{dt} = -N_1\left(Q_{12} + B_{12}I_v\right) + N_2\left(Q_{21} + A_{21} + B_{21}I_v\right) \tag{3.42}$$

$$\frac{dN_2}{dt} = N_1\left(Q_{12} + B_{12}I_v\right) - N_2\left(Q_{21} + B_{21}I_v + A_{21} + Q_{ion} + Q_{pre}\right) \tag{3.43}$$

Where B_{12}, B_{21} are the Einstein coefficients for stimulated absorption and emission, A_{21} is the Einstein coefficient for spontaneous emission, and I_v (W cm^{-2} Hz^{-1}) is the laser spectral intensity. The two energy levels are normally separated by a few electron-volts, thus collisional excitation Q_{12} can be omitted in most combustion measurements. Most excited states are not predissociative (Q_{pre}), unless specifically chosen, and photo-ionization (Q_{ion}) can often be ignored. Equations 3.42 and 3.43 become

$$dN_1\big/dt = -N_1 B_{12} I_v + N_2 \left(Q_{21} + A_{21} + B_{21} I_v \right) \tag{3.44}$$

$$dN_2\big/dt = N_1 B_{12} I_v - N_2 \left(Q_{21} + A_{21} + B_{21} I_v \right) \tag{3.45}$$

As the upper level typically has a negligible population prior to the laser excitation, so the initial condition, $N_2\big|_{t=0} = 0$, is applied to Eq. 3.45. In addition, the total population of the molecule should be conserved, that is, no chemical reactions occur during the measurement

$$N_1 = N_1 + N_2 = constant = N_1^0 \tag{3.46}$$

Where N_1^0 is the total population prior to laser excitation. The solution to the two-level system is given by

$$N_2 = N_1^0 B_{12} I_v \, \tau \left(1 - e^{-t\big/\tau} \right) \tag{3.47}$$

Where the time constant τ is $(Q_{21} + A_{21} + B_{21} I_v + B_{12} I_v)^{-1}$. For laser pulses long compared to τ, the system reaches the steady-state value of

$$N_2 = N_1^0 B_{12} I_v \, \tau \tag{3.48}$$

Or more conveniently be written as

$$N_2 = B_{12} \, I_v \, \frac{A_{21}}{A_{21} + Q_{21}} \, \frac{1}{1 + I_v\big/I_v^{sat}} \tag{3.49}$$

Where the saturation intensity I_v^{sat} is defined as

$$I_v^{sat} = \frac{A_{21} + Q_{21}}{B_{21} + B_{12}}$$ (3.50)

Assuming the fluorescence is emitted equally into 4π steradians, the total number of photons N_p striking on to a photodetector from a collection volume V_c is given by

$$N_p = \eta \frac{\Omega}{4\pi} N_1^0 \ V_c \ B_{12} \ E_v \frac{A_{21}}{A_{21} + Q_{21}} \frac{1}{1 + I_v/I_v^{sat}}$$ (3.51)

where

ε = transmission efficiency of the collection optics

Ω = the collection solid angle (sr)

E_v = the spectral fluence of the laser (J cm⁻² Hz⁻¹)

V_c = the sampling volume, defined by the cross-sectional area of the laser beam A and the axial extent along the beam l from which the fluorescence is detected

For a particular experimental setup, the fluorescence signal can be expressed as

$$P_{flu} = \eta_c \ \Omega V_c f_1(T) x_m \ N \ I_v \frac{A_{21}}{A_{21} + Q_{21}} \frac{B_{12}}{1 + I_v/I_v^{sat}}$$ (3.52)

where

$f_1(T)$ = the fractional population of the lower electronic state

x_m = the mole fraction of the absorbing species

N = the total gas number density (cm⁻³)

I_v = the spectral power of incident laser light (W cm⁻² Hz⁻¹)

$f_1(T)$ represents the fraction of molecules of the absorbing species which are in the specific electronic-vibrational-rotational energy level excited by the laser and is given by the Boltzmann distribution. Therefore, the grouping $[f_1(T)\chi_m N]$ represents the number density of the absorbing species at ground state, N^0_1.

Equation 3.52 can be further simplified for two limiting cases, corresponding to either low or high laser intensity. At low laser intensity, that is, $I_v << I_v^{sat}$, it simplifies to

$$P_{flu} = \eta_c \, \Omega V_c \, f_1(T) \chi_m \, N \, I_v \, B_{12} \frac{A_{21}}{A_{21} + Q_{21}} \tag{3.53}$$

This is the so-called linear fluorescence equation, as the fluorescence is linearly proportional to laser intensity. The term $A_{21}/(A_{21} + Q_{21})$ is termed the *fluorescence quantum yield* and, since $A_{21} << Q_{21}$ at near or above atmospheric pressure, the fluorescence quantum yield is generally much smaller than unity, significantly reducing species detectivity.

In the linear regime, one normally needs to evaluate the quenching rate constant to make quantitative measurements of the species of interest. As the quenching rate is sensitive to pressure, temperature, and composition, it is very difficult to make quantitative LIF measurements in atmospheric and higher-pressure combustion environment.

If the laser intensity is increased until $I_v >> I_v^{sat}$, Eq. 3.52 then becomes

$$P_{flu} = \eta_c V_c \, f_1(T) \chi_m \, N \, B_{12} \frac{A_{21}}{B_{12} + B_{21}} \tag{3.54}$$

and the fluorescence is said to be in the saturation limit. In the saturation regime, the fluorescence signal is independent of both the laser irradiance and the quenching obviating the need to measure or evaluate either one. However, complete saturation is not easily achieved due to either the specific wavelength region of the absorption or the magnitude of the saturation intensity. Saturation is not achieved at the outer edges of a laser beam. In addition, it is not possible to saturate during the entire duration of the laser pulse because of the temporal variation of the laser pulse. It is much more difficult to reach the saturation regime when planar laser-induced fluorescence (PLIF) is used, as is the case for most LIF applications in IC engines.

To perform quantitative fluorescence measurements on a given species, there are several requirements that must be satisfied. The first of these is that the molecule must have a known absorption and emission spectrum. Second, the molecule must have an absorption wavelength accessible to a laser source. Third, the rate of radiative decay of the excited state must be known, which affects the fluorescence power directly. Fourth, excited state losses due to collisions, photo-ionization and predissociation have to be taken into account. The excited state loss due to collisions, known as quenching, may be increased considerably in the presence of other molecules especially at high pressures as encountered in IC engine applications.

3.5 Comparison of LRS, SRS, and LIF

In the previous sections, the fundamental principles of major laser-based diagnostics are described, including laser Rayleigh scattering (LRS), spontaneous Raman scattering (SRS), and laser-induced fluorescence (LIF). They can be compared on an energy-level diagram as shown in Fig. 3.8 in which ground and first excited electronic states are shown with their vibrational and rotational energy levels. Letters v and J represent the vibrational and rotational quantum numbers, respectively. The Raman processes shown are for a diatomic molecule, to which the transition rules $\Delta v = \pm 1$ and $\Delta J = 0, \pm 2$ apply.

As shown in Fig. 3.8, Rayleigh or Raman scattering can occur upon excitation at any visible or ultraviolet wavelength. Fluorescence, however, requires a precise excitation frequency, since molecules in a particular rovibronic energy level of the ground electronic state must be excited to a specific rovibronic energy level in an excited electronic state.

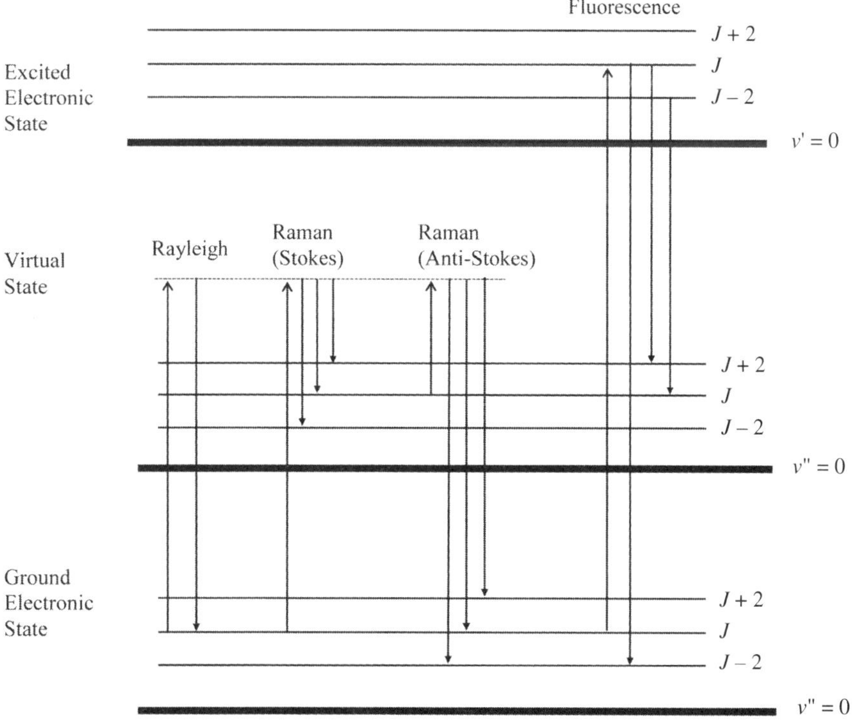

Fig. 3.8 Schematic of the energy-level diagram showing Rayleigh scattering, vibrational Raman scattering, and LIF.

The signal for the three processes can be collated from

$$P = \eta \Omega V_c \left(\frac{d\sigma}{d\Omega} \right) N_{vJ} \, I_i \qquad (3.55)$$

where

P = the power of the measured signal (W)

η = the detection efficiency

Ω = the solid angle of the collection optics (sr)

V_c = the collection volume (cm^3)

$(d\sigma/d\Omega)$ = the differential cross section (cm^2/sr)

N_{vJ} = the number density in the initial rovibronic level ($V,'' \, J''$) of the ground electronic state (cm^{-3})

I_i = the irradiance of the laser beam (W/cm^2)

Eq. 3.55 indicates that for the same experimental conditions, the relative signal for each method depends mainly on the differential cross section. Approximate differential cross section for each scattering method at ambient pressure and temperature are as follows:

Vibrational Raman (N_2)	$\sim 10^{-29}$
Rayleigh Scattering (N_2)	$\sim 10^{-27}$
Fluorescence (tracer molecule)	$\sim 10^{-25}$–10^{-20}
Mie scattering (particles)	10^{-27}–10^{-8}

The principal advantage of Rayleigh scattering is that it is the strongest of the molecular light scattering techniques and well suited as a density measurement technique at low gas density. When compared with Raman scattering, the scattering cross sections for Rayleigh scattering are typically three orders of magnitude larger than the corresponding vibrational Raman cross sections. The simplicity and ease of implementing a Rayleigh scattering system make it very convenient to set up the system and interpret the data. Rayleigh scattering is an elastic scattering, so it is not species specific. Because the Rayleigh signal has the same wavelength as the incident laser, it suffers seriously from Mie scattering of particles and spurious scattered laser light from the surrounding surfaces. As Mie scattering is 10–20 orders of magnitude stronger than

Rayleigh scattering, the environment must be virtually free of particles, (e.g., dust in air, liquid droplets) so that the Rayleigh signal can be detected.

Spectral measurement of Raman scattered light serves to identify the molecular species, while the intensity of Raman scattered light depends on the number of molecules that are thermally populated at particular rovibronic energy levels. Since Raman peaks are spectrally separated from the incident light, spurious scattered light can be discriminated. Raman scattering has the advantage that the interaction time is very short so that quenching effects are not normally encountered. Raman scattering occurs with any laser excitation wavelength. It is capable of providing simultaneous multiple-species concentration measurements. Because SRS does not involve electronic state transitions, it can measure the concentration of stable species, such as N_2, O_2, CO_2, and so forth, in the visible spectrum, which are difficult to measure with the normal LIF technique because these species require ultra-UV light to produce fluorescence. However, SRS is a very weak process and requires careful signal-to-noise consideration, background noise from the scattering of unshifted laser wavelength can be a problem.

LIF is an electronic absorption and emission process that produces relatively strong signal intensity with high spatial resolution. The fluorescence signal is normally red-shifted from the excitation wavelength and hence easier to discriminate than Rayleigh scattering from stray scattered light. Fluorescence signals are directly proportional to the molecular density and can therefore be used to measure species concentration. The strong fluorescence signals allow two-dimensional measurements by using a sheet of incident light. However, quenching effects can reduce fluorescence intensities, especially by oxygen at elevated pressures, in a way that depends on the pressure and temperature, making quantitative measurements difficult in the cylinder.

3.6 Summary

In this chapter, the fundamentals of the three major laser spectroscopic measurement techniques are described. The nature and characteristics of laser Rayleigh scattering (LRS), spontaneous Raman scattering (SRS), and laser-induced fluorescence (LIF) techniques are presented and compared. The implementation and application of LIF to in-cylinder fuel and combustion species measurements will be discussed in Chapter 5. The experimental system and performance of LRS and SRS for in-cylinder fuel and major species measurements will be presented in Chapter 6. The principle of operation and implementation of LRS, SRS, and LIF for in-cylinder gas temperature measurements will then be covered in Chapter 9.

References

Eckbreth, A. C. (1996). *Laser Diagnostics for Combustion Temperature and Species,* 2nd Ed., Amsterdam: Gordon and Breach.

Herzberg, G. (1971). *The Spectra and Structures of Simple Free Radicals.* Ithaca, NY: Cornell University Press.

Inaba, H., and Kobayasi, T. (1972). "Laser Raman Radar—Laser Raman Scattering Methods for Remote Detection and Analysis of Atmospheric Pollution." *Opto-Electronics,* Vol. 4, pp. 101–123.

Long, D. A. (1977). *Raman Spectroscopy.* New York: McGraw-Hill.

Piepmeier, E. H. (1972). "Theory of Laser-Saturated Atomic Resonance Fluorescence." *Spectrochimica Acta,* Vol. 27B, pp. 431–443.

Schrotter, H. W., and Klockner, H. W. (1979). In *Raman Spectroscopy of Gases and Liquids,* edited by A. Weber, pp. 123–166. Berlin: Springer-Verlag.

Further Reading

Lederman, S. (1977). "The Use of Laser Raman Diagnostics in Flow Fields and Combustion." *Prog. Energy Combust. Sec.,* Vol. 3, pp. 1–34.

Linne, M. (2002). *Spectroscopic Measurement: An Introduction to the Fundamentals.* San Diego: Academic Press.

Kohse-Höinghaus, K., and Jeffries, J. B. (2002). *Applied Combustion Diagnostics.* New York: Taylor and Francis.

chapter 4

Principle and Application of LDA and PIV for In-Cylinder Flow Measurements

4.1 Introduction

The details of fluid flow within the cylinders of internal combustion (IC) engines are known to profoundly affect engine performance and emissions. Both large-scale fluid motion, such as swirl and tumble flows, and small-scale turbulent flowfield are employed in the optimization of combustion and emissions. However, the development of three-dimensional (3-D) computational fluid dynamics (CFD) models has offered an attractive tool to engine designers to reduce engine testing and engine design cycle time. Because of its limitations in modeling complex and chemical reaction flows, CFD predictions always require experimental validation in new applications. Therefore, in-cylinder flowfield measurements are important for the validation of CFD models, the verification of design intent, and the study of fundamental in-cylinder flow processes and their interaction with combustion.

Laser Doppler anemometry/velocimetry (LDA/LDV) measurements have provided insight into the in-cylinder flow process. LDA/LDV measures the instantaneous velocity at a single point in the flowfield, from which statistical analysis of the temporal variance in velocity can be performed. In comparison, particle imaging velocimetry (PIV) produces images of flowfield and velocity vectors in an area, from which the flow structure can be readily identified and analyzed. The PIV technique measures the average velocity of flow over a finite period between the two particle images, and the results are suited for spatial correlation analysis in the flowfield. LDA measurement can be performed with a single small optical access point into the engine, whereas PIV measurements typically mandate two orthogonal optical access points to record images of particles in the cylinder.

4.2 Laser Doppler Anemometry

4.2.1 The Principle of Laser Doppler Anemometry

Strictly speaking the operation of LDA is based on the measurement of the Doppler shift of laser light scattered from small particles carried along with the moving fluid. A more lucid explanation can be given for a typical LDA in terms of the fringe pattern that exists in the crossover region because of interference between two laser beams (Fig. 4.1). The spacing of the interference fringe is given by

$$\delta_f = \frac{\lambda}{2\sin\left(\theta/2\right)} \tag{4.1}$$

A particle moving at a velocity v perpendicular to the fringes will thus experience a modulation of light intensity of frequency:

$$f = \frac{2v\sin\left(\theta/2\right)}{\lambda} \tag{4.2}$$

which is also known as the differential Doppler frequency.

As can be seen, there is a low-frequency component of intensity variation as well as the high-frequency component (Doppler frequency). The low-frequency component is often called the *pedestal* and is caused by the Gaussian intensity distribution of a laser beam.

A laser Doppler anemometer may be designed to meet differing requirements depending upon the number of velocity components to be measured, the spatial resolution, and accuracy sought. The overriding consideration is

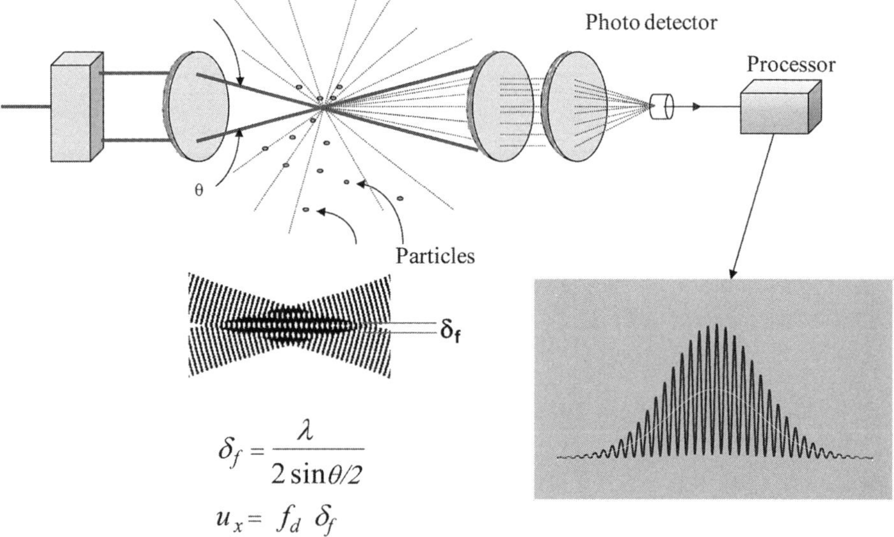

Fig. 4.1 Principle of LDA.

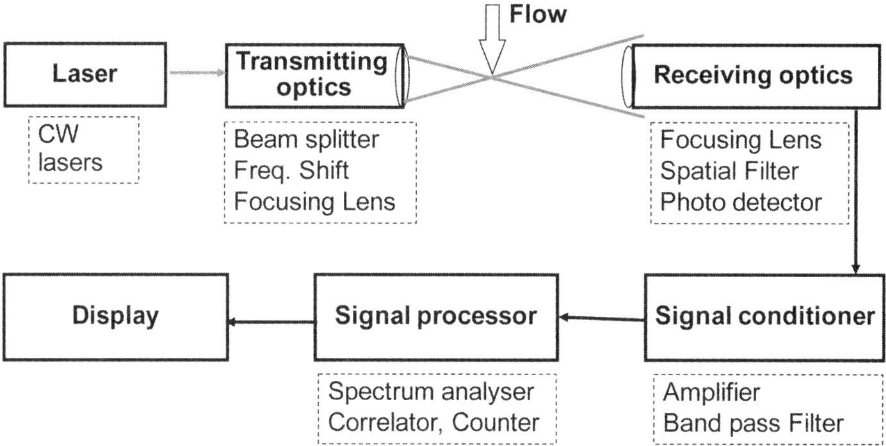

Fig. 4.2 Block diagram of an LDA system.

always the necessity of obtaining an adequate signal-to-noise ratio for signal processing.

An LDA system comprises of four main parts (Fig. 4.2): the laser and transmitting optics, light collection and detection system, the particle generation and seeding system, and signal processing equipment.

4.2.2 Laser and Transmitting Optics

In general, two types of lasers have been incorporated in most LDA systems. A He-Ne laser is characterized by its portability and low cost but low-power output. Recent advances in laser diodes have resulted in the replacement of He-Ne lasers in portable LDA systems. An argon-ion laser is preferred, particularly in engine and combustion systems where higher signal-to-noise ratio is necessary. In addition, its multiwavelength output is ideally suited to a multicomponent LDA system.

The purpose of the transmitting optics is to split the laser output into two parallel beams of equal intensity and then focus them to form the probe volume. Beam splitting can be achieved by the use of a partial reflection beam splitter or polarizing beam splitter (Fig. 4.3). The geometry of the transmitting optics determines the dimensions of the probe volume that is created by the intersection of the two beams. Thus, the beam-expanding optics of Fig. 4.4 are often built into the transmitting optics so that the beam diameter D_L and spacing of the beams D can be adjusted to obtain an optimized probe volume and fringe pattern within. With Gaussian laser beams the probe volume is an ellipsoid, and its dimension can be defined in terms of the dimensions in the coordinate system (Fig. 4.5) as follows:

$$\delta_z = \frac{4\,F\,\lambda}{\pi\,E\,D_L\,\sin\!\left(\dfrac{\theta}{2}\right)} \tag{4.3}$$

$$\delta_y = \frac{4\,F\,\lambda}{\pi\,E\,D_L} \tag{4.4}$$

$$\delta_x = \frac{4\,F\,\lambda}{\pi\,E\,D_L\,\cos\!\left(\dfrac{\theta}{2}\right)} \tag{4.5}$$

The beam waist of the focused laser Gaussian beam d_f is given by

$$d_f = \frac{4\,\lambda}{\pi\,\Delta\theta} = \frac{4F\lambda}{\pi E D_L} \tag{4.6}$$

where $\Delta\theta$ is the convergence angle of the focused laser beam.

Equation 4.2 states that, for a given wavelength of radiation, the signal frequency is proportional to the sine of the intersection angle. Thus, the intersection angle should be chosen so that the signal frequency over the expected velocity range is within the measuring range of the photodetector and

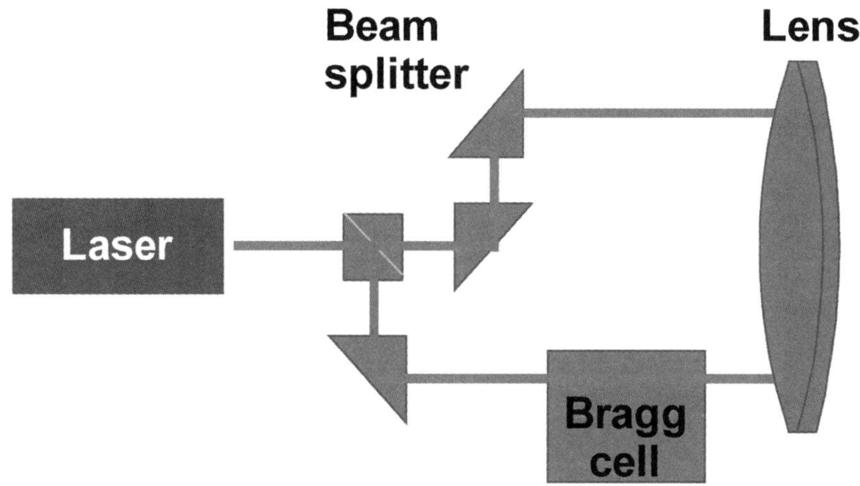

Fig. 4.3 Transmitting optics.

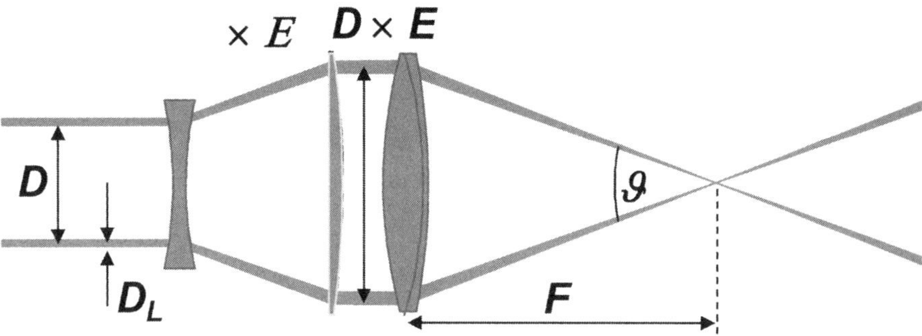

Fig. 4.4 Beam-expanding optics.

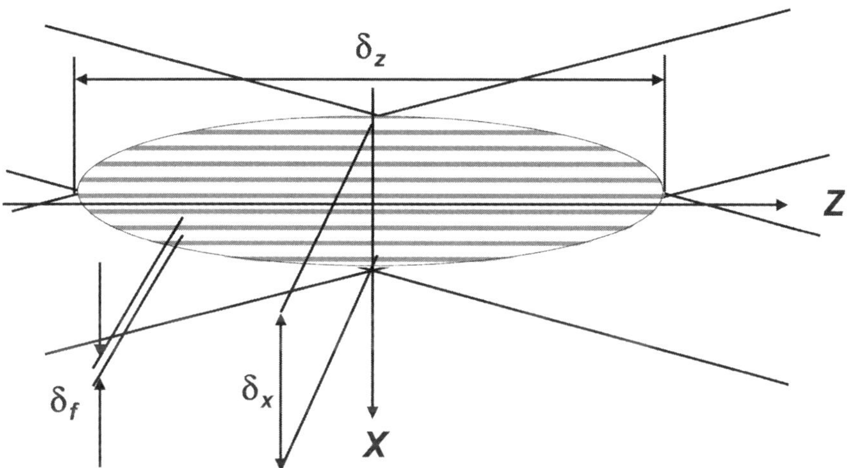

Fig. 4.5 Geometry of the probe volume.

signal processors, while providing an adequate number of fringes for better measurement accuracy.

4.2.3 Light Collection System

Collection of scattered light in the forward direction, as shown in Fig. 4.1, is preferred whenever possible, since light scattering from particles of microns is greatest in this direction. With a backward scattering configuration (Fig. 4.6), a reduction of 2 or 3 orders of magnitude in the received scattered light is expected. In this case, an adequate signal can only be achieved by the use of higher laser power. Sometimes a back-scattering arrangement is unavoidable because of the available optical access to the engine.

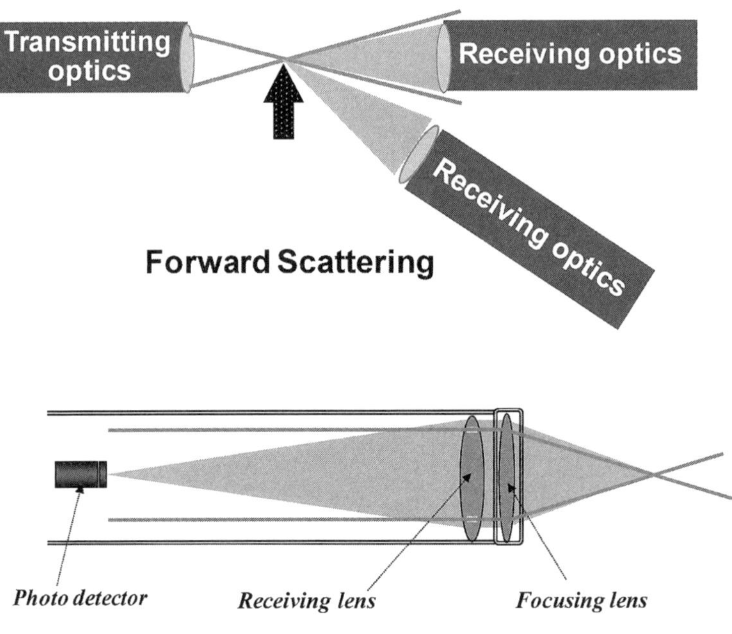

Fig. 4.6 LDA system with different collection configurations.

The function of the light collection system is to ensure that the scattered light originating from the probe volume is collected and imaged onto a photodetector. At the same time, it must reject any scattered light from particles other than those in the probe volume and any reflections from windows or other solid surfaces. The principal discrimination against unwanted light is usually achieved by focusing light scattered from the probe volume onto an accurately positioned aperture (i.e., space filter) in front of the photodetector.

Since the light intensity distribution across the fringe pattern is normally Gaussian, the fringes formed in the outer region of the probe volume are of a poorer quality than the remainder. It is, therefore, desirable to limit the detection of scattered light to the central part of the probe volume, which is defined as the measuring volume. In the forward detection configuration shown in Fig. 4.1, the measuring volume is determined by the diameter of the aperture (spatial filter) and the position of the focusing lens,

$$d_x = \frac{d_a}{M}; \quad d_y = \frac{d_a}{M}; \quad d_z = \frac{d_f}{\sin\left(\theta/2\right)} \tag{4.7}$$

where d_a is the diameter of the aperture, and $M = b/a$ is the magnification of the optical system.

The number of fringes in the measuring volume is given as

$$N_f = \frac{d_a}{\delta_f M}$$ (4.8)

Where δ_f is the fringe separation given by Eq. 4.1.

If LDA measurements are carried out in the presence of combustion, flame luminosity can be suppressed by placing an interference filter in front of the detector, so that only the scattered laser light is detected.

4.2.4 Signal Processing

The primary output of an LDA measurement is a current from the photodetector, which is often a photomultiplier tube. As discussed previously, the photomultiplier output consists of a DC and an AC component. The signal frequency is found in the high-frequency part of the AC component.

Depending upon the particle density in the fluid, it can be quasi-continuous or signal bursts interspersed with periods of low-power shot noise. Burst signals occur when the concentration of seeding particles is low, as it is often the case in engine flow measurements.

The burst signal from the preamplifier is input to the processors' signal conditioner (Fig. 4.2). The conditioner contains low- and high-pass filters for removing noise and the Gaussian pedestal from the Doppler signal. It also contains an amplifier that varies the amplitude of the filtered signal. The filtered signal can then be processed to determine the Doppler shift frequency by different methods. Counters were used in many earlier versions of commercial LDA systems. However, because of the superior quality of a spectrum analyzer, the frequency domain signal processor using fast Fourier transform (FFT) has replaced the counter as the commercially adopted signal processor in LDA experiments. This is particularly true in engine applications when optical access is often limited, and backward scattering detection is sometimes the only choice available. The concentration of seeding particles is limited as well. Consequently, signals tend to have poor signal-to-noise ratio and small amplitude.

4.2.5 Discrimination of Flow Direction by Frequency Shifting

A fundamental problem of the LDA system shown in Fig. 4.1 is that of discriminating the direction of flow. The Doppler frequency given by Eq. 4.2 provides no information about whether the particle is traveling up or down the fringes in the probe volume. The most common technique of discriminating flow direction involves frequency shifting to produce a frequency difference Δv between the two beams. This may be done by frequency-shifting one beam or (usually) both beams (one up and the other down), so that a moving system of fringes is obtained. The speed at which the fringes move is given by

$$v_f = \frac{\Delta v \cdot \lambda}{2 \cdot \sin\left(\theta/2\right)} \tag{4.9}$$

This is illustrated in Fig. 4.7. A stationary particle within the probe volume produces a signal of frequency equal to the shift frequency. Particles moving in the same direction as the fringes lower the Doppler signal frequency; particles moving in the opposite direction raise it, that is,

$$v_D = \frac{\Delta v \cdot \lambda}{2 \cdot \sin\left(\theta/2\right)} \pm \frac{2v \sin\left(\theta/2\right)}{\lambda} \tag{4.10}$$

Thus, to provide direction discrimination, the frequency shift applied must be at least equal to the Doppler shift frequency corresponding to the maximum velocity occurring in a series of measurements. A larger frequency shift is recommended to minimize the fringe biasing. The fringe biasing arises from the fact that signals from particles, which cross the fringe region at large

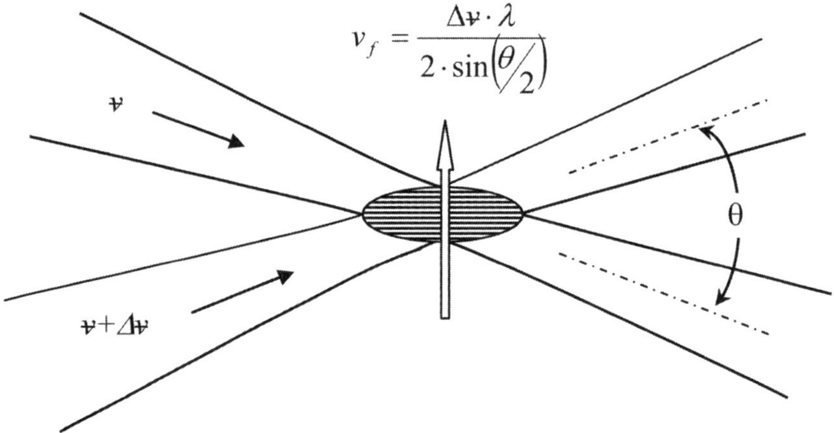

Fig. 4.7 The formation of a moving fringe system by the crossing of two beams of different frequency.

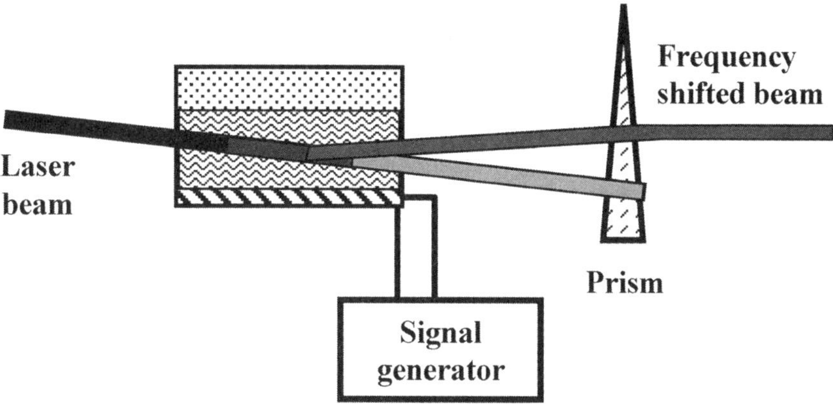

Fig. 4.8 Frequency shifting by means of an acousto-optic Bragg cell.

angles, cross a smaller number of fringes so as to have less probability of being validated by the signal processors.

The most popular frequency shifting technique is obtained with diffraction from an acousto-optic cell (Bragg cell). The principal feature for frequency shifting with a Bragg cell is illustrated in Fig. 4.8. The laser beam is passed through a cell in which acoustic waves, excited by a piezoelectric transducer, are traveling. If the angle between the laser beam and acoustic waves satisfy a certain condition (Bragg condition), diffracted beams are produced as from a moving diffraction grating. The frequency shift of the diffracted beam is equal to the frequency of the acoustic waves, which is controlled by a radio frequency generator. The deviated diffraction (shift) beam is brought back to its original direction by a prism incorporated into the acoustic cell unit. The angle of the prism required is a function of the wavelength and shift frequency. In practice, it is preferable to placing one Bragg cell in each beam to achieve a larger frequency shift without compromising the performance of the photodetector and signal processor.

4.2.6 Seeding and Particles

LDA measurements depend on light scattering by particles present in the fluid. Some suitable particles are usually present naturally in liquid flows. However, in the study of gaseous flows, the artificial addition of particles or seeding is normally required. Particle-seeding requirements can be summarized as follows:

- Particle flow following ability
- Particle light scattering characteristics
- Particle number density in the flow

- Tendency of particles to foul optical windows
- Tendency of particles to abrade cylinder surfaces
- Ability of particles to withstand high-temperature environment in a fired engine

The first requirement places limitations on particle size and the density ratio between particles and fluid. The limits are specified by considering the ability of the particles to follow the turbulent fluctuations up to an upper frequency within certain accuracy. It is estimated that for a particle to fluid density ratio of 1.8×10^4, particles must have diameters less than 2.7 µm to follow fluctuations up to 1 kHz and less than 0.8 µm for frequencies up to 10 kHz. A lower limit of particle size may be defined for very low velocities and density ratios in respect of Brownian motion. Thus, considering particle dynamics alone, 1 µm seems to be the optimum size, provided that the density ratio is sufficiently small.

The second requirement concerns the scattering properties of the particles as a function of size distribution, refractive index, particle shape and concentration, as well as the incident laser power. In the size range of concern, Mie theory defines the scattering intensity as a function of refractive index, particle diameter (D), and scattering angle. The scattering cross section, which is the ratio of the total scattered light to the intensity of the incident light, has a maximum value for the size parameter $x = \pi D / \lambda$, corresponding to a diameter of 0.8 µm for $\lambda = 0.5$ µm or 1 µm for $\lambda = 0.633$ µm. And the directional scattering intensity is strongest in the forward direction.

Optimum signal-to-noise ratio can only be obtained if the particle size is less than the fringe spacing. It is recommended that the ratio should be about 0.25 for the LDA measurements (Drain, 1980).

Several techniques have been employed for generating dispersions of fine particles. These include atomization of liquids, fluidization of powders, chemical reactions, and combustion-based techniques, the essential requirement being that they should supply a uniform concentration of particles into the flow.

Liquid drops are the most satisfactory forms of seeding, because they are nonabrasive, noncontaminating, and, with suitable liquids, noncorrosive and nontoxic. Commercial atomizers are available. Vegetable oil can be used because it does not pose any environmental problem. However, Reeves et al. (1994) found that evaporation loss of vegetable oil droplets near top dead center (TDC) was problematic at partly throttled motoring conditions. They found that Dow Corning® 550 high-temperature silicone fluid remained stable throughout the complete engine cycle.

When measurements are to be carried out in a high-temperature environment, the inert refractory solid particles should be used. The particles can be made of MgO, Al_2O_3, or TiO_2, and Al_2O_3 is the most stable under flame conditions. The most commonly used technique of seeding with solid particles involves the use of a fluidized bed. Window fouling and abrasion damage to windows by solid particles may be severe in the engine measurement.

LDA measurements in firing engines require special attention because of the additional difficulties imposed by the combustion process. The reduction in data rate is one of the most serious problems under firing conditions. In addition to the optimization of optics and electronics, Lorenz and Prescher (1990) found that the particle-seeding system was critical to obtaining high data rates. Completely dry air and preparation of the seeding particles with a hydrophobic material improved significantly the particle generation process.

In LDA measurements, a small quantity of seeded air should be introduced into the main gas flow well before it enters the engine where the flow is being investigated so that the seeding can be dispersed uniformly across the flow.

The level of seeding needs to be controlled as well. Too low a concentration of seeding reduces the data rate. Too much seeding results in more than one particle present in the measuring volume at a time, which causes frequency measurement subject to an error as a result of random phase fluctuation. In addition, flow characteristics may be affected by too much seeding.

4.2.7 Accuracy

LDA measurements in gas flows are characterized by burst signals. There is higher probability of signals from high-velocity particles than from low-velocity particles, because the particles distribute randomly in space, not in time. The probability depends on the instantaneous velocity. Consequently, the ensemble-averaged velocity becomes larger than the real value. This is called *velocity bias*, and the higher the turbulence intensity, the larger the velocity bias becomes.

Various schemes were proposed to correct the velocity bias. McLaughlin and Tiederman (1973) showed that it could be eliminated by using the inverse of the instantaneous velocity vector as the weighting function ($1/|u_i|$). Considering that the inverse of the instantaneous velocity proportional to the residence time of the particle in the measuring volume, Hoesel and Rodi (1977) suggested using this time as the weighting function. The interarrival time was also considered as the weighting function for velocity bias correction (Durao et al., 1980). The general bias correction formulae can therefore be written as

$$\bar{u} = \sum_{1}^{n} u_i \cdot G_i \bigg/ \sum_{1}^{n} G_i \qquad (4.11)$$

$$u' = \left[\sum_{1}^{n} (\bar{u} - u_i)^2 \cdot G_i \bigg/ \sum_{1}^{n} G_i \right]^{\frac{1}{2}} \qquad (4.12)$$

where the weighting function G_i of the ith measurement is given by

$$G_i = 1/|u_i|, \qquad \text{inverse of velocity method}$$

$$G_i = T_i, \qquad \text{residence time weighting}$$

$$G_i = \Delta T_i, \qquad \text{interarrival time weighting}$$

During the signal processing stage, the register of the residence time corresponding to each measurement is thus valuable to obtain more reliable values of mean velocity and the velocity distribution function in highly turbulent flows.

It should be noted that velocity biasing occurs only if the interval between velocity measurements is large compared with the time scale of the velocity fluctuations.

In the counter-type processors, there is fringe bias as well. This arises from the fact that signals from particles that cross the fringe region at large angles with respect to the fringes cross a smaller number of fringes and so have less probability of being validated. This can be corrected for by decreasing the number of cycles required for validation by the counter or by frequency shifting.

4.2.8 Advanced LDA Systems

The modular approach to LDA of the earlier days has been replaced by the integrated systems approach, where the entire system runs under a PC (personal computer). This minimizes the time of setup and ensures that the measurement is accurate. Windows®-based acquisition packages with online help are available to simplify further the measurement process. In many commercial LDA systems, receiving optics are incorporated into the transmitting optics so that the whole system is very compact and self-aligned.

In the simple LDA system shown in Fig. 4.1, any component of velocity perpendicular to the fringes may be measured by suitable rotation of the beam splitter, but only one component can be recorded at a time. Measurements made successively in this way provide considerable information, but there are circumstances where it is necessary to make simultaneous measurements of more than one component of velocity. These may be needed not only to speed up the acquisition of data, but also enable correlation between the components of velocity to be studied, for example, the Reynolds stress term, $\overline{u' \cdot v'}$.

Commercial LDA systems capable of simultaneous two- or three-component velocity measurements are available. To measure two components of velocity in the same probe volume, it is necessary to form two superimposed sets of fringes, which may be discriminated by polarization or more satisfactorily by color. Blue (488-nm) and green (514.5-nm) beams can be conveniently obtained simultaneously from an argon-ion laser operating in its multiline mode. To measure the two components of velocity transverse to the direction of illumination, orthogonal blue and green fringes are formed in a single probe volume. If the third component of the velocity vector is to be measured as well, independent transmitting optics have to be added to deliver the pair of beams at a right angle to the other two pairs of beams.

Fiber-optics are providing a degree of flexibility that is of particular interest in IC engines. Most of the optical system can be located away from the engine and in-cylinder measurements can be accomplished through one optical access (Fig. 4.9). In addition, the difficulty associated with engine vibration can be overcome by mounting the optical fiber probe within a windowed probe holder designed to allow both angular and axial movement of the measurement volume within the cylinder. The drawback with one optical window is the low signal level from the backward light-scattering arrangement, which requires a more powerful laser to compensate for the reduced signal-to-noise ratio.

Automatic data analyses, using the output from the signal processor of a LDA systems, are provided by commercially available software packages. Raw data are processed to provide individual velocity components and their projection in three orthogonal planes, with statistical properties including mean, turbulence intensity, skewness, and flatness of the velocity components. For multicomponent LDA systems, correlations, Reynolds stresses, velocity and flow angle histograms, and vectors expressed in terms of magnitude and angle can be readily displayed or printed. Velocity biasing is corrected by the software package using the particle residence time in the measuring volume. Software packages are also available to control the automatic transversing of

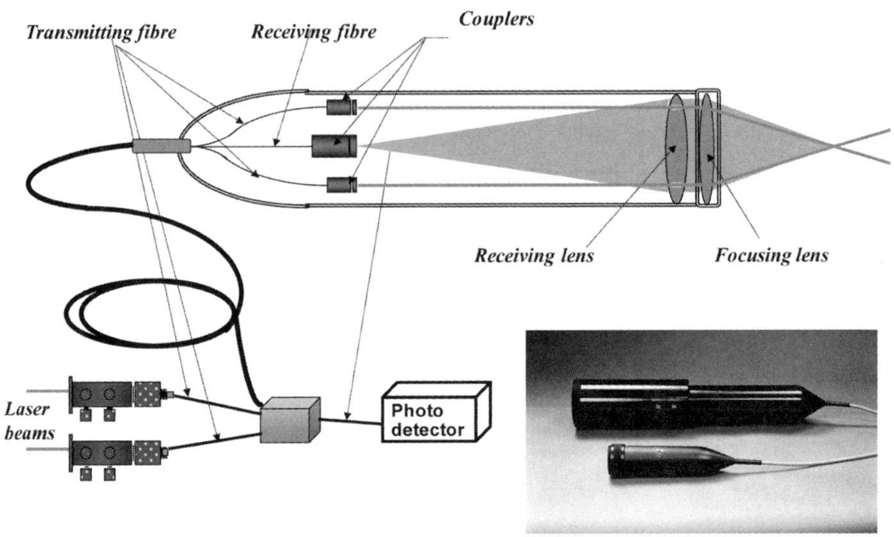

Fig. 4.9 Optical fiber LDA setup and probes.

the table with three axes of translation. These features are greatly increasing the ease and convenience of making accurate LDA measurements.

4.2.9 Data Analysis of In-Cylinder Flows

4.2.9.1 Ensemble-Averaged Analysis

In stationary turbulent flows, the conventional approach is to resolve the instantaneous flow velocity into (time-independent) mean and (time-dependent) randomly fluctuating components. For the nonstationary in-cylinder flows, the analogous decomposition is into a (time-varying) ensemble-mean velocity and random fluctuations about the ensemble mean:

$$U(\theta,i) = \bar{U}(\theta) + u(\theta,i) \tag{4.13}$$

Where $U(\theta,i)$ is the velocity at a certain crank angle (CA) θ in engine cycle i, and the ensemble-averaged velocity is given explicitly by

$$\bar{U}(\theta) = \frac{1}{N(\theta)} \sum_{i}^{N} U(\theta,i) \tag{4.14}$$

Where $N(\theta)$ is the total number of valid velocity estimates available at CA θ over all cycles in the ensemble, and it is assumed that the Doppler signal is sampled uniformly to produce one velocity estimate per crank angle.

A measure of the intensity of the velocity fluctuations is provided by the rms (root mean square) or standard deviation of the instantaneous velocity about the ensemble-mean velocity:

$$u'(\theta) = \sqrt{\frac{1}{N(\theta)} \sum_{i=1}^{N} \left[U(\theta,i) - \bar{U}(\theta) \right]^2} \qquad (4.15)$$

which is also known as the ensemble-rms velocity fluctuation intensity as it is ensemble-averaged.

Since an LDA signal is only produced when a particle moves through the measuring volume, it is necessary to perform an ensemble-averaging over a finite CA window $\Delta\theta$ around the specific CA of interest θ to obtain data as velocity-crank-angle pairs. The ensemble-averaged velocity equation thus becomes

$$\bar{U}(\theta) = \frac{1}{N_t} \sum_{i}^{N_c} \sum_{j=1}^{N_i} U(\theta \pm \frac{\Delta\theta}{2}, i, j) \qquad (4.16)$$

where N_i is the number of velocity measurements recorded in the window during the ith cycle, N_c is the number of cycles, and N_t the total number of measurements. The corresponding equation for the ensemble-averaged rms velocity fluctuation is

$$u'(\theta) = \sqrt{\frac{1}{N_t} \sum_{i}^{N_c} \sum_{j=1}^{N_i} \left[u(\theta \pm \frac{\Delta\theta}{2}, i, j) \right]^2} \qquad (4.17)$$

This need to ensemble-average over a finite CA window introduces an error called crank-angle broadening, resulting from the change in the mean velocity across the window. This error depends on the velocity gradient. The major influence of crank-angle broadening is on the rms velocity fluctuation rather than on the mean value (Rask, 1981).

Alternatively, the smoothed-ensemble data-reduction technique can be used (Rask, 1979). Instead of a single value for the mean velocity, a cubic-spline smoothing routine is used to put a smooth curve through data that has been averaged into 1° CA windows. As a result, the ensemble-averaged mean velocity is evaluated at the exact crank angles of all data points, eliminating the crank-angle broadening of window-based data reduction.

In practice, if an LDA system samples the Doppler signal uniformly with contiguous sampling windows and yields one velocity estimate per crank angle ($\Delta\theta = 1°$), $\Delta\theta$ can be omitted from the calculations; that is, Eqs. 4.14 and 4.15 can be applied directly to the measured data. This will be assumed to be the case in the subsequent analysis in the following sections.

4.2.9.2 Cycle-Resolved-Velocity Analysis

An important implication of Eq. 4.14 is that $\bar{U}(\theta)$ becomes a function of crank angle alone, which implies that the mean flow at a particular crank angle does not vary from one cycle to the next. Flow visualization experiments, however, have shown that this may not always be true. The cyclic variation may be due to the dynamically changing boundary conditions imposed by the piston and valve motions, not to mention fuel-injection and combustion in fired engines. If there are cycle-to-cycle variations, then in the conventional ensemble-averaged analysis, any variations will be include in $u'(\theta)$, resulting in an overestimation of the true in-cycle turbulence intensity.

Two basic approaches have been used for extracting quantitative information from cycle-resolved velocity data. Both are aimed at distinguishing between "true" turbulent velocity fluctuations and cycle-to-cycle variations in the mean velocity.

In the first approach, some form of low-pass filtering is used to estimate the mean velocity in each engine cycle. The rms deviations of the instantaneous velocities from the in-cycle mean velocities are then computed over all the cycles, and they are usually interpreted as a measure of the intensity of the true turbulent velocity fluctuations. We call this velocity-filtering-based analysis.

A second approach assumes that turbulent fluctuations are stochastic or random in nature, whereas large-scale fluctuations or cyclic variations in the mean velocity exhibit an appreciable degree of phase stability or coherence within individual engine cycles. The analysis attempts to identify, parameterize, and extract any phase-coherent contribution to the ensemble-rms fluctuation intensity, leaving only the contribution of the stochastic or turbulent fluctuations.

Velocity-filtering analysis **is** the filtering approach splits the fluctuations into low and high-frequency components relative to a filter cutoff frequency,

$$u(\theta,i) = u_{low}(\theta,i) + u_{high}(\theta,i) \qquad (4.18)$$

This is equivalent to the decomposition of the instantaneous velocity into three components:

$$U(\theta,i) = \bar{U}(\theta) + u_{low}(\theta,i) + u_{high}(\theta,i) \tag{4.19}$$

The in-cycle mean velocity is, therefore, the sum of the ensemble-mean velocity and the low-frequency fluctuation term:

$$\hat{U}(\theta,i) = \bar{U}(\theta) + u_{low}(\theta,i) \tag{4.20}$$

Over which the ensemble-averaging is expected to be equivalent to the conventional ensemble-averaged mean velocity:

$$\overline{\hat{U}(\theta,i)} = \bar{U}(\theta)$$

The intensities of the low-frequency fluctuation is characterized by its rms deviations of the in-cycle mean velocity about the ensemble-mean velocity:

$$u'_{low}(\theta) = \sqrt{\overline{u_{low}^2(\theta,i)}} = \sqrt{\frac{1}{N(\theta)} \sum_{i=1}^{N} \left[\hat{U}(\theta,i) - \bar{U}(\theta) \right]^2} \tag{4.21}$$

Similarly, the high-frequency velocity fluctuation intensity is evaluated as the rms departure of the instantaneous velocity from the in-cycle mean velocity:

$$u'_{high}(\theta) = \sqrt{\overline{u_{high}^2(\theta,i)}} = \sqrt{\frac{1}{N(\theta)} \sum_{i=1}^{N} \left[U(\theta,i) - \hat{U}(\theta,i) \right]^2} \tag{4.22}$$

The conventional ensemble-rms fluctuation intensity is related to the low-and high-frequency intensities by

$$u'(\theta) = \sqrt{\overline{\left[u_{low}(\theta,i) + u_{high}(\theta,i) \right]^2}} = \sqrt{u'^2_{low}(\theta) + u'^2_{high}(\theta) + 2 \cdot \overline{u_{low}(\theta,i) \cdot u_{high}(\theta,i)}} \tag{4.23}$$

The low- and high-frequency fluctuations are usually assumed to be statistically independent, so that the last term in Eq. 6.27 becomes zero, and the conventional ensemble-rms fluctuation intensity is given by

$$u'(\theta) = \sqrt{u'^2_{low}(\theta) + u'^2_{high}(\theta)}$$

(4.24)

The low- and high-frequency fluctuation intensities u'_{low} and u'_{high} are sometimes referred to as the cyclic variation in the mean velocity and the turbulence intensity, respectively.

To determine the low- and high-frequency fluctuation intensities, the in-cycle mean velocity $\hat{U}(\theta,i)$ has to be calculated. The first approach is a moving window average in the time (crank-angle) domain (Ball et al., 1983). In this method, a crank-angle window is selected on the basis of being larger than the turbulent microscale but smaller than the characteristic period of piston motion (e.g., 10° CA). By using heavy seeding and fast sampling, high data rates can be obtained to resolve most of the turbulence frequencies and scales. Because of the burst Doppler signals, the evaluation of the time-varying mean velocity within the selected window requires some sort of data interpolation between individual burst signals, particularly when there are missing data points within the selected window. The interpolation or smoothing can be done by sample-and-hold scheme, in which a missing point is replaced by the intermediately proceeding valid measurement.

The in-cycle mean velocity can also be obtained by polynomial curve fitting, which involves putting a smooth curve through data that has been averaged into 1° CA windows by using cubic-spline curve fitting routine (Rask, 1981). This is equivalent to a low-pass filtering process but its cutoff frequency is not well defined.

The third approach involves the use of a frequency domain filter (Liou and Santavicca, 1985). Fourier transforming the instantaneous velocity-crank-angle record $U(\theta,i)$ produces a spectrum of various frequencies. The low-pass filter is then applied in the frequency domain by setting all the Fourier coefficients above the selected cutoff frequency f_{low} to zero and retaining the coefficients below f_{low} unaltered. Finally, the inverse Fourier transform of this truncated series back to time (crank-angle) domain generates the desired estimate of the in-cycle mean velocity $\hat{U}(\theta,i)$. Since the method is based on the selection of a cutoff frequency that is crank-angle dependent, it involves some arbitrariness and requires physical justification. Daneshyar and Fuller (1986) proposed the use of the inverse stroke duration as the cutoff frequency. Fansler and French (1988) assigned the value of f_{low} to the highest frequency present in the conventional ensemble-averaged velocity $\bar{U}(\theta)$.

Autocorrelation based analysis considers the frequency content of the in-cylinder velocity fluctuations. Nothing has been said about the degree to which the fluctuations are random (chaotic, in-coherent) or deterministic (organized, coherent). The classic statistical tool for assessing the relative contributions of random and deterministic processes is the autocorrelation function.

The nonstationary velocity autocorrelation function of the random, intermittent LDA data can be evaluated by (Glover, 1986; Fansler and French, 1988):

$$R(\theta,\tau) = \frac{\dfrac{1}{N_p(\theta,\tau)}\displaystyle\sum_{i=1}^{N} u(\theta,i)\cdot u(\theta+\tau,i)}{u'(\theta)\cdot u'(\theta+\tau)} \tag{4.25}$$

where $u(\theta,i) = U(\theta,i) - \bar{U}(\theta)$ is the fluctuation about the ensemble-averaged velocity, $u'(\theta)$ is the corresponding ensemble-rms fluctuation intensity, θ is referred to as the reference crank angle, and τ is the delay or lag. The quantity $N_p(\theta,\tau)$ is the total number of valid cross-product pairs for each (θ,τ) slot over the entire ensemble of N engine cycles. A series of discrete windows of width $\Delta\tau$ (e.g., 1° CA) in the delay τ is used to estimate $R(\theta,\tau)$.

To improve the statistical quality of the autocorrelation estimate, the result of Eq. 4.25 can be averaged over a reference range (Glover, 1986) that is somewhat wider than the slot width $\Delta\tau$, assuming that the velocity is locally stationary over the duration of the reference range. The final expression is therefore given by

$$R(\theta,\tau) = \frac{1}{N_{ref}}\sum_{j=1}^{N_{ref}}\left[\frac{\displaystyle\sum_{i=1}^{N} u(\theta_j,i)\cdot u(\theta_j+\tau,i)}{N_p(\theta_j,\tau)\cdot u'(\theta_j)\cdot u'(\theta_j+\tau)}\right] \tag{4.26}$$

Fansler and French (1988) compared three reference ranges (1, 6, 12) and found that the 12° reference range used did not affect $R(\theta,\tau)$ profiles when 1200 cycles were analyzed.

Finally, an alternative approach based on conditional sampling is used to identify groups of engine cycles of similar behavior that can then be further analyzed using ensemble-averaging. This is particularly useful to resolve the relationship between flow characteristics before ignition near the spark plug or in the unburned region ahead of the propagating flame, with some combustion parameters during individual engine cycles (Swords et al., 1982; Witze and

Martin, 1986). The combustion parameters investigated include peak cylinder pressure, flame speed, and flame arrival time.

4.2.10 Applications of LDA to IC Engines

There have been many applications of LDA to IC engines since the 1970s. Because of the complexity of the engine flow processes, model engines with simplified geometry and running at slow speed were used in early studies of the effect of individual engine parameters (e.g., compression ratio, valve shape, etc.) on the in-cylinder flowfield using LDA (e.g., Arcoumanis et al., 1983; Morse et al., 1980; Tindal et al., 1988). During the same period, measurements of the steady flows through helical and directed ports were conducted by Wigley and Hawkins (1978) and Coghe et al., (1988).

Several researchers have reported LDA measurements in motor engines (e.g., Bopp et al., 1986; Liou et al., 1984; Suen et al., 1987). A consistent conclusion emerging from these studies is that the turbulence intensity at TDC, with open combustion chambers in the absence of swirl, has a maximum value equal to about half the mean piston speed:

$$u'(TDC) \approx 0.5 \cdot \overline{S}_p$$

In recent years, multivalve SI (spark-ignition) engines have been designed with pent-roof combustion chambers. For these engines, tumble has been found to accelerate the combustion process. *Tumble* is defined as a rotational air motion around an axis perpendicular to the cylinder axis, and *swirl* is the rotation around an axis parallel to the cylinder axis. When tumble was generated by appropriate intake port design in studies, it was found that the peak turbulence intensity occurred much closer to TDC and was significantly higher than that in the absence of tumble motion, because of the collapse of the tumble flow in the early part of compression stroke (Arcoumanis et al., 1990; Hadded and Denbratt, 1991). The effects of the intake port design (Omori et al., 1991) and the piston shape (Baby and Floch, 1997) on the formation and breakdown of tumble have also been reported.

Cycle-resolved LDA measurements have been used to investigate the cycle-to-cycle variations. Fansler and French (1988), for example, obtained cycle-resolved measurements in a motored engine, and the velocity fluctuations were analyzed by the frequency-domain filtering method and the evaluation of nonstationary velocity autocorrelation functions.

Turbulence length scales have been measured using primarily two kinds of LDA techniques. The integral length scale is defined as the integral of the autocorrelation coefficient $R(\theta,x)$ of the fluctuating velocity at two adjacent points, x and $x + \Delta x$, in the flow with respect to the variable distance between the points:

$$l_l(\theta) = \int_0^\infty R(\theta, \Delta x)\, dx$$

and

$$R(\theta, \Delta x) = \frac{1}{N-1} \sum_{i=1}^{N} \frac{u(\theta, x, i) \cdot u(\theta, x + \Delta x, i)}{u'(\theta, x) \cdot u'(\theta, x + \Delta x)}$$

where

$u(\theta, x, i) = U(\theta, x, i) - \bar{U}(\theta, x)$ is the fluctuation velocity

$u'(\theta, x) = \sqrt{\overline{\left[u(\theta, x, i)\right]^2}}$, that is, the rms or standard deviation of the instantaneous velocity about the ensemble-mean velocity

Fraser and Bracco (1988) determined the lateral integral length scales using a two-point, single-probe volume LDA system in a motored single-cylinder engine. In this approach, a single elongated probe volume was generated. The measurement locations within the elongated probe volume were determined by imaging the entire probe volume onto a slit aperture, on the other side of which were the input ends of two optical fibers. Photomultipliers mounted on the other ends of the two optical fibers were connected to two counter processors, where Doppler signals were processed.

Turbulence length scale measurements by two-probe-volume LDA technique was carried out in a motored direct-injection diesel engine by Corcione and Valentino (1990). The technique was realized by forming two probe volumes using separate transmitting optics for the green and blue beams of an argon-ion laser. The probe volume of the two green beams was fixed, and the blue beams could be moved by means of a rotating optics. The distance between the two probe volumes could be altered between 0 and 10 mm.

In another approach, the scanning LDA was used for the measurement of turbulence length scale (Glover et al., 1988). The fixed mirror in the barrel extension of a single optical cylinder engine was replaced with a rotating mirror driven from the crankshaft at twice the engine speed by a toothed belt and gear arrangement. This resulted in a scan duration across a window of about 6° CA, and its measuring volume velocity was much higher than the turbulence intensity. Therefore, the probe volume was traversed rapidly through the flow, freezing the turbulence.

A number of investigations into the correlation between the flowfield and combustion was reported in fired engines (Hadded and Denbratt, 1991; Hall et al., 1986; Lorenz and Prescher, 1990; Park et al., 2004). A consensus emerging from these studies that there is a strong correlation between the turbulence intensity at the spark timing and the rate of early combustion. To resolve the relationship between flow characteristics before ignition near the spark plug, or in the unburned region ahead of the propagating flame, with some combustion parameter during individual engine cycles, conditional sampling of groups of engine cycles of similar behavior were used (Swords et al., 1982; Witze and Martin, 1986). The conditional sampling parameters include peak cylinder pressure, flame speed, and flame arrival time.

In recent years, CFD engine simulation has been developed to model the in-cylinder flow and combustion processes. LDV has proved to be a useful tool as well, because it provides time-resolved quantitative velocity and detailed turbulence measurements (Galindo et al., 2001; Obokata et al., 2009), which are complementary to the PIV results (Ishima et al., 2008; Malcolm et al., 2011).

4.3 Particle Image Velocimetry (PIV)

4.3.1 Principle of Particle Image Velocimetry

The principle of the PIV technique is simple: The velocity vector is calculated from the displacement of an element of fluid in a known time interval. In its simplest form, the PIV technique requires a thin slice of the flowfield being illuminated by a laser light sheet. Particles in the flow within the light sheet scatter the light. The scattered light is detected by a camera placed at a right angle to the light sheet (Fig. 4.10). The laser light sheet is double-pulsed (switched on and off very quickly, twice) at a known time interval, Δt. The first pulse of the laser freezes images of the initial positions of these particles onto the first frame of the camera. The camera frame is advanced, and the second frame of the camera is exposed to the light scattered by the particles from the second laser pulse. There are, thus, two images, the first showing the initial positions of the particles and the second their final positions due to the movement of the flowfield. Alternatively, the double exposed particle images can be recorded on the same photographic film.

The intersection volume between the region illuminated by the laser light sheet and the field of view of the imaging optics determines the region of the flow to be measured, which is denoted the measurement region (Fig. 4.11). The smallest resolvable area of the camera, the so-called interrogation area (IA), projected back onto the measurement region, determines the interrogation volume (IV) that constitutes a single velocity vector.

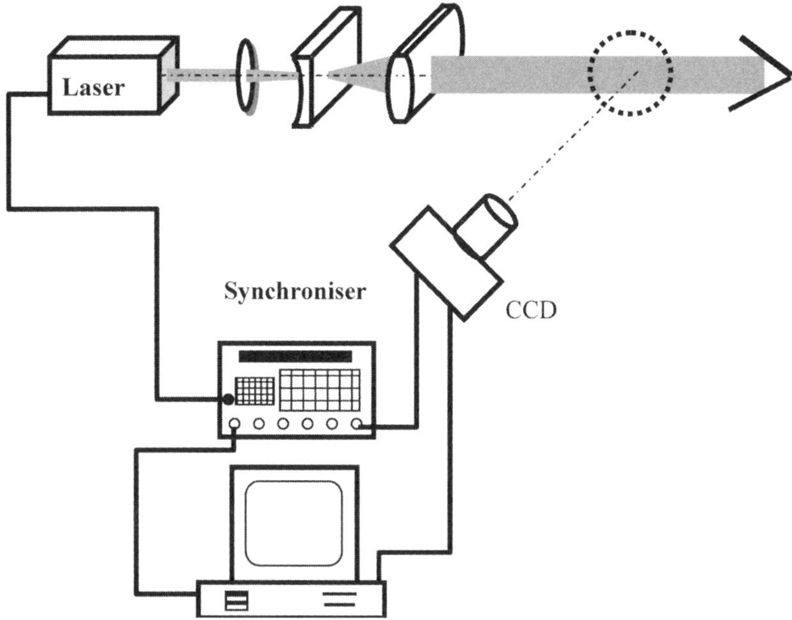

Fig. 4.10 A typical PIV experimental setup.

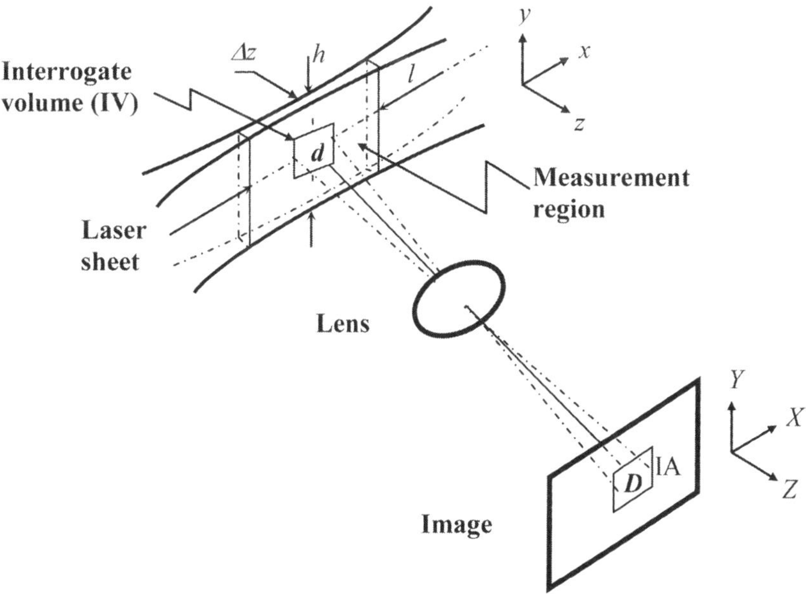

Fig. 4.11 Definition of measurement region and measurement volume.

The particle displacement in the object plane, Δx and Δy, are found from the displacement in the image plane, ΔX and ΔY, on the recording medium:

$$\Delta x = \frac{1}{M}\Delta X \quad and \quad \Delta y = \frac{1}{M}\Delta Y \tag{4.27}$$

Knowing the magnification factor of the imaging optics M and the time separation Δt between the two laser pulses, the velocity projections on the object plane, u and v, may be found from Eq. 4.28:

$$u = \frac{\Delta x}{\Delta t} \quad and \quad v = \frac{\Delta y}{\Delta t} \tag{4.28}$$

4.3.2 Operation of the Digital PIV System

The photographic film recordings used in the early PIV system have been completely replaced by solid-state imaging devices. In particular, cross-correlation CCD (charge-coupled device) cameras permit the particle images of the two consecutive laser pulses to be recorded onto two separate frames, offering significant improvement in terms of velocity gradient tolerance, dynamic range, and signal-to-noise ratio, as well as automatic detection of flow direction. CCD imaging arrays of up to 16 million pixels are commercially available.

The cross-correlation CCD camera is a full-frame, progressive scan, interline transfer CCD camera. In this type of camera, each active pixel has its own storage site, and its charge is read out sequentially (progressive scan). This permits very fast transfer of the entire exposed image into the adjoining storage sites within a few microseconds or less in conjunction with high resolution. The fast transfer, then, allows two single exposed PIV images to be captured at a time delay slightly longer than the transfer time.

The advent of cross-correlation CCD camera enables the frame-straddling method to be used in the modern digital PIV system. In the frame-straddling technique, two laser pulses are timed so that they are straddled between two consecutive frames of one-half of the total pixels (Fig. 4.12). To achieve this, the laser pulses are timed such that the first pulse occurs in the first camera frame (frame n), and the second pulse occurs during the second frame (frame $n + 1$). The maximum PIV image rate is half of the camera's frame rate (normally 30 Hz) with pulse delays as short as 0.5 µs. Although the frame separation time is fixed for a given camera, the time delay between the two images can be adjusted by the timing of the two laser pulses. For very-high-speed measurement, the laser pulses are timed such that the first pulse occurs immediately before the frame transfer takes place (frame n), while the second pulse occurs immediately thereafter (frame $n + 1$). For slow flows, the first and

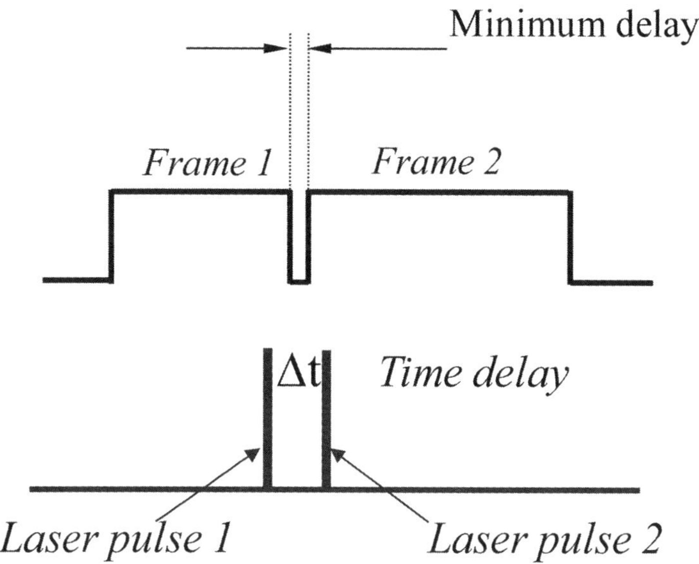

Fig. 4.12 Principle of the frame-straddling technique.

second laser pulses should be placed at the start of the first frame and the end of the second frame, respectively.

Because these cameras also often have asynchronous reset options, they can be triggered externally. This allows PIV data capture to be synchronized with the experiment and is therefore particularly useful in rotating, periodic, or transient flows, such as those in engines.

The digitized particle images can then be transferred to a PC for subsequent analysis or to a dedicated digital correlator for real-time measurements. The analysis is then carried out by means of the cross-correlation method.

Frame straddling, particularly if using short straddle times, requires very precise synchronization and control of the PIV lasers, CCD camera, and frame grabber. As with any PIV system, this control is provided by a synchronizer, typically made up of a digital delay generator and controller.

4.3.3 Lasers and Light Sheets

Lasers provide light in a conveniently collimated and intense form, and so they are ideally suited for the light source for the PIV measurement. The choice of laser is controlled by a number of criteria. Central in the choice of an illumination source is its high power, because the PIV measurement requires illumination over an extended volume of flow and the amount of light scattered from particles must be detectable by the camera. Pulse width must be short enough to freeze the flow.

The frequency-doubled Nd:YAG pulsed laser is the most commonly used laser for PIV measurement. There are two types of Nd:YAG lasers used in PIV experiments: the single-cavity Nd:YAG laser with the double-pulse option and the twin-oscillator, twin-amplifier frequency-doubled Nd:YAG laser. The simple one is a single-cavity Nd:YAG laser with the double-pulse option. Two Marx banks are used to open the laser's Pockels cell (Q-switching device) twice during one flashlamp discharge. This divides the available energy into two pulses. The first pulse is triggered externally, and a potentiometer on the double-pulse option panel adjusts the timing of the second pulse. However, this reduces the energy of each pulse to much less than that of pulses in single-pulse mode. Pulse separations are limited to typically 20–200 µs. The pulse energy and width vary with the pulse separation. This leads to particle images of uneven intensity that must be accommodated by the sensitivity dynamic range of the recording media.

The laser used in most commercial PIV systems is a twin-oscillator, twin-amplifier frequency-doubled Nd:YAG laser. Each of the two oscillators and amplifiers can be triggered separately, thus allowing infinite control of the laser pulse separations and simple control of the individual pulse energy. Alternatively, two separate lasers may be employed. These systems, however, require accurate coalignment of the two separate outputs.

The illumination of flow in PIV measurements requires the formation of a thin laser light sheet within the flow. The laser light sheet is typically achieved with an arrangement of cylindrical and spherical lenses. Spherical lenses are used to focus the laser beam to the required beam waist diameter, giving the light sheet thickness. Cylindrical lenses are used to expand and collimate the beam into a sheet. The width of the laser light sheet depends upon the area of interest and the recording optics. The thickness of the laser light sheet should be selected taking into account the magnitude of the out-of-plane velocities present (the velocity component normal to the light sheet) and the pulse separation for recording the flow. Large out-of-plane components of velocity require a thick light sheet, reducing the intensity of the illumination and the amount of light scattered. According to Keane and Adrian (1990), the minimization of data loss caused by movement normal to the light sheet is achieved when

$$w \cdot \Delta t \leq \Delta z / 4 \tag{4.29}$$

where w is the out-of-plane velocity component, Δt is the time between exposures, and Δz is the light-sheet thickness.

A further consideration in illumination source is beam delivery. Many PIV measurements in engines can be accomplished using a set of lenses on an optical bench. In some applications the limited space and harsh environment require some means by which the light energy can be delivered to the measurement volume. Fiber-optic delivery system is difficult to apply for PIV, because the requirement for very short pulses and high-output beam quality limits the high-energy transmission. Though less elegant and flexible than a fiber-optic system, flexible mirror arm systems in commercial PIV can be used for beam delivery, in which mirrors and optical lenses are incorporated into tubes mechanically linked together.

4.3.4 Imaging Optics and Perspective Errors

Several important optical imaging parameters need to be considered when estimating the dynamic range of a PIV system. Sharp and small particle images minimize the uncertainty in locating the image centroid or correlation peak centroid and hence the error in velocity measurement. Smaller particle images permit more closely spaced particle pairs of higher velocity to be resolved. Sharp and small particle images lead to high particle image intensity, since for the same amount of scattered light by the seeding particle, the light energy per unit area increases quadratically with decreasing particle images.

However, the particle image diameter is often diffraction-limited by the focusing lens. The diffraction-limited imaging occurs because of the finite aperture of the focusing lens. The diffraction-limited minimum image diameter is given by

$$d_{diff} = 2.44\, f_{\#}\, (M+1)\, \lambda \tag{4.30}$$

where $f_{\#}$ is the f-number of the focusing lens, defined as the ratio between the focal length f and the aperture of the lens A, and M is the magnification factor given by

$$M = \frac{S}{s} \tag{4.31}$$

where S is the distance between the image plane and lens, and s is the distance between the lens and the object plane.

If lens aberrations are absent, the true particle image d_i can be estimated from

$$d_i = \sqrt{M^2 d_p^2 + d_{diff}^2} \tag{4.32}$$

Table 4.1 shows the calculated values of the images of a small particles (1 µm) in conjunction with the depth of field δ_z, calculated from

$$\delta_z = 2\,f_\#\,d_{diff}\,(M+1)\big/M^2 \tag{4.33}$$

It can be seen from Table 4.1 that a large aperture leads to sharp particle images, as more light is collected. However, to obtain a sharp image, the seeding particles need to fall within the depth of field of the camera lens. For maximum use of the available scattered light, the depth of field should be approximately equal to the light sheet thickness. Therefore, the minimum f-number, or maximum aperture for a fixed focal length, is determined by the required depth of field.

As in most cases, the displacement vector measured by a typical PIV system is in a two-dimensional (2-D) plane. The 2-D displacement vector in the image plane $\mathbf{D} = \mathbf{X}_i' - \mathbf{X}_i$, (Fig. 4.13) and the particle displacement vector $\mathbf{d}$ is related by

$$\mathbf{D} = -M \cdot \mathbf{d} = -M \cdot (u\boldsymbol{i} + v\mathbf{j}) \cdot \Delta t \tag{4.34}$$

However, most practical flows are 3-D, and the laser light sheet is of finite thickness. In such cases, Eq. 4.34 becomes

$$\mathbf{D} = -M \cdot \left[\left(d_x\mathbf{i} + d_y\mathbf{j}\right) + d_z\,\frac{(x\mathbf{i} + y\mathbf{j})}{s} \right] = -M \cdot \Delta t \cdot \left[\left(u\mathbf{i} + v\mathbf{j}\right) + w \cdot \frac{(x\mathbf{i} + y\mathbf{j})}{s} \right] \tag{4.35}$$

Table 4.1 Calculated Values of the Particle Images of a 1-µm Particle
($\lambda = 532$ nm, $M = 0.25$)

$f_\#$	D_i (µm)	δ_z (mm)
1.4	2.3	0.25
2.8	4.6	0.5
4.0	6.5	1.1
5.6	9.1	2.0
8.0	13.0	4.2
11	17.9	7.8
16	26.0	16.6

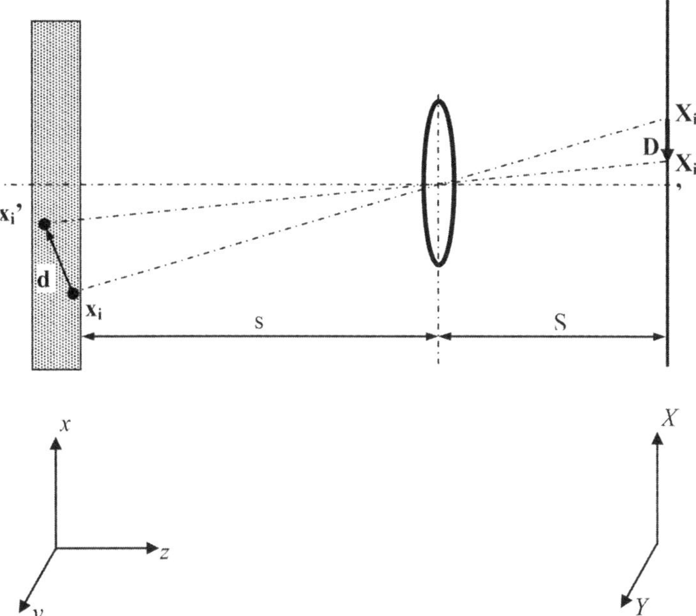

Fig. 4.13 Image formation of particle movement on the recording plane.

The additional third term due to perspective effects means that a particle displacement normal to the light sheet affects the particle image displacement. This effect introduces an uncertainty in measuring the in-plane velocity components (u, v), especially at the edges of the measurement volume where (x, y) values are larger. Equation 4.35 shows that this so-called perspective error can be reduced by using a small magnification factor and shorter pulse separation, and by placing the imaging system further away from the measurement volume. If it is not properly addressed in the design of PIV measurements, the perspective error can easily go up to more than 10% of the mean flow velocity.

4.3.5 Evaluation of Particle Image Displacement Vectors

To obtain the velocity vector map, particle images need to be evaluated by locally correlating the two images of particles on two separate frames. Figure 4.14 shows a schematic representation of the evaluation process to determine the displacement vector. Each PIV recording is subdivided into a number of square interrogation regions; then the velocity vector in each interrogation region is evaluated by the cross-correlation method.

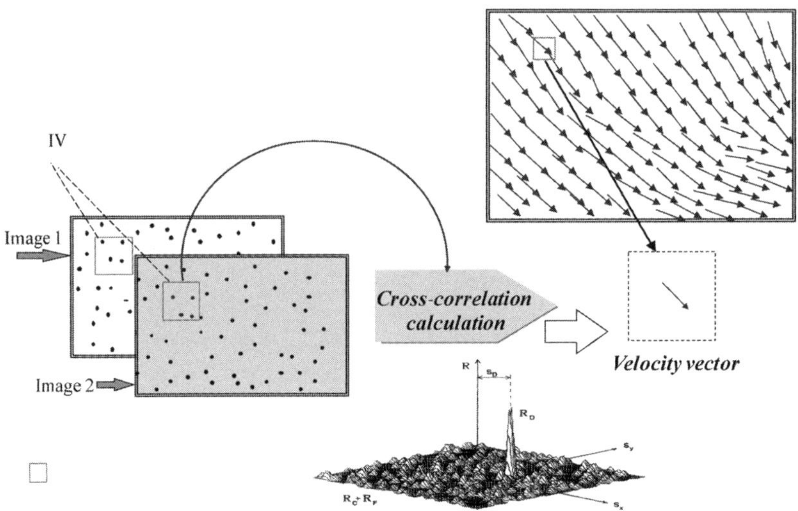

Fig. 4.14 Cross-correlation analysis used in modern digital PIV system.

Each interrogation region in each image consists of a random distribution of particle images, which correspond to certain pattern of N particles in the flow:

$$\Psi = \begin{pmatrix} x_1 \\ x_2 \\ \bullet \\ \bullet \\ x_N \end{pmatrix} \qquad (4.36)$$

$$\text{with} \quad x_i = \begin{pmatrix} x_i \\ y_i \\ z_i \end{pmatrix}$$

where Ψ represents the state of the particle ensemble at a given time t, and x_i is the position vector of the particle i at time t, in the interrogation volume of the laser sheet.

In the following analysis, capital letters refer to the coordinates in the image plane (Fig. 4.13) so that

$$X = \begin{pmatrix} X \\ Y \end{pmatrix}$$

Lowercase letters are used to represent the coordinates in the object plane. Assuming that the magnification factor is M, then

$$x = \frac{X}{M} \qquad and \qquad y = \frac{Y}{M}$$

The image intensity field of a single exposure can be written as (Raffel et al., 2007):

$$I(X,\Psi) = \sum_{i=1}^{N} P(x_i) \cdot \tau(X - X_i) \tag{4.37}$$

Where $\tau(X - X_i)$ is the point spread function of the imaging lens, and it describes the impulse response of the imaging lens. $P(x_i)$ is the system transfer function, giving the light energy of the image of an individual particle i inside the interrogation volume and its conversion into an electronic signal.

Assuming that there is a constant displacement d of all particles inside the interrogation volume (in the measurement region). Thus, the particle locations at the second exposure are given by

$$x_i' = x_i + d = \begin{pmatrix} x_i + d_x \\ y_i + d_y \\ z_i + d_z \end{pmatrix} \tag{4.38}$$

The corresponding particle image displacement is given by

$$\mathbf{D} = \begin{pmatrix} M \cdot d_x \\ M \cdot d_y \end{pmatrix} \tag{4.39}$$

The intensity distribution of the interrogation region for the second exposure can be modeled as:

$$I'(X,\Gamma) = \sum_{j=1}^{N} P'(x_j + d) \cdot \tau(X - X_j - D) \tag{4.40}$$

The particle image intensity distribution of the corresponding interrogation region for the first exposure is given by

$$I(X,\Gamma) = \sum_{i=1}^{N} P(x_i) \cdot \tau(X - X_i) \tag{4.41}$$

The cross-correlation function of two interrogation areas can then be written as

$$R_{II}(s,\Psi,d) = \langle I(X,\Gamma) \cdot I'(X+s,\Gamma) \rangle$$
$$= \frac{1}{A_I} \sum_{i,j}^{N} P(x_i) \cdot P(x_j + d) \int_{A_I} \tau(X - X_i) \cdot \tau(X - X_j + s - D) \cdot dX \tag{4.42}$$

Where **s** is the separation vector in the correlation plane. By distinguishing the $i \neq j$ terms, which represent the correlation of different particle images and therefore randomly distributed noise in the correlation plane, and the $i = j$ terms, which contain the displacement information desired, we come to the following expression:

$$R_{II}(s,\Psi,d) = \frac{1}{A_I} \sum_{i \neq j} P(x_i) \cdot P(x_j + d) \int_{A_I} \tau(X - X_i) \cdot \tau(X - X_j + s - D) \cdot dX$$
$$+ \frac{1}{A_I} \sum_{i=j} P(x_i) \cdot P(x_i + d) \int_{A_I} \tau(X - X_i) \cdot \tau(X - X_i + s - D) \cdot dX$$

or

$$R_{II}(s,\Psi,d) = \sum_{i \neq j} P(x_i) \cdot P(x_j + d) \cdot R_\tau(X_i - X_j + s - D)$$
$$+ R_\tau(s - D) \sum_{i=j} P(x_i) \cdot P(x_i + d) \tag{4.43}$$

which can be decomposed into three parts:

$$R_{II}(s,\Psi,d) = R_C(s,\Psi,d) + R_F(s,\Psi,d) + R_D(s,\Psi,d) \tag{4.44}$$

where the first two terms R_C and R_F contribute to the background noise in the correlation plane (Fig. 4.14), both resulting from the $i \neq j$ terms. $R_D(s,\Psi,d)$ represents the component of the cross-correlation function that corresponds to the correlation of images of particles obtained from the first exposure with images of identical particles obtained from the second exposure ($i = j$ terms), that is,

$$R_{D}\left(s,\Psi,d\right)= R_{\tau}(\mathbf{s}- D)\sum_{i=j} P(x_{i})\cdot P(x_{i}+d) \tag{4.45}$$

Hence, this displacement correlation distribution reaches its peak at $s = D$, which means that the average particle image displacement vector D can be determined from the maximum of the displacement correlation distribution in the correlation plane. The sign of D uniquely defines the direction of flow in the interrogation.

Although correlation may be calculated directly using Eq. 4.43. In practice, the most efficient way of calculating correlation is carried out by means of FFT algorithms. The correlation theorem states that the cross-correlation of two functions is equivalent to a complex-conjugate multiplication of their Fourier transforms:

$$R_{II} \Leftrightarrow \hat{I} \bullet \hat{I}'^{*} \tag{4.46}$$

Where $\hat{I}$ and $\hat{I}'$ are the Fourier transforms of the functions I and I', respectively. In practice, the Fourier transform is efficiently performed using the FFT algorithm. The calculation of the cross-correlation function can therefore be performed by computing two 2-D FFTs on equal-sized samples of the image followed by a complex-conjugate multiplication of the resulting Fourier coefficients. The most common FFT implementation requires the input data (size of the interrogation window) to have a binary-based dimension (e.g., 32×32 or 64×64 pixel samples).

4.3.6 Seeding Particles

The seeding requirements described in Section 4.2.6 are also applicable to in-cylinder PIV measurements. The requirements for extended illumination in the PIV measurement and sufficient light energy of scattered particle images mandate the use of high-power lasers. However, at higher laser energies, flare or spurious scattering from surrounding surfaces increases, causing significant image noise on the recording medium. To reduce the spurious scattering, the use of slightly larger seeding particles may be advantageous.

The seeding density needs to be sufficiently high. As more particle image pairs enter into the correlation calculation, the probability of a valid displacement detection increases. The actual number of image pairs captured in an interrogation area depends on three factors: the overall seeding density, the amount of in-plane displacement, and the amount of out-of-plane displacement. Keane and Adrian (1990) recommended that more than five particle pairs should be used in single-exposure/double-frame PIV per interrogation region, as in a modern digital PIV system. The second effect, which the particle image density has for the evaluation of PIV images, is that it can reduce the measurement uncertainty substantially. Therefore, if a flow can be densely seeded, then both a high valid detection rate, as well as a low measurement uncertainty, can be achieved using small interrogation areas, which in turn allows for high spatial resolutions. An upper limit of seed density is reached above which decorrelation occurs due to speckle formation.

4.3.7 Optimization of a PIV System

As noted earlier, several experimental parameters are involved in successful PIV measurement. Having discussed the operation of a PIV system in detail, we now consider how to optimize PIV measurements.

A PIV system may be described in terms of the following properties:

- Spatial/temporal resolution
- Measurement dynamic range
- Accuracy
- Speed of data processing

Spatial resolution is important because it determines the ability of the system to measure small spatial scales of flow structures. Spatial resolution is determined by the thickness of the laser sheet and the quality of the imaging optics. As noted previously, laser sheet thickness is determined by sheet-forming optics, and its thickness Δz is limited by the data drop-out rate due to the out-of-plane velocity component, $w \cdot \Delta t \leq \Delta z/4$. Therefore, a thicker laser sheet should be used so long as there is sufficient scattered energy density on the recording medium.

Temporal resolution is limited by the speed at which one can advance the film or transfer the image electronically. Standard film cameras are limited to a few frames per second. High-speed time-resolved PIV recordings at 100 image pairs per second have been reported using a high-repetition Cu vapor laser and a high-speed film drum camera (Stolz et al., 1992). Typical cross-correlation cameras are capable of capturing up to 15 PIV image pairs per second, provided that the laser can be operated at such a high frequency.

The measurement dynamic range is determined by the pulse separation, imaging conditions, and dimensions of interrogation regions. The minimum measurable displacement in the image plane is limited either by the effective image diameter given by Eq. 4.32,

$$d_i = \sqrt{M^2 d_p^2 + d_{diff}^2}$$

or by the spatial resolution of the recording medium, (e.g., the pixel size of a CCD camera, d_{pixel}).

Hence, the minimum velocity that can be resolved is given by

$$V_{min} = \Delta t \cdot d_i / M \qquad or \quad V_{min} = \Delta t \cdot d_{pixel} / M \qquad (4.47)$$

To optimize correlation process, an upper limit for the particle displacement in cross-correlation images should be half of the length L of the interrogation region. Therefore, the maximum velocity that can be measured given this restriction is

$$V_{max} = 0.5 \Delta t \cdot L / M \qquad (4.48)$$

The dynamic range can then be expressed as a single ratio of these two quantities:

$$D_v = V_{max} / V_{min} = 0.50\, L/d_i \quad or \quad D_v = V_{max} / V_{min} = 0.50\, L/d_{pixel} \qquad (4.49)$$

Typical values of the velocity dynamic range for cross-correlation PIV measurements can reach 100. In engine flows, the limited measurement range will prevent the simultaneous measurement of high- and low-velocity regions. Therefore, erroneous velocity measurements can be expected in regions of the flow that fall outside the measurement range, and these will have to be removed in a systematic validation procedure. In practice, the limited velocity dynamic range may be further restricted at the lower end of the measurement dynamic range by enlarged particle images caused by optical aberration of imaging optics and noise in the recording medium.

It is interesting to analyze the measurement dynamic range further when the particle image is diffraction limited, that is, $d_i = d_{diff}$. Assuming that the depth of field is equal to the light sheet thickness, then for a cubic interrogation volume, we have

$$L/M = \delta_z = l = \Delta z \qquad (4.50)$$

Substitution of Eqs. 4.30 and 4.33 into 4.49 gives the following result:

$$D_v \propto \frac{L}{d_i} = \frac{1}{1.22}\sqrt{\frac{l}{\lambda}} \qquad (4.51)$$

Thus, the measurement dynamic range is a function of the interrogation volume size and the laser wavelength. Equation 4.51 also shows that the dynamic range is independent of the imaging system magnification, assuming diffraction-limited imaging. This has several implications for the design of a PIV experiment. First, magnification can be reduced to allow a larger area of flow to be recorded without compromising the dynamic range, as long as the recording medium is able to sufficiently resolve the particle image. Second, with sufficient scattering, a lower magnification allows the imaging system to be placed further from the measurement section, limiting perspective errors and relaxing restrictions on experimental design.

The pulse separation Δt should be selected so that

$$\frac{\left(D_{max}\big/M\right)}{V_{max}} \leq \Delta t \leq \frac{\left(D_{min}\big/M\right)}{V_{min}} \qquad (4.52)$$

As noted previously, the minimum measurable displacement D_{min} in the image plane is either limited by the spatial resolution of the recording medium (pixel size of a CCD camera) or the effective image diameter d_i. The upper limit for the particle displacement D_{max} should be one quarter of the interrogation region length L.

The overall measurement accuracy in PIV is affected by a number of factors extending from the recording process all the way to the methods of evaluation, including recording medium noise, velocity gradients, out-of-plane particle motion, and particle motion outside the measurement dynamic range. The perspective error can be significantly reduced by placing the imaging system further away from the measurement plane. As the perspective error diminishes in proportion to the out-of-plane displacement, it can be reduced to a great extent if the measurement plane can be oriented with the predominant flow direction. The particle-seeding density should be sufficiently high so that there will be eight pairs of particles in each interrogation region, which will increase the valid data rate of PIV evaluation.

The measurement accuracy in PIV evaluation can be assessed experimentally by using actual PIV recordings for which the displacement data are known reliably. For example, PIV recordings of a quiescent flow can be used in determining the measurement accuracy.

In summary, successful PIV measurement demands the optimization of several parameters, ranging from the illumination source and imaging optics all the way to the evaluation of PIV recordings. The important parameters involved in the optimization of PIV measurement are summarized in Table 4.2. The basic procedures involved are as follows:

1. The design of a PIV experiment can begin with the selection of laser light sheet thickness, which should be sufficiently thick so that maximum out-of-plane displacement is less than a quarter of the light sheet thickness.

2. The required laser pulse separation should be chosen so that the maximum particle image displacement is approximately one-half of the interrogation region size for the cross-correlation analysis.

3. The imaging optics should have a magnification factor much less than 1, and its *f*-number setting selected so that the light sheet width falls within its depth of field. A longer distance between the imaging system and the object plane is preferred, but there should be enough light energy of scattered particle image for recording.

4. The interrogation process bears important consequences on the measurement dynamic range and accuracy. A larger interrogation region increases dynamic range, but it compromises accuracy and spatial resolution. During interrogation, the interrogation windows corresponding to the two separately exposed particle images should be offset by the estimated mean particle image displacement vector for improved data yield.

In practice, the pulse separation can be optimized by repeated measurements using different pulse separations. The interrogation region size therefore has to be adjusted accordingly at the interrogation stage to accommodate the maximum particle displacements that are present. As a consequence, the offset required in cross-correlation stage needs to be optimized as well.

4.3.8 Postprocessing of PIV Data

After evaluation of PIV recordings, sometimes a number of incorrectly determined velocity vectors are present as a result of noise or artefacts in the particle images. These spurious velocity vectors are referred to as outliers (Fig. 4.15). In the simplest case, they can be detected by visual inspection of the magnitude and direction of velocity vectors and their deviation from their nearest neighbors. In practice, data validation algorithms, such as the global

Table 4.2 Summary of the PIV System Variables and Their Optimization

Laser Source	Light sheet thickness(Δz):	$\Delta z \geq 4 \cdot w \cdot \Delta t$; thicker sheet with sufficient spatial resolution.
	Pulse separation (Δt):	Cross-correlation: $D_{max} \leq 0.5\,L$
Seeding Particles	Particle size (d_p):	1 μm–10 μm; small particle to follow turbulent flow, large particle to increase scattered light.
	Seeding density:	≈6 particles in the interrogation region for improved valid data rate (signal-to-noise ratio).
Imaging Optics	Magnification (M)	<1.0; smaller magnification for increased measurement area and reduced perspective error.[1]
	f-number ($f\#$)	selected so that depth-of-field $\delta_z \geq \Delta z$ and particle image diameter $D_i \leq d_{pixel}$.[1]
Cross-Correlation CCD Camera	Pixel size and No.:	The format of camera determines the spatial resolution of PIV recordings.
	Frame rate (30 fps):	The number of PIV image pairs per second is half the frame rate.
Interrogation Regions	Size ($L \times L$):	Larger L for increased dynamic range, smaller L for improved accuracy.
	Number:	Limited to a square matrix of $2^n \times 2^n$ by FFT algorithms used in digital correlation calculations.
Evaluation Scheme	Image shift between the two frames:	The two separately exposed images should be shifted by the estimated mean particle image displacement vector to reduce the in-plane loss of correlation.

Note: 1. So long as there is sufficient light energy of scattered particle images on the recording medium.

histogram operator and the dynamic mean value operator, can be implemented to remove these outliers (Raffel et al., 2007).

The removal of certain spurious velocity vectors leaves gaps in the vector field that need to be filled. This normally involves interpolation using a weighted average of the surrounding data. If required, the continuity of the data can be further improved by smoothing using numerical low-pass filters.

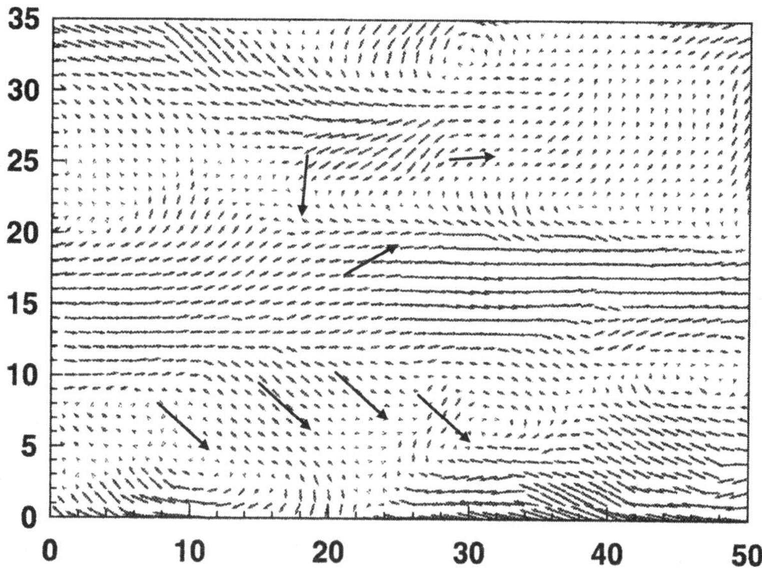

Fig. 4.15 Raw velocity vector map with outliers.

Results of PIV measurements are normally presented as velocity vector maps. Ensemble-averaging of cyclic flows and vector field operations (e.g., vorticity to detect tumble or swirl in the flow) are needed to understand fluid mechanical features. PIV data animation is particularly useful for better understanding of time series of PIV recordings.

4.3.9 Advanced PIV Systems

With the availability of high-speed solid state lasers and cameras, high-speed time-resolved PIV measurements can be performed. Diode-pumped solid-state lasers can provide a burst of short laser pulses at tens of kHz. In conjunction with high-speed video cameras, high-speed PIV measurements at several kHz can be obtained by performing a cross-correlation analysis of the adjacent frames.

Three-dimensional or stereoscopic PIV systems have also been developed for extracting the third, out-of-plane, velocity component from two cameras placed at 90° to each other to observe the light sheet plane from two different angles (Fig. 4.16). The camera and lens are aligned with an angle to the light sheet, and hence distortion of image fields occurs. Therefore the angle is kept so small that defocusing is acceptable. To overcome focusing problems in angular displacement, tilting the camera sensor relative to the lens is required and the Scheimpflug condition should be satisfied. In this arrangement image plane, the lens's principal axis plane and object plane (light sheet) meet at the same point (Fig. 4.16). The Scheimpflug arrangement allows the plane of best

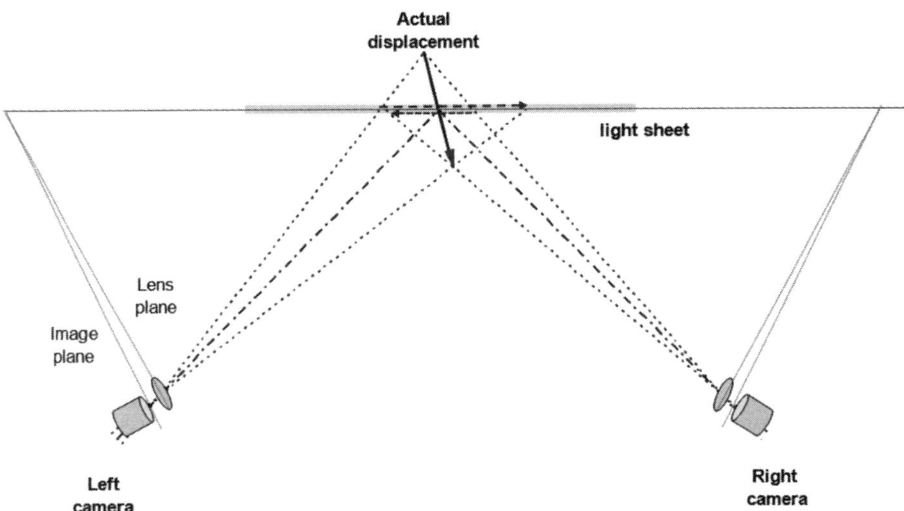

Fig. 4.16 Schematic of stereoscopic PIV setup.

focus to be located in the plane of light sheet while having the camera view the light sheet from an off-axis angle. The Scheimpflug configuration introduces perspective distortion to the images, causing a rectangle in light sheet plane to be imaged as a trapezoid on the image sensor (Fig. 4.17). Stereoscopic evaluation is possible only within the area covered by both cameras. Perspective distortion is corrected by using a second- order mapping function of a matrix that transforms the pixel coordinate system into real coordinate systems. This is included in the 3-D PIV system as a numerical model describing how objects in space are mapped onto the sensor of each camera.

The correct focus and tilt angles are achieved when all the particles within the camera field of view are in good focus. The real-time display of the images captured by the camera makes this setup process straightforward. The commercial stereoscopic PIV stereo-camera systems allow user selection of the geometry of the camera setup. The user has full control of the distance and angle between the cameras as well as fine control of the angle between the CCD array and camera lens to focus correctly.

Parameters for the numerical model are determined through camera calibration in order to map the coordinates in images to those in the object plane. Camera calibration is performed by recording images of a target with calibration markers at known positions and comparing known marker positions with corresponding marker positions on each camera image. In this way, model parameters can be adjusted to give the best possible fit, by using 3-D PIV software that includes camera calibration routines to measure and account for perspective distortion arising from the skewed orientation of the cameras, routines to compute the third velocity component from the two camera datasets, and presentation of three-component velocity information.

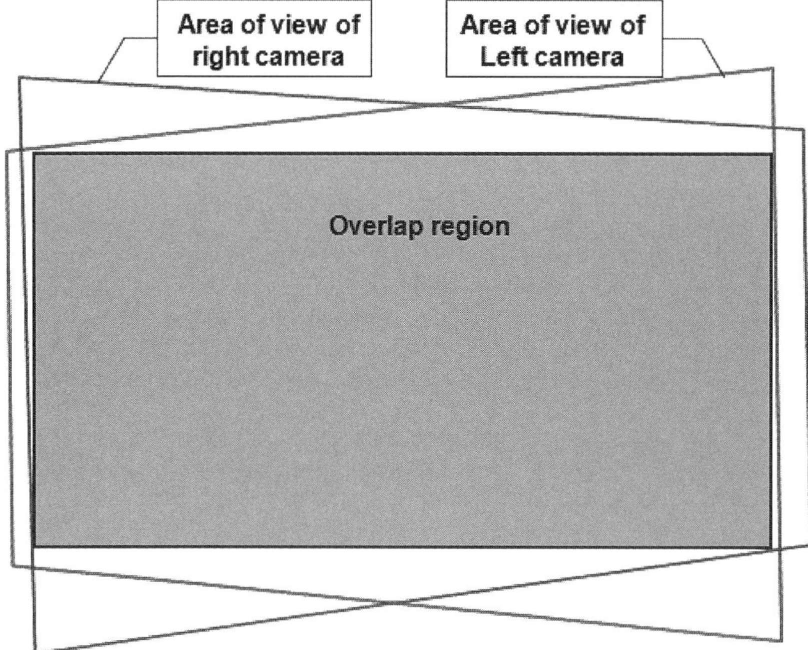

Fig. 4.17 Images of stereoscopic PIV.

The application of stereoscopic PIV to in-cylinder flow measurement is complicated by the requirement of large optical access and the additional distortion caused by the cylindrical optical liner.

4.3.10 PIV Measurements in IC Engines

The first and one of the most comprehensive in-cylinder flow measurements using PIV was carried out in the late 1980s by Reuss et al., (1989). The PIV experiments were performed in a specially constructed single-cylinder optical engine. The engine was equipped with a single poppet valve to maximize the size of a window in the cylinder head, through which the particle images were recorded. Images of TiO_2 particles from a single double-pulsed Nd:YAG laser were photographed with a 35-mm camera. The presence of strong axial swirl within the laser sheet allowed the velocity direction to be determined unambiguously using autocorrelation analysis. In addition to instantaneous velocity vector maps, large- and small-scale flow structures were analyzed by low-pass and high-pass filtering, respectively. Reuss and his team then applied the same PIV setup to a skip-fired engine so that the velocity, vorticity, and strain measurements ahead of a propagating flame could be made (Reuss et al., 1990).

As digital recording and processing become standard in modern PIV systems, PIV can be readily applied to obtain in-cylinder flow measurements. The most recent examples include the investigation of the structure and evolution of the velocity field in the optically accessible diesel engine (Deslandes et al., 2004; Petersen and Miles, 2011); the in-cylinder flow characterization and CFD validation by PIV and LDV measurements (Ishima et al., 2008; Malcolm et al., 2011); investigation of the intake port design and cylinder head geometry on in-cylinder flow structures in SI engines (Karhoff et al., 2011; Heim and Ghandhi, 2011; Li et al., 2004). With renewed interest in direct-injection spark-ignition engines, high-speed time-resolved PIV measurements were carried out to understand the evolution of flow structures in such engines (Jarvis et al., 2006; Müller et al., 2011). An overview of in-cylinder flow velocity measurements carried out since 1980s can be found in a review paper by Farrell (2005).

In the following section, the application of PIV to in-cylinder flow measurements carried out in a laboratory at Brunel University are presented to illustrate the capability of the PIV techniques and additional flowfield information that can be extracted by specially constructed data-processing algorithms (Li et al., 2004).

4.3.10.1 Experimental Setup

The experiment was conducted on a single-cylinder optical engine with an extended cylinder block. Figure 4.18 shows the complete engine and PIV system setup. A 4-cylinder head with four valves per cylinder and a pent-roof combustion chamber was mounted on the extended cylinder block. The engine had two optical accesses: an extended transparent piston with a 55-mm diameter quartz window and a 30-mm long quartz ring between the cylinder block and the cylinder head. Swirl motion was measured on the horizontal plane 12 mm below the cylinder head. Tumble motion could be measured on both the vertical symmetric plane of the combustion chamber and the vertical plane through the centers of the intake and exhaust valves. Measurements were taken under motored conditions at engine speed of 1200 rpm.

The second harmonic output (532 nm) from two Nd:YAG lasers was used as the light source for the PIV measurement. The beams from the two lasers were combined into a collinear beam in a beam combination optics box. A pulse-delay generator was used to control the laser timing and laser energy. When it received an external trigger signal, it sent out four delayed signals according to the delay times that were required to fire the flash lamps and the Q-switches of the two lasers, respectively. The trigger signal was generated by a shaft encoder and a counter at a preset crank angle. The shaft encode used in the experiment had a resolution of 1° CA.

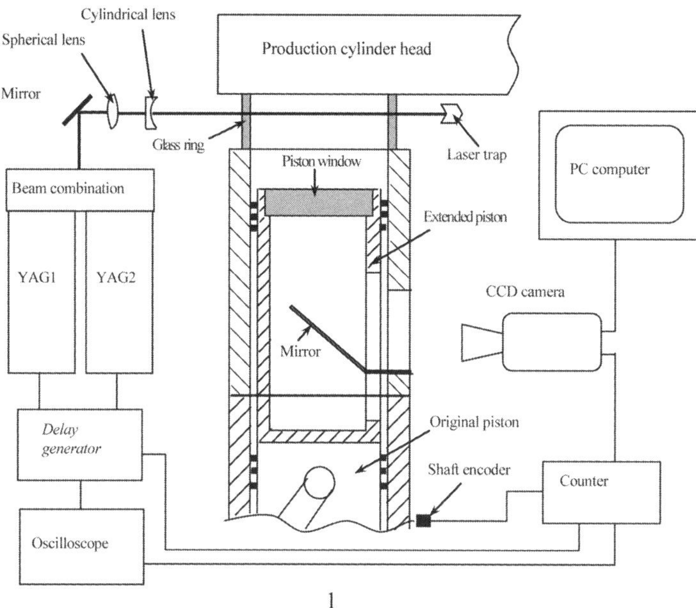

Fig. 4.18 PIV and engine setup for in-cylinder flow measurements.

A cross-correlation CCD camera of 1008×1018 pixels and 9-μm pixel spacing was used in the experiment. As shown in Fig. 4.19 , after receiving the trigger signal, the camera generates a strobe trigger output to start two exposure sequences in 20 μs. The first exposure lasts 255 μs, and the electronic shutter closes in 2° CA after receiving the trigger signal at 1200-rpm engine speed. The second exposure lasts about 33 ms, ending at about 243° CA after the trigger signal. In the meantime, the delay generator controls when the two lasers will fire after it receives the trigger signal. The delay times between the two laser pulses was set so that the first laser pulse was located in the last part of the first exposure period and the second laser pulse at the start of the second exposure period. Thus a frame-straddle pair of images could be captured in a short interval or laser pulse separation, ΔT. The first image from the first laser pulse was first stored in a vertical transfer registers on the CCD chip before it was transferred to a frame grabber in the PC, during which the second image was captured. The second image of particles in the flowfield by the second laser pulse was then transferred to the frame grabber in another 33 ms. The frame grabber in the computer immediately passed the digitized image information to the computer memory. The whole process of recording a pair of images and transferring them to the computer took about 483° CA at an engine speed of 1200 rpm. The next pair of particle images would start to be recorded only when the next trigger signal arrived.

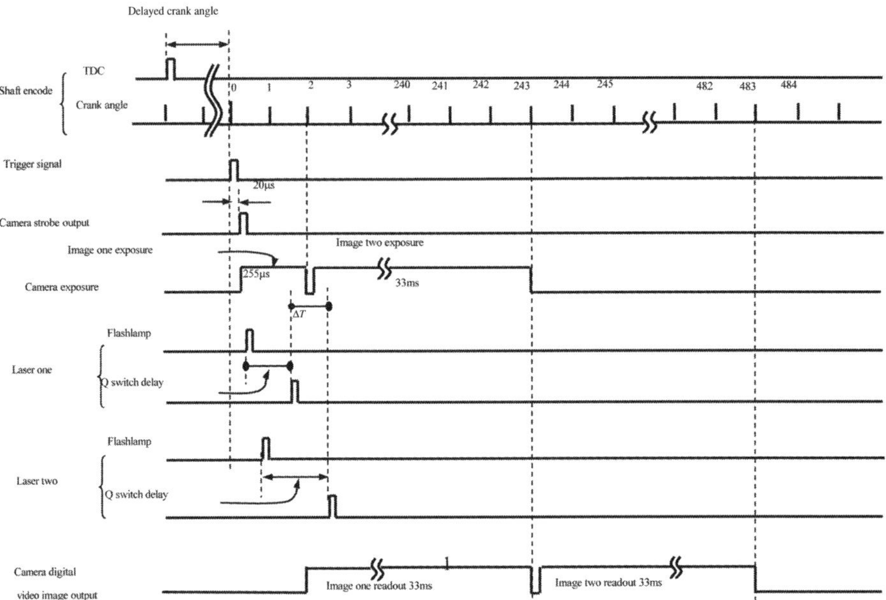

Fig. 4.19 Schematic timing diagram of the PIV measurement.

A jet atomizer was used to generate a fine mist of 0.5- to 5-µm silicone oil droplets. The particle mist was introduced into the cylinder through a flexible pipe connected to the intake port. A precision low-pressure regulator in the atomizer was used to control the seeding density.

The light sheet was produced by a spherical lens with a 1000-mm focal length and a concave cylindrical lens with a –76-mm focal length. The beam waist thickness was estimated to be 500 µm, and the divergence of the laser sheet thickness was about 90 µm within the measured flowfield. During recording, the CCD camera used a small *f*-number of 2.8 for achieving sharper particle images. With the small *f*-number, lower laser power was needed, leading to lower background light noises. The magnification factor was determined to be 1/7.8 so that the complete flowfield within the laser sheet could be recorded. The depth of field was estimated to be about 1.5 mm, larger than the thickness of the light sheet.

The laser pulse separation is an important parameter for obtaining better PIV results. The choice of laser pulse separation ΔT depends on the interrogation spot size, measurable velocity range, the number of particle pairs, and particle displacement in the interrogation area. To get a correct vector with high signal-to-noise ratio in the selected interrogation area, the maximum particle in-plane displacement should be limited to one-third of the spot size and maximum out-of-plane displacement should be less than one-fourth of the light sheet thickness. Moreover, the minimum in-plane displacement should be two particle-image diameters. Since the velocity range in the cylinder was from 1 to 50 m/s at an engine speed of 1200 rpm, the typical pulse separation ΔT was varied from 10 µs during the induction stroke to 30 µs during the compression stroke.

To eliminate the influence of background light noise caused by surface reflection, the combustion chamber was painted black and an optical band-pass filter with a central wavelength of 532 nm was mounted in front of the camera. It was also found that the overlapping of light sheets from the two lasers was crucial to obtain good cross-correlation images.

For each measurement, 120 PIV realizations were acquired from consecutive engine cycles at a particular crank angle so as to compute ensemble-averaged flow parameters.

During the postprocessing process, the image pairs were divided into small interrogation regions (32 × 32 pixels), corresponding to a spatial resolution of 2.2 × 2.2 mm. Although the spatial resolution selected was not small enough to describe the fine turbulent structures such as Kolmogorov length scale or even Taylor microscale (less than 2 mm), it was still small enough to be able to describe the integral length scale (approximately 5–12 mm) and larger-scale flow structures with scale size greater than the integral length scale. A cross-correlation algorithm was applied to the particle images within each region to determine the particle displacement. The Gaussian algorithm was used to detect the correlation peak in the current study. About 5% spurious vectors were present during the data process owing to the loss of particle pairs or low seeding density, particularly in the intake process large out-of-plane velocity presents. These spurious vectors were removed using a standard deviation operator and replaced by a mean velocity of the adjacent values.

Some degree of random measurement error hides in the original velocity vectors because of the combined effects of turbulence structure, velocity gradients, particle size and number, and even background light noise. These errors may be increased by an order of magnitude in computing turbulence parameters like rms fluctuating velocity, integral length scale, vorticity, and strain rate. Consequently, the smoothing technique using convolution of the two-component vector with an axisymmetric Gaussian kernel $w(k,m)$ was used to attenuate the high-frequency velocity fluctuation caused by these errors, as described by the following equations:

$$U(x_i, y_i) = \sum_{m=-M}^{M} \sum_{k=-K}^{K} w(k,m) U(x_{i-k}, y_{j-m}) / W \qquad (4.53)$$

$$V(x_{i,}y_i) = \sum_{m=-M}^{M} \sum_{k=-K}^{K} w(k,m)V(x_{i-k},y_{j-m})/W \qquad (4.54)$$

where

$$w(k,m) = \exp[-2(k^2 + m^2)/p^2]$$

$$W = \sum_{m=-M}^{M} \sum_{k=-K}^{K} w(k,m)$$

$U(x_i,y_i)$ and $V(x_i,y_i)$ are two components of velocity vectors at certain point. The quantity p determines the kernel size and therefore the smallest scale, which is to be retained in the smoothed velocity distribution. To prevent real high-frequency velocity fluctuation from being removed, a small value $p = 1.1$ was chosen.

4.3.10.2 Flowfield Analysis

Determination of ensemble-averaged mean velocity and fluctuating velocities. The in-cylinder flowfield prior to combustion is manifestly nonstationary, as a result of the dynamically changing boundary conditions imposed by the piston and the valve motions. The nonstationary aspect is responsible for the cycle-to-cycle variation in the in-cylinder flowfield. Furthermore, the in-cylinder flowfield is highly turbulent in the individual cycles. To identify the cyclic variations and in-cycle turbulence levels, the instantaneous velocity vector $\vec{U}_{(x,y,\theta,i)}$ may be decomposed into an ensemble-averaged mean velocity and two fluctuating components as follows:

$$\vec{U}_{(x,y,\theta,i)} = \vec{U}_{EA(x,y,\theta)} + \vec{u}_{LF(x,y,\theta,i)} + \vec{u}_{HF(x,y,\theta,i)} \qquad (4.55)$$

where

superscript $\rightarrow$ donates velocity vector

(x,y) denote the coordinates in the 2-D planar flowfield

θ is the crank angle at which the measurement is performed

i represents a particle cycle

The first term on the right-hand side of Eq. 4.55, $\vec{U}_{EA(x,y,\theta)}$, is the ensemble-averaged mean velocity over many cycles. The second and third terms, $\vec{u}_{LF(x,y,\theta,i)}$ and $\vec{u}_{HF(x,y,\theta,i)}$, are related to low-frequency velocity fluctuation (cyclic variation) and high-frequency velocity fluctuation (in-cycle turbulent velocity fluctuation), respectively. Both are composed of the all-frequency velocity fluctuation $\vec{u}_{F(x,y,\theta,i)}$, which is simplified as the fluctuating velocity in the following equation:

$$\vec{u}_{F(x,y,\theta,i)} = \vec{u}_{LF(x,y,\theta,i)} + \vec{u}_{HF(x,y,\theta,i)} \qquad (4.56)$$

In addition, the sum of the two terms on the right-hand side of Eq. 4.55 is defined as the in-cycle mean velocity, that is,

$$\vec{U}_{B(x,y,\theta,i)} = \vec{U}_{EA(x,y,\theta)} + \vec{u}_{LF(x,y,\theta,i)} \qquad (4.57)$$

Two basic approaches have been previously used for extracting cyclic variation and in-cycle turbulent velocity fluctuation from LDA measurements, as described in Section 4.2.9. In the first approach, some form of low-pass filtering in the temporal frequency domain is used to estimate the in-cycle mean velocity for a measured point (x,y). The rms deviations of the instantaneous velocities from the in-cycle mean velocities are then computed over all the cycles $u'_{HF,EA}$, and usually interpreted as a measure of the intensity of the true turbulent velocity fluctuations. The second approach separates the two fluctuation components by assuming the turbulent fluctuations are stochastic or random in nature and the cyclic variations should have an appreciable degree of phase stability or coherence. The mathematic tool used in second approach is the classical statistical tool based on the autocorrelation function.

In this study, a low-pass filtering scheme in the spatial frequency domain will be introduced. The analysis begins with the conventional ensemble-averaging analysis. The ensemble-averaged mean velocity is given explicitly by

$$\vec{U}_{EA(x,y,\theta)} = \frac{1}{N} \sum_{i=1}^{N} \vec{U}_{(x,y,\theta,i)} \qquad (4.58)$$

Where N is the total number of cycles in which the velocity vectors at a particular crank angle θ are measured. For simplicity, we will omit θ from this and subsequent formulae.

To obtain the fluctuation components in Eq. 4.55, a 2-D spatial filtering technique can be used. First, the 2-D instantaneous velocity data $\vec{U}_{(x,y,\theta,i)}$ are transformed into the spatial frequency domain using 2-D FFT. Then a spatial cutoff frequency in mm^{-1} is selected, above which all Fourier coefficients were set to zero. finally, the in-cycle mean velocity $\vec{U}_{B(x,y,i)}$ is obtained by performing an inverse FFT.

Low-frequency fluctuating velocity $\vec{u}_{LF(x,y,i)}$ can be achieved by subtracting ensemble-averaged velocity $\vec{U}_{EA(x,y)}$ from the in-cycle mean velocity, according to Eq. 4.57. High-frequency fluctuating velocity $\vec{u}_{HF_{(x,y,i)}}$ and the fluctuating velocity $\vec{u}_{F_{(x,y,i)}}$ are easily calculated from Eqs. 4.55 and 4.56.

The choice of a correct spatial cutoff frequency, no doubt, is important during the cycle-resolved procedure because the cutoff frequency provides a criterion for separating low- and high-frequency fluctuating velocities. In previous cycle-resolved LDV analysis, different temporal cutoff frequencies had been selected between 300 and 650 Hz at similar engine speed. Based on some preliminary studies, a spatial cutoff frequency 0.05 mm^{-1} was found to be appropriate in the cycle-resolved analysis.

After obtaining the fluctuating velocity components, the ensemble-averaged rms fluctuating velocity can be calculated as follows:

$$u'_{EA(x,y)} = \sqrt{\frac{1}{N}\sum_{i=1}^{N} u^2_{(x,y,i)}} \tag{4.59}$$

where $u_{(x,y,i)}$ may be the low-frequency, high-frequency, and all-frequency fluctuating velocities. Sequentially, the kinetic energy of fluctuating velocity can be derived from rms fluctuating velocities:

$$E_{EA(x,y)} = \frac{1}{2}[u'^2_{EA(x,y)} + v'^2_{EA(x,y)}] \tag{4.60}$$

where $u_{EA(x,y)}$ and $v_{EA(x,y)}$ is rms fluctuating velocity components in x and y coordinate directions, respectively. They may be rms fluctuating velocities of low-frequency, high-frequency, and all-frequency.

Figure 4.20(a)–(f) shows the cycle-resolved velocity fields at 60° CA before TDC of the compression stroke. The spatial cutoff frequency used in low-pass filtering is 0.05 mm^{-1}, corresponding to 20 mm in spatial domain. The ensemble-averaged mean velocity field shows the presence of solid-body-like swirl vortex. The instantaneous velocity field is rather different. It displays a 3-D flow characteristic, especially in the region around the point (0,−10). Although a swirl vortex is visible in the flowfield, it is far from a solid-body vortex.

The difference between the instantaneous velocity and the ensemble-averaged mean velocity can be readily seen in the fluctuating velocity field. It can be seen from Fig. 4.20(c) that though the fluctuating velocity is randomly distributed, there are some larger-scale structures, such as a vortex, for example, at point (4,8), and source-like flow, for example, at point (−12,−20), where the velocity vectors diverge outward from a point.

The low-pass filtered velocity distribution, shown in Fig. 4.20(d), that is, an in-cycle mean velocity field, is similar to its instantaneous velocity field in Fig. 4.20(a). It contains a large-scale vortex structure-swirl. And not only swirl

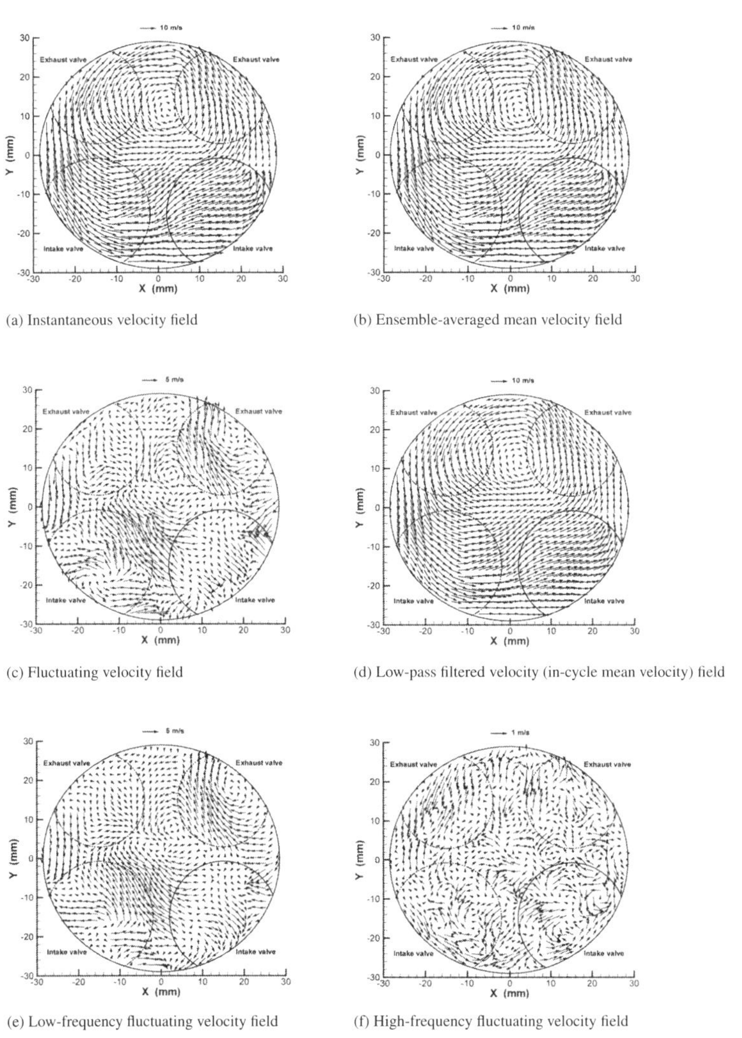

(a) Instantaneous velocity field

(b) Ensemble-averaged mean velocity field

(c) Fluctuating velocity field

(d) Low-pass filtered velocity (in-cycle mean velocity) field

(e) Low-frequency fluctuating velocity field

(f) High-frequency fluctuating velocity field

Fig. 4.20 Cycle-resolved velocity fields (100th cycle, 60° CA BTDC, 1200 rpm, spatial cutoff frequency 0.05 mm⁻¹).

center and velocity distribution are identical, but even velocity value is nearly same. However, the low-pass velocity field is smoother than the instantaneous velocity field because high-frequency velocity fluctuation has been removed.

The distribution of low-frequency fluctuating velocity shown in Fig. 4.20(e), that is, cyclic variation of bulk flow, is analogous to that of the fluctuating velocity field in Fig. 4.20(c). The larger-scale vortex structure at point (4,8) and source-like flow structure at point (−12,−20) are still visible. As the high-frequency random fluctuation has been removed from it, the low-frequency fluctuating velocity field is smoother than the total fluctuating velocity field in Fig. 4.20(c), and some larger-scale structures such as vortex at point (18,18) and source-like flow at point (−2,2) are more readily identifiable.

In the high-frequency fluctuating velocity field in Fig. 4.20(f), the velocity value is typically less than 1 m/s, much lower than the total fluctuating velocity value, and hence it becomes more random than the total fluctuating velocity field. But some smaller vortices can still be seen at positions (11,1), (21,−10), (0,10), (−16,−8), and so on. The size of these vortex structure is about or less than 8 mm.

Further comparisons among Fig. 4.20(c), (e), and (f) show that not only the velocity distribution but also the velocity value are nearly the same between the low-pass filtered velocity field and the instantaneous velocity field, and between the low-frequency fluctuating velocity field and the total fluctuating velocity field. The high-frequency fluctuating velocity value is very small. This result indicates that flow structure with frequency lower than 0.05 mm^{-1} (or with spatial scale larger than 20 mm) is absolutely dominant in the in-cylinder flowfield. The cyclic variation is mostly responsible for the spatial turbulence present in the in-cylinder flowfield.

Figure 4.21 shows the corresponding distributions of cycle-resolved rms fluctuating velocities: rms low-frequency fluctuating velocity fields in plots (a) and (b); rms high-pass fluctuating velocity in plots (c) and (d); and rms fluctuation velocities in the x and y directions in plots (e) and (f).

Figure 4.22 shows the related distributions of 2-D cycle-resolved fluctuation kinetic energy. The relationship among total fluctuation kinetic energy, low-frequency kinetic energy, and high-frequency kinetic energy is analogous to the relationship among cycle-resolved rms fluctuating velocities shown in Fig. 4.21. The low-frequency fluctuation kinetic energy field has a distribution similar to the total kinetic energy field. Both fields have nearly identical crests and troughs, but the low-frequency kinetic energy distribution shows more clearly crests and troughs than the total kinetic energy field.

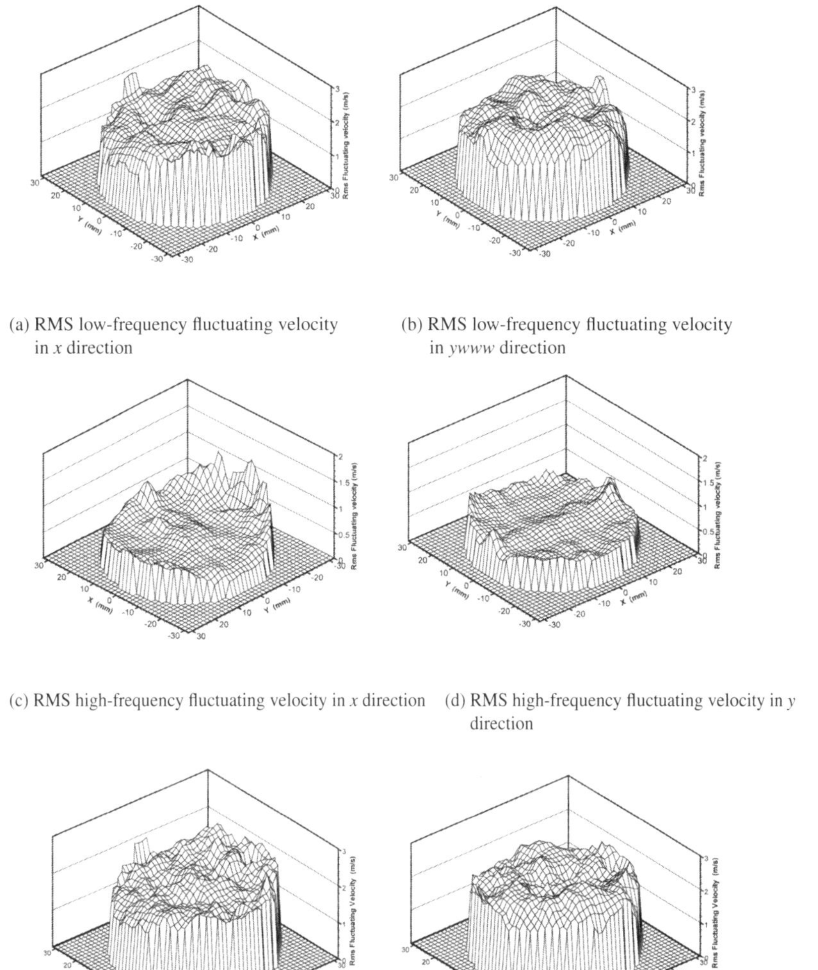

(a) RMS low-frequency fluctuating velocity in x direction

(b) RMS low-frequency fluctuating velocity in *ywww* direction

(c) RMS high-frequency fluctuating velocity in x direction

(d) RMS high-frequency fluctuating velocity in y direction

(e) RMS fluctuating velocity in x direction

(f) RMS fluctuating velocity in y direction

Fig. 4.21 Distributions of cycle-resolved rms fluctuating velocities in x and y directions at 60° CA BTDC (before top dead center).

Estimation of integral length scale is a measure of the size of the largest eddies, which contain most of the turbulence kinetic energy. It is defined as the integral of the autocorrelation coefficient of fluctuating velocity at two adjacent points in the flowfield with respect to the variable distance between the points. In LDA measurements, the integral length scale was usually estimated from integral time scale multiplied by mean velocity in terms of the Taylor hypothesis. But this method usually leads to great uncertainty because the mean velocity in the cylinder may be greater than the turbulence intensity,

121

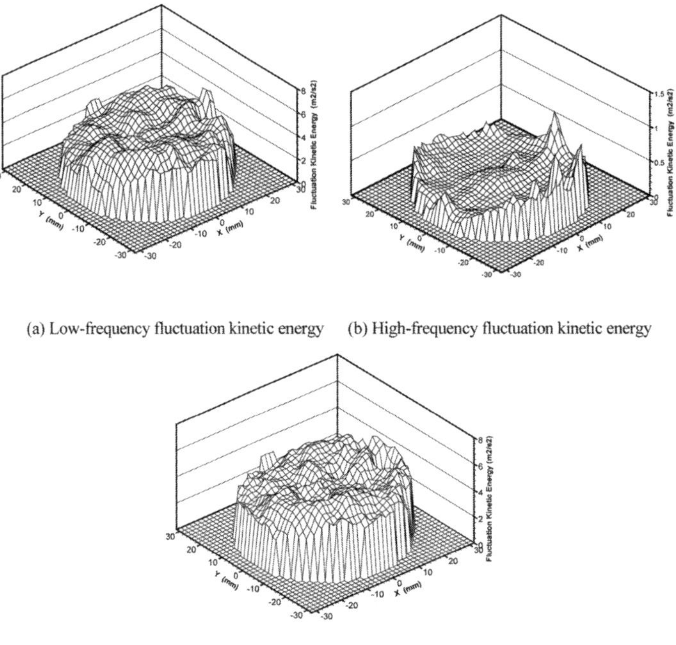

(a) Low-frequency fluctuation kinetic energy   (b) High-frequency fluctuation kinetic energy

(c) Total fluctuation kinetic energy

Fig. 4.22 Distributions of cycle-resolved fluctuation kinetic energy at 60° BTDC.

and the turbulent velocity distribution is normally nonisotropic. These flow characteristics do not fit with the conditions of Taylor hypothesis. In planar PIV measurement, however, the integral length scale can be directly evaluated from spatial correlation functions of fluctuating velocity:

$$L_{x(x,y)} = \int_0^\infty R(x,y,\xi,0)\,d\xi \tag{4.61}$$

$$L_{y(x,y)} = \int_0^\infty R(x,y,0,\eta)\,d\eta \tag{4.62}$$

where $L_{x(x,y)}$ and $L_{y(x,y)}$ are the integral length scales at point (x,y) along the x and y directions, respectively. $R_{(x,y,\xi,\eta)}$ is the spatial autocorrelation function of the fluctuating velocity at point (x,y), given by

$$R_{(x,y,\xi,\eta)} = \frac{\dfrac{1}{N}\sum_{i=1}^{N} u_{F(x,y,i)} \cdot u_{F(x+\xi,y+\eta,i)}}{u'_{F,EA(x,y)} \cdot u'_{F,EA(x+\xi,y+\eta)}} \tag{4.63}$$

where (ξ, η) are the displacement coordinates along the x and y directions in the correlation plane.

Similar to the calculation of the ensemble-averaged flow parameters, a series of discrete spatial windows with sizes of 2.2×2.2 mm in the spatial delay coordinate (ξ, η) is used to estimate the autocorrelation function.

Because the fluctuating velocity $u_{F(x,y,i)}$ contains the cyclic variability of large-scale flow structure, the integral length scale of turbulence may be overestimated by using $u_{F(x,y,i)}$. It is desirable to use the high-frequency fluctuating velocity component $u_{HF(x,y,i)}$ to calculate the integral length scale in Eq. 4.63. But greater error may be present during calculation of the autocorrelation function owing to the smaller value and irregular distribution of the high-frequency velocity field. In addition, the value of the high-frequency fluctuating velocity component is sensitive to the cutoff frequency. Therefore as a compromise, $u_{F(x,y,i)}$ was used to calculate the integral length scale. If $u_{F(x,y,i)}$ in Eq. 4.63 is substituted by $v_{F(x,y,i)}$, the integral length scale of the other fluctuating velocity component is obtained.

Table 4.3 shows the calculated integral length scales at some selected spatial locations. It is seen that the integral length scale in x direction is different from that in y direction at each point of measurement region. In addition, it can also be shown that the integral length scale of each velocity component (u_F, v_F) is different from each other. Furthermore, the result shows that the value of the integral length scale varies with the points in the flowfield.

Vorticity and strain rate play fundamental roles in turbulent flows, for example, in energy transfer by vortex stretching and the production of the Reynolds stress and turbulent kinetic energy. The strain rate affects the turbulent combustion because of its involvement in the flame stretching, wrinkling, and extinction in the reaction sheet. In the present PIV measurements, the spatial resolution is limited to 2.2×2.2 mm. This is not small enough to identify the smallest scale vorticity and stain rate directly responsible for the flame structure within the reaction sheet. But the calculated vorticity and strain rate may still be significant for understanding the fuel and air mixing process and hence the subsequent turbulent combustion process.

From the 2-D PIV velocity data, the following parameters can be determined: the out-of-plane component of vorticity ω_z, the in-plane components of shearing strain ε_{xy} and the extensional strain ε_{xx}, ε_{yy}:

$$\omega_z = \frac{\partial V}{\partial x} - \frac{\partial U}{\partial y} \tag{4.64}$$

Table 4.3 Integral Length Scale of Fluctuating Velocity

Points in the Field	Fluctuating Velocity Component	Lx(mm)	Ly(mm)
(0,0)	u_F	10.8	7.8
	v_F	5.7	6.7
(2,12)	u_F	10.7	7.5
	v_F	5.2	9.1
(−10,−10)	u_F	9.62	6.17
	v_F	8.89	9.20
(12,−10)	u_F	8.17	5.85
	v_F	6.69	9.76
(10,10)	u_F	8.16	7.32
	v_F	6.41	9.66

$$\varepsilon_{xy} = \frac{\partial U}{\partial y} + \frac{\partial V}{\partial x} \tag{4.65}$$

$$\varepsilon_{xx} = \frac{\partial U}{\partial x} \tag{4.66}$$

$$\varepsilon_{yy} = \frac{\partial V}{\partial y} \tag{4.67}$$

Since the density change with time is much smaller than the velocity gradient in the engine cylinder, the out-of-plane strain ε_{zz} can be approximately estimated from the continuity equation of impressible fluids:

$$\varepsilon_{zz} = -(\varepsilon_{xx} + \varepsilon_{yy}) = -(\frac{\partial U}{\partial x} + \frac{\partial V}{\partial y}) \tag{4.68}$$

In a PIV velocity data map, two extensional strains ε_{xx} and ε_{yy} can be evaluated from the first-order differences at each point. The vorticity at each point of the vector map is usually evaluated in terms of the Stokes theorem by computing the circulation around a closed contour surrounding the point and then reducing the closed area to zero, that is,

$$(\omega_z)_{i,j} = \lim_{A \to 0} \frac{\Gamma_{i,j}}{A} = \lim_{A \to 0} \frac{1}{A} \oint \vec{U} \cdot d\vec{l} \tag{4.69}$$

where $\vec{l}$ and A are the path and the area of the closed contour, respectively. Instead of an infinitely small area, a 3×3 small rectangular contour is chosen to calculate the averaged vorticity within the enclosed area. The circulation $\Gamma_{i,j}$ at central point (i,j) can be calculated using a standard integration scheme such as the trapezoidal rule around the neighboring eight points. So Eq. 4.69 can be written as

$$(\omega_z)_{i,j} \cong \frac{\Gamma_{i,j}}{4\Delta x \cdot \Delta y}$$

with

$$\Gamma_{i,j} = \frac{1}{2}\Delta x(U_{i-1,j-1} + 2U_{i,j-1} + U_{i+1,j-1})$$

$$+ \frac{1}{2}\Delta y(V_{i+1,j-1} + 2V_{i+1,j} + V_{i+1,j+1})$$

$$\tag{4.70}$$

$$- \frac{1}{2}\Delta x(U_{i+1,j+1} + 2U_{i,j+1} + U_{i-1,j+1})$$

$$- \frac{1}{2}\Delta y(V_{i-1,j+1} + 2U_{i-1,j} + U_{i-1,j-1})$$

where Δx, Δy is the grid spacing in two coordinate directions, respectively. For the out-of-plane or extensional strain, an analogy to the relation between vorticity and circulation can be given: In place of the circulation, the net flow across the boundaries of the contour is calculated as follows:

$$
\begin{aligned}
(\varepsilon_{zz})_{i,j} \cong\ & \frac{V_{i-1,j-1} + 2V_{i,j-1} + V_{i+1,j-1}}{8\Delta y} \\[2ex]
& -\frac{V_{i+1,j+1} + 2V_{i,j+1} + V_{i-1,j+1}}{8\Delta y} \\[2ex]
& +\frac{U_{i+1,j-1} + 2U_{i+1,j} + U_{i+1,j+1}}{8\Delta x} \\[2ex]
& -\frac{U_{i-1,j+1} + 2U_{i-1,j} + U_{i-1,j-1}}{8\Delta x}
\end{aligned}
\tag{4.71}
$$

The shear strain can also be obtained in an analogy to the vorticity and the out-of-plane strain, although no such similar variables like circulation or net flow exists for the shear strain.

$$
\begin{aligned}
(\varepsilon_{xy})_{i,j} \cong\ & -\frac{U_{i-1,j-1} + 2U_{i,j-1} + U_{i+1,j-1}}{8\Delta y} \\[2ex]
& +\frac{U_{i+1,j+1} + 2U_{i,j+1} + U_{i-1,j+1}}{8\Delta y} \\[2ex]
& -\frac{V_{i-1,j+1} + 2V_{i-1,j} + U_{i-1,j+1}}{8\Delta x} \\[2ex]
& +\frac{V_{i+1,j-1} + 2V_{i+1,j} + U_{i+1,j+1}}{8\Delta x}
\end{aligned}
\tag{4.72}
$$

Fig. 4.23 is a contour map of the instantaneous out-of-plane vorticity for the same cycle as described in Fig. 4.22. Because vorticity and high-frequency randomly fluctuating velocity are both the fundamental characteristic of small-scale turbulence, the high-frequency fluctuating velocity distribution is superimposed, as shown in Fig. 4.21(f), onto the correspondent vorticity contour map to identify any interactions between them. It can be seen that the vorticity is not uniformly distributed within the flowfield. Several small areas with identical signs and a similar magnitude of vorticity are irregularly distributed in the contour map. These areas of intense vorticity are linked

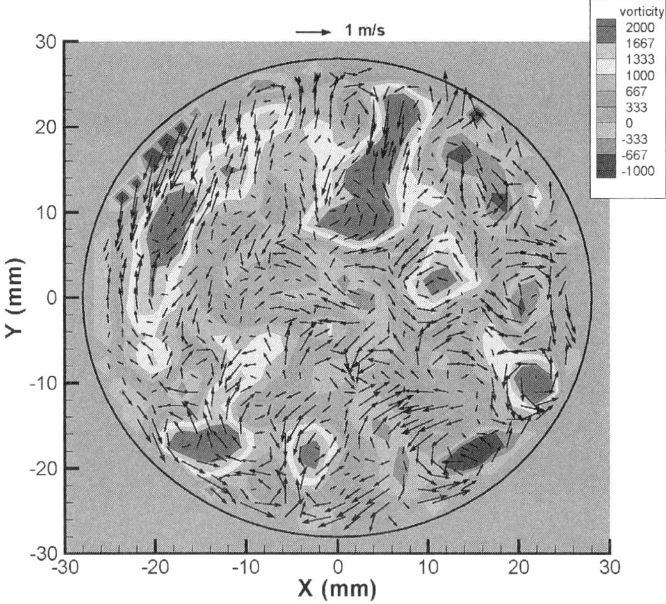

Fig. 4.23 Out-of-plane vorticity contour map for 100th cycle at 60° CA BTDC, 1200 rpm.

to small-scale vortex-like structures. Counterclockwise small vortices, for example, at points (21,–10), (4,10), (–3,–19), and (12,2), correspond to positive intense vorticity (red in the figure) and clockwise vortices, for example, at points (21,–2) and (3,0), correspond to negative intense vorticity. It should be noticed that the vorticity is determined directly from the basic instantaneous velocity data, whereas the high-frequency fluctuating velocity is obtained using a spatial-filtering technique. The close correspondence of the vorticity maps with the high-pass filtered velocity fields indicates that the results of whole data processing are satisfying.

Fig. 4.24(a)–(d) show contour maps of shearing strain and extensional strains with the identical color scale obtained from the same PIV realization as above. High-pass filtered velocity vectors are drawn on the same figure to identify the relationship between these strains and high-frequency fluctuating velocity field. Similar to the vorticity contour distribution, these shearing strain and extensional strains are distributed nonuniformly. A number of small patches with high values of strain rate are seen in the contour maps. The structure and magnitude of these instantaneous strain-rate fields may have important implications for flame development in premixed-charge engines. Many studies have demonstrated that strain rate on the order of a few hundred to a few thousand s^{-1} can significantly reduce the laminar burning velocity or even lead to flamelet extinction.

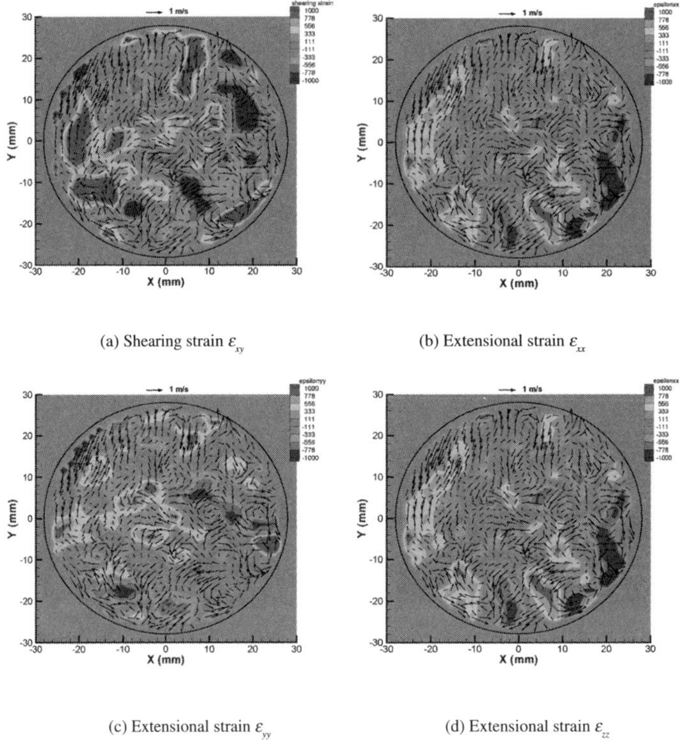

(a) Shearing strain ε_{xy} (b) Extensional strain ε_{xx}

(c) Extensional strain ε_{yy} (d) Extensional strain ε_{zz}

Fig. 4.24 Contour maps of the shearing strain and extensional strains for 100th cycle PIV data at 60° CA BTDC, 1200 rpm.

4.4 Summary

The fundamentals of LDA and PIV and their applications to IC engines have been discussed in this chapter. LDA has been applied to engine research for over three decades and provided much useful insight into the flow processes and their interaction with combustion within the cylinder of IC engines. The advent of PIV systems has provided the ability to record instantaneous, cycle-resolved, 2-D maps of velocity within extended areas in the cylinder. In particular, the modern cross-correlation digital PIV system allows not only quickly acquiring a great number of particle images in consecutive engine cycles but also processing them and displaying velocity vectors on line. This allows PIV experiment to be easily optimized. In addition, hundreds of PIV realizations can be obtained in consecutive engine cycles so as to analyze ensemble-averaged flow behaviors and cycle-resolved flow structures. Furthermore, the recent advancement in the high-speed cameras and diode-pumped solid-state lasers has made it possible to obtain crank-angle-based high-speed PIV measurements. In addition to stereoscopic PIV, alternative 3-D PIV techniques are being developed for whole volume 3-D flow measurements, as discussed by Raffel et al.(2007)

References

Arcoumanis, C., Bicen, A. F., and Whitelaw, J. H. (1983). "Squish and Swirl-Squish Interaction in Motored Model Engines." *J. Fluids Engng*, Vol. 105, pp. 105–112.

Arcoumanis, C., Hu Z., Vafidis, C., and Whitelaw, J. H. (1990). "Tumbling Motion: A Mechanism for Turbulence Enhancement in Spark-Ignition Engines." SAE Paper No. 900060, SAE International, Warrendale, PA.

Baby, X., and Floch, A. (1997). " Investigation of the In-Cylinder Tumble Motion in a Multi-Valve Engine: Effect of the Piston Shape." SAE Paper No. 971643, SAE International, Warrendale, PA.

Ball, W. F., Pettifer, H. F., and Waterhouse, C. N. (1983). "Laser Doppler Velocimeter Measurements of Turbulence in a Direct Injection Diesel Combustion Chamber." Paper No. C52/83, ImeChE Publications.

Bopp, S., Vafidis, C. and Whitelaw, J. H. (1986). "The Effect of Engine Speed on the TDC Flowfield in a Motored Reciprocating Engine." SAE Paper No. 860023, SAE International, Warrendale, PA.

Coghe, A., Brunello, G., and Tassi, E. (1988). "Effects of Intake Ports on the In-cylinder air Motion Under Steady Flow Conditions." SAE Paper No. 880384, SAE International, Warrendale, PA.

Corcione, F. E., and Valentino, G. (1990). "Turbulence Length Scale Measurements by Two-probe Volume LDA Technique in a Diesel Engine." SAE Paper No. 902080, SAE International, Warrendale, PA.

Daneshyar, H., and Fuller, D. E. (1986). "Definition and Measurement of turbulence Parameters in Reciprocating IC Engines." SAE Paper No. 861529, SAE International, Warrendale, PA.

Deslandes, W., Dumont, P., Dupont, A., Baby, X., et al. (2004), "Airflow Cyclic Variations Analysis in Diesel Combustion Chamber by PIV Measurements." SAE Paper No. 2004–01–1410, SAE International, Warrendale, PA.

Drain, L. E. (1980). *The Laser Doppler Technique*. New York: Wiley.

Durao, D. F., Lake, J., and Whitelaw, J. H. (1980). "Bias Effects in Laser Doppler Anemometry." *J. Phy. E: Scientific Instruments*, Vol. 14, pp. 442–445.

Fansler, T. D., and French, D. (1988). "Cycle-Resolved Laser Velocimetry Measurements in a Reentrant-Bowl-in-Piston Engine." SAE Paper No. 880377, SAE International, Warrendale, PA.

Farrell, F. V. (2005). "Examples of In-Cylinder Velocity Measurements for Internal Combustion Engines." Vol. 221, *Proc. IMechE Part D: J. Automobile Engineering*.

Fraser, R. A., and Bracco, F. V. (1988). "Cycle-Resolved LDA Integral Length Scale Measurements in an IC Engine." SAE Paper No. 880381, SAE International, Warrendale, PA.

Galindo, J., Pastor, J., Serrano, J., Pastor, J., et al. (2001). "Multidimensional Modeling of the Scavenging and Injection Processes of a Small Two-Stroke Engine Compared to LDV Measurements," SAE Paper No. 2001–01–3233, SAE International, Warrendale, PA.

Glover, A. R. (1986). "Towards Bias-Free Estimates of Turbulence in Engines." *Proc. of 3rd Intl. Symp. On Applications of Laser Anemometry to Fluid Mechanics.* Lisbon.

Glover, A. R., Hundleby, G. E., and Hadded, O. (1988). "The Development of Scanning LDA for the Measurement of Turbulence in Engines." SAE Paper No. 889379, SAE International, Warrendale, PA.

Hadded, O., and Denbratt, I. (1991). "Turbulence Characteristics of Tumbling Air Motion in Four-Valve SI Engines and Their Correlation with Combustion Parameters." SAE Paper No. 910478, SAE International, Warrendale, PA.

Hall, M. J., Bracco, F. V., and Santavicca, D. A. (1986). "Cycle-resolved Velocity and Turbulence Measurements in an IC Engine with Combustion." SAE Paper No. 860320, SAE International, Warrendale, PA.

Heim, D., and Ghandhi, J. (2011). "A Detailed Study of In-Cylinder Flow and Turbulence using PIV." *SAE International Journal of Engines*, Vol. 4, No. 1. pp. 1642–1668.

Hoesel, W., and Rodi, W. (1977). "New Biasing Elimination Method for laser Doppler Velocimeter Counter Processing." *Review of Scientific Instruments*, Vol. 48, pp. 910–919.

Ishima, T., Obokata, T., Nomura, T., and Takahashi, Y. (2008). "Analysis on In-Cylinder Flow by Means of LDA, PIV and Numerical Simulation under Steady State Flow Condition." SAE Technical Paper No. 2008–01–1063, SAE International, Warrendale, PA.

Jarvis, S., Justham, T., Clarke, A., Garner, C., et al. (2006). "Motored SI IC Engine In-Cylinder Flow Field Measurement Using Time Resolved Digital PIV for Characterisation of Cyclic Variation." SAE Paper No. 2006–01–1044, SAE International, Warrendale, PA.

Karhoff, D., Bücker, I., Dannemann, J., Klaas, M., et al. (2011), "Experimental Analysis of Three-Dimensional Flow Structures in Two Four-Valve Combustion Engines." SAE Technical Paper No. 2011–24–0044, SAE International, Warrendale, PA.

Keane, R. D., and Adrian, R. J. (1990). "Optimization of Particle Image Velocimeters. Part I: Double Pulsed Systems." *Meas. Sci. Tech.* Vol. 2, pp. 963–974.

Li, Y., Zhao, H., Leach, B., Ma, T., and Ladommatos, N. (2004). "Characterization of an in-Cylinder Flow Structure in a Highly Tumble Spark-Ignition Engine." *Int. J. Engines Research*, Vol. 5, pp. 375–400.

Liou, T. M., Hall, D. A., Santavicca, D. A., and Bracco, F. V. (1984). "Laser Doppler Velocimetry Measurements in Valved and Ported Engines." SAE Paper No. 840375, SAE International, Warrendale, PA.

Liou, T. M., and Santavicca, D. A. (1985). "Cycle Resolved LDV Measurements in a Motored IC Engine." Transactions of the ASME, Vol.108, pp.232-240.

Lorenz, M., and Prescher, K. (1990). "Cycle Resolved LDV Measurements on a Fired SI Engine at High Data Rates Using a Conventional Modular LDV-System." SAE Paper No. 900054, SAE International, Warrendale, PA.

Malcolm, J., Behringer, M., Aleiferis, P., Mitcalf, J., et al. (2011). "Characterisation of Flow Structures in a Direct-Injection Spark-Ignition Engine Using PIV, LDV and CFD." SAE Technical Paper No. 2011–01–1290, SAE International, Warrendale, PA.

McLaughlin, D. K., and Tiederman, W. G. (1973). "Biasing Correcting for Individual Realization of Laser Anemometer Measurements in Turbulent Flows." *Phys. Fluids*, Vol. 16, pp. 2082–2088.

Morse, A. P., Whitelaw, J. H., and Yianneskis, M. (1980). "The Influence of Swirl on the Flow Characteristics of a Reciprocating Piston-Cylinder Assembly." *J. Fluids Engng*, Vol. 194, pp. 291–299.

Müller, S., Arndt, S., and Dreizler, A. (2011). "Analysis of the In-Cylinder Flow Field / Spray Injection Interaction within a DISI IC Engine Using High-Speed PIV." SAE Paper No. 2011–01–1288, SAE International, Warrendale, PA.

Obokata, T., Kato, M., Ishima, T., and Kaneko, M. (2009). "Database Constructions by LDA and PIV to Verify the Numerical Simulation of Gas Flows in the Cylinder of a Motored Engine," SAE Technical Paper No. 2009–28–0010, SAE International, Warrendale, PA.

Omori, S., Iwachido, K., Motomochi, M., and Hirako, O. (1991). "Effect of Intake Port Flow Pattern on the In-Cylinder Tumbling Air Flow in Multi-Valve SI Engines." SAE Paper No. 910477, SAE International, Warrendale, PA.

Petersen, B., and Miles, P. (2011). "PIV Measurements in the Swirl-Plane of a Motored Light-Duty Diesel Engine." *SAE International Journal of Engines*, Vol. 4, No. 1, pp. 1623–1641.

Park, D., Sullivan, P., and Wallace, J. (2004). "Different Velocity Data Analysis for Flows Near a Spark Plug in the Combustion Chamber of a Spark-Ignition Engine," SAE Paper No. 2004–01–1351, SAE International, Warrendale, PA.

Raffel, M., Willert, C. E., Wereley, W., and Kompenhans, J. (2007). *Particle Image Velocimetry: A Practical Guide*, 2nd Ed. Springer. Heidelberg.

Rask, R. B. (1979). "Laser Doppler Anemometer Measurements in an Internal Combustion Engine." SAE paper No. 790094, SAE International, Warrendale, PA.

Rask, R. B. (1981). "Comparison of Window, Smoothed-Ensemble, and Cycle-by-Cycle Data Reduction Techniques for Laser Doppler Anemometer Measurements of In-Cylinder Velocity." In *Fluid Mechanics of Combustion Systems*, edited by T. Morel, R. P. Lohmann, and J. M. Rackley, pp. 11–20. ASME.

Reeves, M., Garner, C. P., Dent, J. C., and Halliwell, N. A. (1994). "Particle Image Velocimetry Measurements of in-Cylinder Flow in a Multi-Valve Internal Combustion Engine." *Pro. Instn. Mech. Engrs*, Part D, Vol. 201, pp. 63–70.

Reuss, D. L., Adrian, R. J., Landreth, C. C., French, D. T., and Fansler, T. D. (1989). "Instantaneous Planar Measurements of Velocity and Large-Scale Vorticity and Strain Rate in an Engine Using Particle-Image Velocimetry." SAE Paper No. 890616, SAE International, Warrendale, PA.

Reuss, D. L., Bardsley, M. E. A., Felton, P. G., Adrian, R. J., and Landreth, C. C. (1990). "Velocity, Vorticity and Strain-Rate Ahead of a Flame Measured in an Engine Using Particle Image Velocimetry." SAE Paper No. 900053, SAE International, Warrendale, PA.

Stolz, W., Kohler, J., Lawrenz, W., Meier, F., Bloss, W. H., Maly, R. R., Herweg, R., and Zahn, M. (1992). "Cycle Resolved Flow Field Measurements Using a PIV Movie Technique in a SI Engine." SAE Paper No. 922354, SAE International, Warrendale, PA.

Suen, K. O., Tindal, M. J., Yianneskis, M., and Paul, G. R. (1987). "The Developments of Swirl in the Cylinder of a Direct Injection Diesel Engine." *Proc. 2nd Int. Conf. On Laser Anemometry-Advances and Applications.*

Swords, M. D., Kalghatgi, G. T., and Watts, A. J. (1982). "An Experimental Study of Ignition and Flame Development in a Spark-Ignition Engine." SAE Paper No. 821220, SAE International, Warrendale, PA.

Tindal, M. J., Cheung, R. S., and Yianneskis, M. (1988). "Velocity Characteristics of Steady Flows Through Engine Inlet Ports and Cylinders." SAE Paper No. 880383, SAE International, Warrendale, PA.

Wigley, G., and Hawkins, M. G. (1978). "Three-Dimensional Velocity Measurements by Laser Anemometry in a Diesel Engine Under Steady Flow Conditions." SAE Paper No. 780060, SAE International, Warrendale, PA.

Witze, P. O., and Martin, J. K. (1986). "Cyclic Variation Bias in Spark-Ignition Engine Turbulence Measurements." In *Laser Anemometry in Fluid Mechanics*, edited by R. J. Adrian et al.

chapter 5

In-Cylinder Fuel and Combustion Specie Measurement by Laser-Induced Fluorescence

5.1 Introduction

As discussed in Chapter 3, laser-induced fluorescence (LIF) is an electronic absorption and emission process that produces relatively strong signal intensity. The fluorescence signal is normally red-shifted from the excitation wavelength, and hence it can be separated from stray scattered light. Fluorescence intensity is directly proportional to the molecular density, and it can be used to measure the concentration. The strong fluorescence signals allow two-dimensional imaging of species to be obtained by using a laser sheet. However, quenching effects can reduce fluorescence intensities, especially by oxygen at elevated pressures, in a way that depends on pressure and temperature, thereby making quantitative measurements in an engine difficult.

The application of LIF to combustion flows and IC (internal combustion) engines is normally carried out by multipoint planar imaging. Two-dimensional planar laser-induced fluorescence (PLIF) may be viewed as a modern form of species visualization, providing spatially resolved information in a plane rather than integrated over a line of sight. The wider use of PLIF in combustion flows and IC engines is due to the strength of its fluorescence process relative to its primary competitors, Rayleigh and Raman scattering, and to its species specificity.

5.2 Fuel Vapor Concentration Measurement by PLIF

5.2.1 Principle of Operation

As discussed in Chapter 3, there are two regimes that LIF can operate, depending on the relative laser intensity. If the laser intensity is strong enough, laser-induced fluorescence will become independent of both the laser irradiance and the quenching. However, in engine applications, it is extremely difficult to reach the saturation regime because of the high-pressure environment inside the combustion chamber. Instead, LIF is in the so-called linear regime and its signal level is given by Eq. 3.53, repeated here as Eq. 5.1:

$$P_{flu} = \eta_c \Omega V_c\, f_1(T) x_{21}\, N\, I_{21} B_{21} \frac{A_{21}}{A_{21} + Q_{21}} \tag{5.1}$$

Eq. 5.1 states that the fluorescence signal is directly proportional to the mole fraction of the species of interest x_m, which forms the basis of LIF for species concentration measurements. The function $f_1(T)$ indicates that the fluorescence signal is affected by the species temperature as a result of the change in species populations at different energy states with temperature. Note that in the linear regime, one normally needs to evaluate the quenching rate constant Q_{21} to make quantitative measurements of the species of interest. The quenching rate is sensitive to pressure, temperature, and composition.

Several requirements must be satisfied to obtain quantitative fluorescence measurements on a given species. First, the molecule must have a known absorption and emission spectrum. Second, the molecule must have an absorption wavelength accessible to a laser source. Third, the rate of radiative decay of the excited state, which affects the fluorescence power directly, must be known. Fourth, excited state losses caused by collisions, photo-ionization, and predissociation have to be taken into account. Some aspects of these requirements will be considered later. For a detailed discussion of quantitative LIF measurements for combustion diagnostics, see the review paper by Schulz and Sick (2005).

In theory, three approaches can be employed for LIF fuel concentration measurements: natural fluorescence from the fuel, fluorescence from a tracer molecule of matching properties to the fuel, and fluorescence from exciplex-forming tracers. The first two strategies are used for fuels existing in either liquid form or gaseous form. A third approach, LIEF (laser-induced exciplex fluorescence) is used to study liquid fuel droplets and fuel vapor concentration simultaneously with a pair of tracer species, as discussed in Chapter 7.

Some of the practical fuels may fluoresce when excited by UV (ultraviolet) light. The use of LIF from the fuel itself simplifies the experiments, but the interpretation of the results can be difficult. Because of the variation in fuel composition from batch to batch, the fluorescence signal may not be reproducible because of the lack of information on the photochemistry of the fluorescent component. Also, the high-boiling-point aromatics in the fuel are likely to fluoresce the strongest, which may not be representative of the fuel as a whole. In most applications, tracers of known absorption and emission properties have been used to mark the fuel concentration in the cylinder.

5.2.2 Selection of Fluorescence Tracers

To obtain quantitative fuel distribution, several properties of the tracer need to be considered:

- Absorption of the available laser wavelength
- Satisfactory fluorescence yield
- Sufficiently low quenching properties
- Nontoxic, stable, and soluble in fuel
- Matching vaporization properties

In addition, the fluorescence should be sufficiently red-shifted to facilitate separation of elastically scattered laser light from the LIF signal.

The first requirement depends on the laser available. Two types of pulsed lasers are commonly employed: excimer lasers and Nd:YAG lasers with third and fourth harmonic output because of their high power outputs and short wavelengths. The last two features directly affect the fluorescence signal level, delectability, and accuracy. The quenching rate of the fluorescence signal reduces the fluorescence signal level. Quenching can be particularly high by the atmospheric oxygen at high pressures in the cylinder and can even prevent the use of air. The last two properties will affect the accuracy of the measurement. To be used as a tracer for the fuel, it must be soluble in the fuel. The matching boiling point helps to mimic the fuel vaporization process. To study the evaporation process in direct-injection gasoline engines, some model fuels have been formulated to match the evaporation properties of the tracer (Stevens et al., 2007).

The required absorption wavelength in the visible and UV region restricts the tracers to two main categories: aromatic and carbonyl compounds. The fluorescence tracers used for LIF measurements are summarized in Table 5.1. The data of relative LIF intensity and the pressure and temperature effects were adapted from Fujikawa et al. (1997). As shown in Table 5.1, aromatic compounds (e.g., toluene) are strongly quenched by oxygen. Carbonyl compounds (compounds containing –C=O bond) are less sensitive to the oxygen quenching, and they can be excited by a variety of light sources, for example, third and fourth harmonics of Nd:YAG (355 nm and 266 nm) and excimer lasers (248 nm and 308 nm). Among the carbonyl compounds used are aldehydes (–C=O bond at one end), ketones (–C=O bond in the middle), and diones (two –C=O bonds). Diones, such as biacetyl (diacetyl, 2,3 butanedione), have longer excitation and fluorescence wavelengths. Acetone and 3-pentanone have higher fluorescence yield than aldehydes. In addition, ketones are preferred over aldehydes as tracers because they are less reactive

135

Table 5.1 Summary of Tracers Used in LIF Measurements

Tracer and Fuel	T_{boil} (°C)	Absorption Band (nm)	Emission Band (nm)	Relative LIF Intensity	Pressure Effect** (%/ MPa)	Temperature Effect (%/1000 K)	Excitation Source And Reference
$C_6H_5CH_3$, **toluene**, * (5.0 vol.% in iso-octane)	111	230–280	270–320	90 @ 248 nm 33 @ 266 nm	-80 @ 248 nm -90 @ 266 nm	-30 @ 248 nm 80 @ 266 nm	KrF excimer laser 248 nm, Reboux et al. (1994, 1996).
$C_6H_5N(CH_3)_2$, **N,N-dimethylaniline** (0.2 vol. % in iso-octane)	193	200–320	310–420				KrF excimer laser 248 nm, Urushihar et al. (1996); Itoh (1995).
$C_{16}H_{10}$, **fluoranthene**, (1,2-benzene) (< 1.0 mass.% in iso-octane)	Solid 384	200–400	400–600				XeCl excimer laser 308 nm, Meyer et al. (1995, 1997).
C_2H_5CHO, **propionaldehyde** C_3H_7CHO, **butyraldehyde** C_5H_9CHO, **N−valerdehyde** $C_6H_{13}CHO$, **heptaldehyde** C_6H_5CHO, **bensaldehyde** (9.1 vol. % in iso-octane)	49 75 103 153 178	220–340 220–340	300–550 300–550	0.14 @ 248 nm 0.11 @ 248 nm			XeCl excimer laser 308 nm, Swindal et al. (1995).
$CH_3CO\ CH_3$, **acetone**	56	200–330	300–600	@ 248 nm 1.0 @ 266 nm	60 @ 248 nm 5.0 @ 266 nm	-25 @ 248 nm -7.0 @ 26 6 nm	XeCl excimer laser 308 nm, Wolff et al. (1994).
$C_2H_5COCH_3$, **2-butanone**, (ethylmethylketone, EMK) (100 %)	80	210–310	300–600				XeCl excimer laser 308 nm, Lawrenz et al. (1992).
$C_2H_5COC_2H_5$, **3-pentanone** (diethylketone) (2.0–25.0 vol.% in iso-octane)	102	220–340	330–600	1.3 @ 248 nm 1.3 @ 266 nm 0.0 @ 308 nm	70 @ 248 nm 3.0 @ 266 nm 30 @ 308 nm	-28 @ 248 nm -11 @ 266 nm	Excimer lasers 248 nm, Johansson et al. (1995). Nd:YAG laser 266 nm, Ghandhi & Bracco (1995). Excimer lasers 308 nm, Berckmüller (1996, 1997).
$CH_3COCOCH_3$, **biacetyl**, (2,3−butanedione) (2.0–6.0 vol.% in iso-octane)	88	200–480	440–510				Nd:YAG laser 355 nm, Baritaud & Heinze (1992); Deschamps et al. (1994).
NO_2, **nitrogen dioxide** (0.2 vol % in nitrogen)	gas	UV-Visible	580–650				Nd:YAG laser 532 nm, Zhao et al. (1994).

Notes: * Toluene was used in FARLIF measurements. ** The pressure effects are valid up to 10 bar.

in a high-pressure and high-temperature environment. Ketones attack rubber or neoprene. Because fuel pumps, regulators, and fuel injectors contain rubber or neoprene O-rings, the fuel supply system of an engine needs to be modified using ketone-resistant components such as stainless steel and Teflon.

5.2.3 Experimental Setup of PLIF

The typical setup of a PLIF system is shown in Fig. 5.1. A pulsed laser, typically a high-energy pulsed UV laser (e.g., excimer, Nd:YAG, and tunable dye lasers) is triggered to generate a laser pulse at a specified crank angle generated from an engine driven shaft encoder. The laser output is formed into a laser sheet using combinations of cylindrical and spherical lenses. The fluorescence signal is imaged at the right angle using a gated intensified CCD (charge-coupled device) camera. A band-pass filter may be placed in front of the camera to reject elastic scattering from the walls. A commercial lens can also act as a filter if it is opaque in the UV region. A delay generator synchronizes the operation of the laser, the ICCD (intensified charge-coupled device) camera, and an image acquisition system via a PC (personal computer). To determine the required settings for the delay generator, a photodetector connected to an oscilloscope can be used to measure the time taken for the laser to produce a laser pulse after being sent the signal from the delay generator, along with the width of that pulse. For an excimer laser, the typical delay time has been found to be around 1.0–2.0 µs. The intensifier should be gated for a short period of time (~200 ns) after the laser pulse to discriminate against background light and combustion luminosity (Fig. 5.2). Furthermore, as the typical capture rate of an ICCD camera and associated

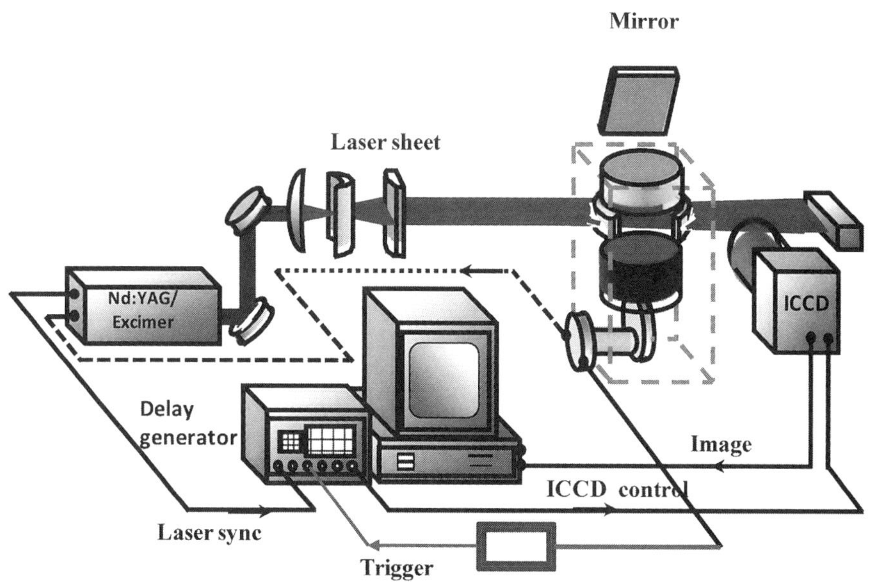

Fig. 5.1 Schematic of the optical setup for PLIF experiments.

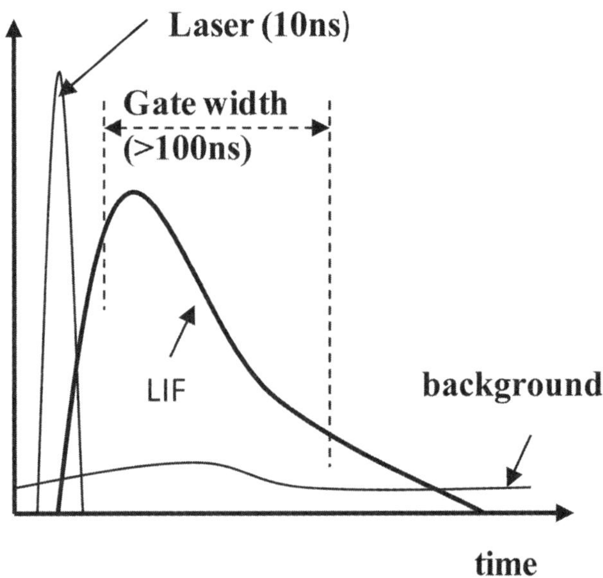

Fig. 5.2 The timing event of LIF signal and laser output.

image acquisition system is limited to less than a few frames per second, a PLIF image can only be captured at every other few cycles. Thus, a divider unit is often employed to generate a trigger so that laser is only fired when the ICCD camera is ready. To minimize the amount of fuel vapor exhausted from the engine and to reduce contamination of the optical access windows, the fuel injection in a direct-injection engine should only take place in cycles when an image is to be captured by means of the same divider unit.

5.2.4 Calibration of PLIF Images

To obtain quantitative results, the raw LIF images need to be postprocessed and calibrated. The first process involves the removal of systematic errors in the imaging system. This includes the removal of background and dark current signals from the recorded images, followed by corrections for nonuniformity in the spectral and spatial response of the pixels and laser sheet intensity. (Image processing and calibration were discussed in Chapter 2.) Other systematic errors, such as beam steering of the laser beam by density gradients, attenuation of the laser sheet across the imaged region, laser-induced perturbations (Eckbreth, 1996), partial saturation, and radiative trapping (absorption of the fluorescence within the combustion gases) are often avoidable or negligible. But these effects need to be considered in some applications (Schulz and Sick, 2005).

The LIF images thus obtained represent the concentration of fluorescing molecules (tracers) at a particular condition. Since LIF intensity depends on temperature and pressure as well as tracer density, the LIF signal needs to be calibrated to obtain the fuel concentration. Two calibration approaches can be adopted. The first involves in situ calibration at each crank angle for different air/fuel ratios of homogeneous mixtures, which means a calibration curve for each crank angle. The second approach is to relate LIF images to a calibration image taken at atmospheric pressure and ambient temperature. It can be shown (Berckmüller et al., 1994, 1996) that effects of pressure and temperature on the LIF signal during the intake and compression strokes can be corrected for by a normalization factor f_N, shown in the following equation:

$$ f_N = \frac{p_1 \cdot T_0 \cdot a + b \cdot T_1}{p_0 \cdot T_1 \cdot a + b \cdot T_0} \tag{5.2}$$

where p_0 and T_0 represent pressure and temperature at the calibration condition, and p_1 and T_1 are the pressure and temperature of the gas in the cylinder when the LIF images are taken. The in-cylinder pressure can be obtained accurately from a pressure transducer. In-cylinder gas temperature is difficult to measure and is normally estimated. During the intake stroke, the gas temperature may be assumed constant. During the compression stroke, polytropic compression can be used to estimate the bulk gas temperature. Constants a and b represent the effect of pressure and temperature on fluorescence yield and may be determined experimentally.

5.2.5 In-Cylinder Fuel Distribution Measurements by PLIF

5.2.5.1 Overview of PLIF Measurements in IC Engines

Over the last two decades, PLIF has been used widely for in-cylinder fuel concentration measurements. Table 5.2 presents some of the typical PLIF measurements in IC engines. Among the varied combination of PLIF setups, excimer (XeCl or KrF) lasers and 3-pentanone combinations are considered to be the most favored configurations, because of higher pulse energy, flexible repetition rate, and minimized quenching effects (e.g., Berckmüller et al., 1996).

In many cases, PLIF has been used as a visualization technique for in-cylinder fuel distribution. Some have attempted more quantitative PLIF measurements. For example, Johansson et al. (1995) obtained quantitative images of the fuel distribution within the engine by performing calibration with homogeneous known fuel/air mixtures in the cylinder. The authors claimed that the pulse-to-pulse spatial variations and the noise from image intensifier yielded an error of about 10%. The background noise contributed another 3% error. The effects of quenching and absorption on PLIF were considered small. Therefore, the total error was estimated being around 15%. A similar experimental setup was employed to measure quantitative fuel distributions in a Honda VTEC single-cylinder engine by Berckmüller et al. (1994; 1996). A mixture of 75% of iso-octane and 25% of 3-pentanone by volume was excited by a XeCl excimer

Table 5.2 Examples of PLIF experiments

Laser Source and Laser Sheet (width × thickness)	Detection System and Spatial Resolution	Dopant and Fuel	Applications and Reference
Nd:YAG 532 nm, 180 mJ, 50 mm × 250 μm	ICCD camera, 60 ns, >580 nm. 0.23 × 0.23 × 0.25 mm	NO_2 (0.2 vol.%) N_2	N_2 injection to simulate fuel injection in a motored engine at low speeds. Zhao et al. (1994).
Nd:YAG 355 nm, 80 mJ, 20 mm × 500 μm	ICCD camera, 10 μs, 440 − 480 nm. 0.25 × 0.2 × 0.5 mm	Biacetyl, (2 − 6 vol.%) iso-octane.	Effect of injection timing and engine speed on the mixture homogeneity in a single cylinder SI engine. Baritaud et al. (1992), Deschamps et al. (1994).
XeCl 308 nm, 150 mJ 27 mm × 1.0 mm	ICCD camera − −	Fluoranthene, (<1 mass%) iso-octane.	Effect of air/fuel ratio, engine speed, injection timing, and charge motion on the mixture distribution in a SI engine Meyer et al. (1995, 1997).
XeCl 308 nm, -, 20 mm × 0.5 mm (10 mm × 0.5 mm)	ICCD camera − −	Acetone (-) 3-pentanone (5 vol.%) iso-octane	Effect of air/fuel ratio, injection timing and EGR on the mixture distribution in an SI engine. Wolff et al. (1994), Knapp et al. (1997).
XeCl 308 nm, 220 mJ 76 mm × 0.3 mm	ICCD camera − −	EMK (100%)	Fluorescence characteristics of carbonyl compounds, imaging of air/fuel ratio in a square piston SI engine. Lawrenz et al. (1992).
KrF 248 nm, 400 mJ 69 mm × 2 mm	ICCD camera − , 280–400 nm. 0.19 × 0.19 × 2.0 mm	DMA, (0.2 vol. %) iso-octane	The effect of swirl and tumble flows on the mixture distribution in a single-cylinder SI engine. Itoh et al. (1995); Urushihara et al. (1996).
KrF 248 nm, 100 mJ 12 mm × 0.1 mm	ICCD camera 100 ns, visible range. 0.19 × 0.19 × 0.1 mm	3-Pentanone (0.2 vol.%) iso-octane	The effect of injection timing and piston geometry on the mixture homogeneity in a large SI engine. Johansson et al. (1995).
XeCl 308 nm, 150 mJ 50 mm × 0.25 mm	ICCD camera 200 ns, visible range. 0.25 × 0.25 × 0.25 mm	3-Pentanone (25.0 vol.%) iso-octane	The effect of intake system and injection timing on the mixture distribution in an SI engine. Berckmüller et al. (1996, 1997).
XeCl 308 nm, 60 mJ 20 mm × 0.2 mm	ICCD camera 2 μs, visible range. −	Aldehydes (9.1 vol.%) iso-octane	The effect of fuel boiling temperature on the mixture homogeneity in an SI engine. Swindal et al. (1995).
Nd:YAG 266 nm, -, −	Intensified CID camera 150 ns, visible range. −	3-Pentanone (5.0 vol.%) iso-octane	Fuel distribution in a special designed two-stroke direct injection gasoline engine. Ghandhi & Bracco (1995).
Nd:YAG 266 nm, 10 mJ 50 mm × 0.5 mm	ICCD camera 200 ns, >345 nm. 0.25 × 0.25 × 0.5 mm	Gasoline fuel	Fuel distribution in a two-stroke direct-injection gasoline engine. Fansler et al. (1995).
KrF 248 nm, 100 mJ 19 mm × 1 mm	ICCD camera 200 ns, >280 nm. 0.19 × 0.19 × 1.0 mm	Toluene (5.0 vol.%) iso-octane	Fuel/air ratio LIF (FARLIF) measurements in a single-cylinder SI engine. Reboux et al. (1994, 1996).

laser. A calibration equation was employed to convert the LIF signals into quantitative fuel/air ratio measurements.

To avoid fluorescence from impurities in commercial gasoline as well as deposit forming on the optical window, iso-octane or a mixture of iso-octane and n-heptane (primary reference fuel, or PRF) is often used, as in the previous two examples. However, iso-octane represents only a small fraction of gasoline components. To carry out PLIF measurements under more realistic engine conditions, a compound fuel of a similar distillation curve to that of gasoline has been formulated and used in PLIF measurements (Itoh et al., 1995, Stevens et al., 2007). In the case of direct injection, significant difference was found in the in-cylinder mixing patterns when iso-octane and a new compound fuel were used as the test fuel.

5.2.5.2 Example of In-Cylinder Fuel Distribution by PLIF

This section describes an application of PLIF to the study of stratified fuel fraction gasoline engines (Li et al., 2004). This research investigated whether two different fuel fractions could be distributed preferentially in the cylinder through two separate intake ports designed for generating high tumble flows. The measurement was performed in a single-cylinder engine with optical accesses via a fused silica piston window and a transparent upper liner section of 30 mm in height.

Two injectors were fitted near the entry of each intake port. Hexane doped with 3-pentanone (25% in weight) and iso-octane doped with DMA (10% in weight) were used to represent the two fuels. When excited at 308 nm by a XeCl laser, DMA produces fluorescence in the range of 300–450 nm with the peak intensity at 335 nm, and 3-pentanone has an emission band from 350 nm to 500 nm with intensity peak at 435 nm. The difference in the two intensity peaks is large enough to be separated by two narrow band optical filters.

Figure 5.3 presents a schematic diagram of the two-tracer LIF measurement system. The output of an XeCl excimer laser was directed to its required position via a series of laser mirrors before it was formed into a laser sheet of 2-mm thickness, via a spherical lens of 1000-mm focus length, to loosely focus the beam in the cylinder and a concave cylindrical lens with a focal length of −76 mm.

An image doubler was set up to capture the two LIF images of DMA and 3-pentanone. In the image doubler setup, a dichroic beam splitter at 45° reflects most fluorescence signals in the UV range (below 400 nm) of the camera and allows the visible light range (over 400 nm) to be detected by the ICCD via a mirror behind it. In addition, two narrow band-pass filters were placed side by side in front of the camera, one with 435 ± 12.5 nm FWHM (full width at half

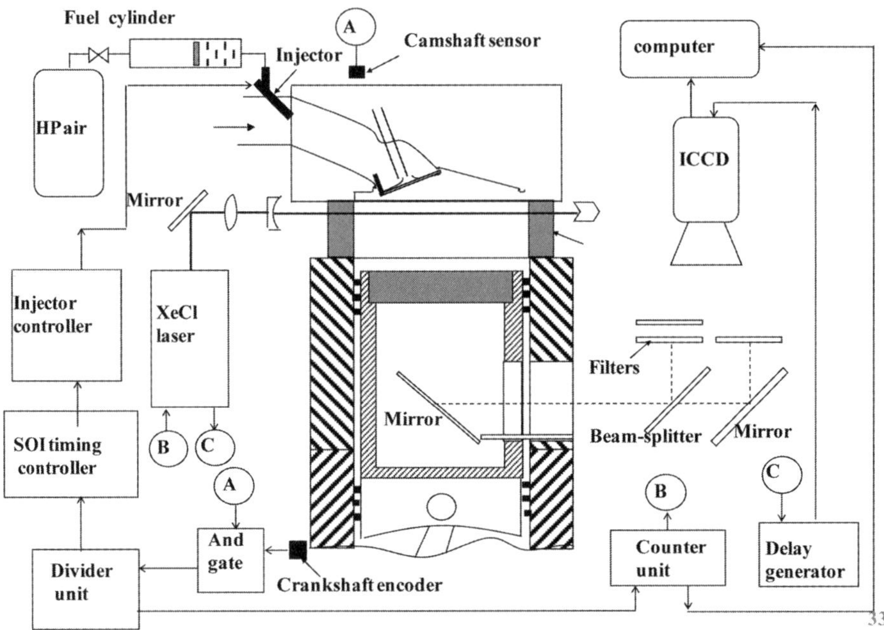

Fig. 5.3 Two-tracer PLIF setup.

maximum) for 3-pentanone and the other filter in the DMA optical path was 350 nm ± 5 nm (FWHM. Because the fluorescence intensity of DMA was much stronger than 3-pentanone, additional neutral density filters were placed in the optical path of DMA. Different neutral density filters were used at each crank angle to balance the two image intensities and optimize the light collection level in conjunction with adjusting the camera gain and aperture settings).

A Nikon® UV lens with a focal length of 105 mm and a maximum aperture of $f/4.5$ was mounted in front of the ICCD. The spatial resolution of the image was determined as 0.2 mm by a ruler placed along the measurement plane. Because of the slow image transfer rate of the ICCD camera, the image capture system was restricted to 0.2 Hz. Therefore, a divider unit was built so that a valid TDC (top dead center) signal was generated for every 75 engine cycles, corresponding to 7.5 seconds at 1200 rpm. The valid TDC signal with the crank angle pulses from the shaft encoder was then used to generate two trigger signals at a preset crank angle through two counters. As shown in Fig. 5.4, one of the trigger signals was used to start the fuel injection process and the other for the laser to start the firing sequence. About 1.1 μs after receiving the trigger signal, a synchronous output signal from the laser was sent to the delay generator, from which another trigger signal was generated (after a short delay) and sent to the ICCD camera. The delay time was determined as 0.74 μs by detecting the laser output using a photodiode detector through a reflective glass plate in the laser beam path. The gate width of the intensifier was set to a relatively long duration of 300 ns to take into account of the pulse-to-pulse variations in the laser firing time when the laser was operated in the externally triggered mode.

During experiments, five background images were taken at each crank angle after the 20 LIF images had been recorded under identical operating conditions. They were taken without fuel injections and averaged to reduce the effect of shot-to-shot variation in the laser output. The resulting mean background image was then subtracted from the corresponding LIF images.

To minimize the effect of shot-to-shot variation of laser and cyclic variation of fuel distribution, 20 background-corrected images were then arithmetically averaged to give the in-cylinder fuel distribution and its variation within 20 cycles.

To determine the spatial distribution across the laser sheet and to correct for laser absorption in the measurement planes, a homogeneous air/fuel mixture was needed within the cylinder. Another injector was installed in the intake system 1 meter upstream of the intake valves. Because of its relatively high volatility, hexane doped with 3-pentanone was used to prepare the homogeneous air/fuel mixture. Intake air was heated to 65°C ± 5°C by an air heater and a temperature controller. As a result, fuel was completely vaporized to form a homogeneous air/fuel mixture in the cylinder. To obtain an averaged distribution of the light intensity in the measurement planes, 20 images and five background images were recorded at the same crank angle. In general, the fluorescence intensity was uniform in both vertical and horizontal measurement planes. However, the fluorescence intensity on the laser beam

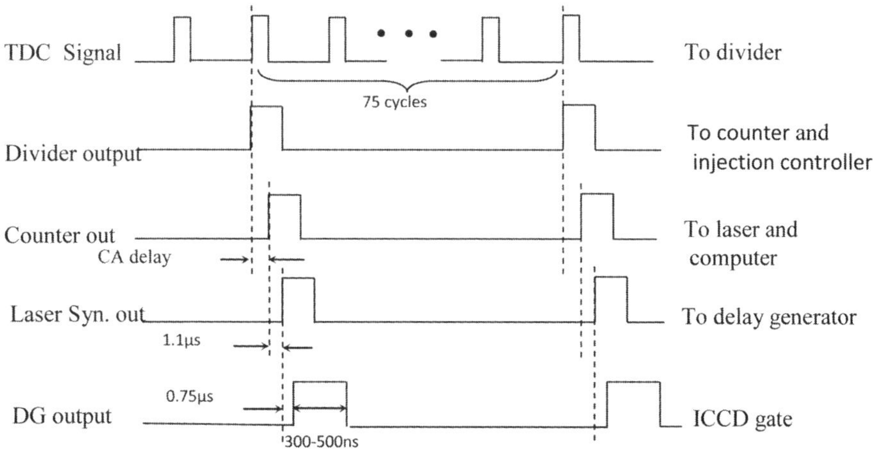

Fig. 5.4 Schematic timing diagram of the two-tracer PLIF experiment.

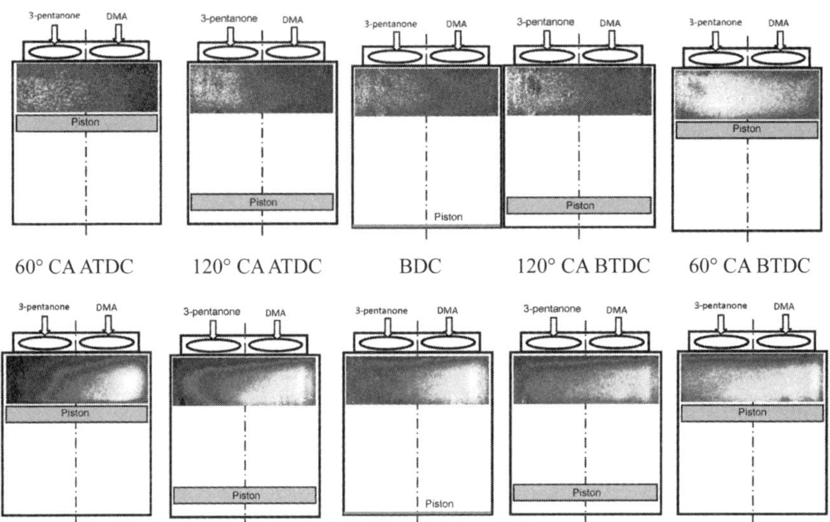

Fig. 5.5 In-cylinder distribution of two fuels.

entry side appeared stronger than that on the laser exit side, as a result of the laser absorption effect, which could be minimized by less concentration of absorbing species (tracers).

Figure 5.5 shows the simultaneously recorded PLIF images of the two fuel components at different crank angle (CA) positions. The upper row shows the iso-octane/3-pentanone distribution, and the lower row of images represents the corresponding hexane/DMA distributions. Because of the strong tumble motion, clear stratification between the two fuels from each intake port was preserved until 60 CA BTDC (before top dead center).

5.3 Visualization of Combustion Species by PLIF

5.3.1 Introduction

The hydroxyl (OH) radical has been the most studied combustion species in combustion and flames using LIF. This is because of its importance as a highly reactive chemical species in combustion, its relative abundance during combustion, and the availability of convenient UV laser system for its excitation.

NO is produced during high-temperature combustion and is one of the major pollutants from IC engines. Detecting NO by LIF in a combustion environment is considerably more difficult than detecting OH because NO has a much smaller absorption cross section and requires a shorter UV excitation wavelength.

Formaldehyde (CH$_2$O) is known to be an intermediate generated during the early stage of the autoignition process. As autoignition proceeds into the high-temperature regime, formaldehyde is rapidly consumed in the degenerate branching reactions that precipitate high-temperature ignition. Formaldehyde is therefore a good natural tracer of autoignition and a marker to indicate the start of high-temperature combustion when it is consumed. Its abundance in the autoignition region and the high sensitivity of its LIF process can therefore be employed to visualize the occurrence of autoignition sites.

5.3.2 Experimental Considerations

The experimental setup for PLIF measurements of these species is very similar to that for PLIF fuel concentration measurements (Fig. 5.1). When LIF from fuel or tracers is detected, as demonstrated in Section 5.2, broadband excitation and detection are used because the detailed vibrational-rotational spectrum cannot be resolved for large polyatomic molecules. The concentration of combustion species is much smaller than that of fluorescing tracers, and hence their LIF signals are weaker. However, the absorption process will be greatly enhanced if the laser output can be tuned to a specific spectral line between the two electronic states. Simple polyatomic molecules, especially diatomic molecules, have well-separated vibrational-rotational energy levels within each electronic energy state. Hence, excitation and emission of specific spectral lines are employed in the LIF measurements of OH and NO. Though there are four atoms in a formaldehyde (CH$_2$O) molecule, the vibrational-rotational lines of formaldehyde are sufficiently separated and can be excited individually. Therefore, narrow band absorption features of these species often demand a tunable laser. One of the most common excitation sources for the LIF measurements of these species is an Nd:YAG pumped dye laser. The pump beam from the Nd:YAG laser is set at either the second (532-nm) or third (355.5-nm) harmonic for pumping the laser generation process in the liquid dye. The pump beam frequency employed depends on the dye being pumped, which, in turn, depends on the final frequency required. The dye laser is typically pumped transversely. To obtain the desired output frequency in the UV region, the dye laser beam is frequency-doubled or sum-mixed in crystals in a frequency conversion unit. The tuning of the output frequency is achieved by rotating a grating in the laser system. Some dye lasers can be pumped by a XeCl excimer laser at 308 nm. For OH and NO, tunable excimer lasers can also be used for direct fluorescence excitation, as discussed later in this chapter.

Fine-tuning the laser frequency to a particular spectral line(s) is accomplished by passing a small laboratory burner through the reaction zone (for OH) or postflame gases (for NO), from which scans of the fluorescence excitation spectrum can be used to determine the excitation wavelength for the maximum LIF signal. The excitation scans are carried out by tuning the laser across the absorbing transitions in a given spectral region. The excited fluorescence is then monitored over a broad spectral region using a wide monochromator slit or a band-pass interference filter.

For qualitative visualization of these species and their distribution by PLIF, which is often the primary objective of many in-cylinder studies, it is often sufficient to employ a single broadband UV laser with wavelength overlapping with the absorption spectra of the species, for example, OH excitation by a XeCl excimer laser and CH_2O by the third harmonics of an Nd:YAG laser.

There are a number of particular issues associated with quantitative LIF measurements of combustion species. This can be understood by looking the LIF signal in the linear regime in Eq. 5.1. which is valid if the laser spectral intensity I_v does not saturate the excitation transition. This is usually the case, particularly in IC engine applications where pressure is high. The laser energy that is to be transferred to the probe volume should remain constant, which requires monitoring the individual laser pulse energy and correction for the change in window transmission. The laser beam may be attenuated by absorption on its way to the probe volume by fuel or combustion products.

The selection of spectral lines for excitation has large effects on the detection and interpretation of LIF signals. The excitation wavelength should be chosen such that the laser is not absorbed by other species in the cylinder. When OH radicals are excited at near UV wavelengths around 300 nm, the beam attenuation is less a problem. However, when shorter wavelengths are employed for OH or NO excitation and detection, absorption by fuel and combustion products may become significant. Additional difficulty arises from the LIF of other species, particularly hot oxygen (O_2), when the excitation of LIF is obtained with shorter wavelengths. The LIF signal depends on the absorption cross section of the measuring species at the excitation wavelength. Excitation of spectral lines that have large absorption cross sections will increase the fluorescence signal.

Pressure broadening and shifting of the rotational lines significantly changes the species absorption spectrum and hence the overlap between the laser and the species absorption profile. Di Rosa and Hanson (1994) suggested that this effect may be taken into account by adding an overlap correction coefficient, $g_\lambda(P)$ to Eq. 5.1, which yields the following equation:

$$I_{flu} = C\, f_1(T)\chi_m\, N\, g_\lambda(P)\, I_v\, B_{12} \frac{A_{21}}{A_{21} + Q_{21}} \tag{5.3}$$

The value of $g_\lambda(P)$ can be determined experimentally (Hildenbrand et al., 1998) or accounted for by performing in situ calibration, which corrects for effects like quenching, system response, and so forth.

In Eq. 5.7 the grouping $(f_1(T)\chi_m\, N)$ represents the number density of fluorescing species at the lower rotational state $N_{j'}$ from which molecules are excited to a higher energy state. It is a function of temperature as described by the Boltzmann distribution:

$$N_J = \frac{N}{Q_{rot}} g_l (2J + 1)\, e^{-BJ(J+1)\,hc/kT} \tag{5.4}$$

where the rotational partition function, $Q_{rot} = kT/hcB$, and g_l represents the degeneracy of the state. The degeneracy indicates the number of molecules which occupy any given energy state. The temperature dependence of the lower state population, therefore, needs to be corrected for quantitative measurements. Alternatively, the lower state should be chosen such that its population is insensitive to temperature variations.

After excitation, the excited species will either fluoresce or lose its energy without radiation as a result of collisions with other molecules, that is, quenching. Because combustion species are normally measured during the combustion period when the pressure and temperature are high, the collisional quenching seriously reduces their fluorescence yield. The quenching rate is proportional to pressure and approximately inversely proportional to the square root of the temperature. Since the gas pressure is virtually uniform throughout the combustion chamber and the temperature gradient is small in the burned gas region, the quenching rate may be considered to be constant throughout the burned gas region. The quantification of quenching rate, however, is particularly difficult since the quenching rate is a strong function of temperature, pressure, and the nature of the colliding species.

The pressure, temperature, and quenching effects described so far, along with system calibration factor C, may be accounted for by performing in situ calibration, where known amounts of NO are seeded to the intake charge, and the fluorescence intensity is measured to generate a correlation between the LIF signal and NO concentration at a particular condition (Hildenbrand et al., 1998). Alternatively, a detailed spectroscopic code (Paul et al., 1993; 1994) has been used to make these corrections to the total NO concentration in the image obtained from a direct-injection diesel combustion engine by Dec and Coy (1996). The calibration of the LIF signal is more difficult to implement for OH radicals, whose concentration cannot be accurately known or determined.

Having survived the quenching process, the fluorescence signal from different transition bands can be detected. Resonant detection, in which the LIF signal is measured at the excitation wavelength, should be avoided because it suffers from interference of Rayleigh scattering and spurious scattering from dust or walls. In addition, radiation trapping should be considered. Radiation trapping takes place when the spontaneous emission (fluorescence) coincides with an absorption band of the same species at different energy states or different species. For example, if OH is excited via the (2,0) $A^2\Sigma$–$X^2\Pi$ transition, the emission from (2,1) band could be absorbed by OH radicals at the first vibrational energy state of the ground electronic state ($X^2\Pi$) if their population is large. Measurement geometry may be used to advantage when permissible, for example, performing measurements near the outer combustion region to minimize the optical path and distance through the medium that the signal travels. Radiation trapping can be avoided by proper selection of excitation and detection schemes as well.

The absorption problem may also be present for the fluorescence light at shorter wavelengths by intermediates formed during combustion, including partially oxidized hydrocarbons and PAHs (polycyclic aromatic hydrocarbons). This needs to be considered particularly for PLIF of NO, which involves low UV fluorescence detection.

The detection of LIF signal requires the separation of the signal light from other sources of radiation at different wavelengths, often at intensity levels well above that of the signal. Since the OH LIF signal lies in the near UV region, a band-pass interference filter can be used to reject spurious scattered light and flame luminescence. Whereas the fluorescence signal of NO lies in the low UV region, dielectric-coated mirrors are preferred because they can be optimized for narrow band reflections with a very high reflectance (>0.995). An UV filter setup consisting of four dielectric-coated mirrors centered around 230 nm has been used to isolate the NO fluorescence from other fluorescence by Schulz et al. (1996).

The use of a gated, intensified CCD camera is necessary because of the weak nature of OH and NO LIF signals. When the laser and the detector are synchronized from a shaft encoder mounted on the engine, the short gating period of the intensifier will minimize the interference from ambient light and flame luminescence. In addition, the removal of systematic errors present in the imaging system should be carried out, as discussed in Section 5.2, before the final results can be obtained.

5.3.3 PLIF Measurements of OH

In the case of LIF measurements of OH, two electronic states are involved: the ground electronic state, $X^2\Pi$, and the first excited electronic state, $A^2\Sigma$, as shown in Fig. 3.2. Three vibrational bands of the $A^2\Sigma$–$X^2\Pi$ transition have been used in OH LIF in IC engines. Table 5.3 summaries experimental PLIF setups for OH measurements in IC engines. The most common setup employs an Nd:YAG pumped dye laser to excite the OH radical around 284 nm, which corresponds to $Q_1(9)$ and $Q_2(8)$ bands of the (1,0) vibrational band of $A^2\Sigma$–$X^2\Pi$ transition, because of their high fluorescence yields and low sensitivity to temperature variations. The OH LIF signal from the (1,1) vibrational band is detected using a narrow band-pass filter centered at 313 nm. The detection of (1,1) vibrational band of OH radical reduced the effect of quenching rate on LIF signals in comparison with the detection of LIF from the (0,0) vibrational band at 310 nm. The detection of (1,1) fluorescence also reduced the effect of radiation trapping since the population of vibrationally excited OH radicals (v''=1) responsible for radiation trapping was much smaller than that of the OH radicals at ground vibrational level (v''=0). However, for qualitative visualization, both (1,1) and (0,0) vibrational bands are detected by a relatively broad band-bass filter. Excitation by a XeCl excimer laser at 308.2 and the subsequent resonant detection of the $Q_1(3)$ rotational band of the (0,0) vibrational band is also possible but suffers from severe interference of elastically scattered light from the walls and dust particles.

Table 5.3 Summary of OH Excitation and Detection Schemes Used in PLIF Measurements

Laser	Excitation Wavelength	Detection Wavelength	Applications	Comment	Reference
Nd:YAG pumped dye laser , 3 mJ	283.92 nm (1,0) $Q_2(8)$, $Q_1(9)$	310.9 ±11 nm (1,1) (0,0)	Diesel combustion in an RCM fueled with low sooting fuel	Strong signal	Kosaka et al. (1996)
Nd:YAG pumped dye laser , 24 mJ	284.01 nm (1,0) $Q_2(8)$, $Q_1(9)$	312 ±16 nm (1,1) (0,0)	Single cylinder DI diesel engine with low sooting fuel	Strong signal	Dec and Coy (1996)
Nd:YAG pumped dye laser	283.92 nm (1,0) $Q_2(8)$, $Q_1(9)$	310 nm (0,0)	Single-cylinder SI engine fueled with gasoline	Strong signal	Tanaka and Tabata (1994)
Nd:YAG pumped dye laser	283.553 nm (1,0) $Q_2(8)$, $Q_1(9)$	305–340 nm (0,0)	DI diesel engine with i-octane and n-tetradecane in O_2 enriched air	Strong signal	Nakagawa et al. (1997)
Nd:YAG pumped dye laser, 10 mJ	281 nm (1,0)	312 ±10 nm (1,1) (0,0)	SI engine with transparent head fueled with propane	Strong signal	Schipperijn et al. (1988)
Nd:YAG pumped dye laser , 9 mJ	284.26 nm (1,0) $Q_2(9)$	313 ±5 nm (1,1)	2-stroke SI engine with transparent head with propane	Reduced radiation trapping	Felton et al. (1988)
XeCl excimer laser, 250 mJ	308.2 nm (0,0) $Q_1(3)$	310 ±10 nm (1,1) (0,0)	Square piston SI engine with methane	Suffers from spurious scattering	Suntz et al. (1988)
XeCl excimer laser, 20 mJ	308.2 nm (0,0) $Q_1(3)$	310 ±10 nm (1,1) (0,0)	Side valve L-head SI engine with propane or iso-octane	Suffers from spurious scattering	Serpenguzel et al. (1993)
Tunable KrF excimer laser, 250 mJ	248.45 nm (3,0) $P_2(8)$	297 ±10 nm (3,2)	Four-cylinder SI engine with iso-octane or gasoline	Insensitive to quenching (LIPF), weak signal	Andresen et al. (1990) Vannobel et al. (1993)

Visualization of OH in a diesel engine is considerably more difficult. There are two main sources of interference in diesel combustion. One is the elastically scattered laser light from soot particles and liquid fuel droplets. The other is the broadband emission from soot particles, LIF of soot particles, and fluorescence from PAHs. Since the LIF signal of OH is collected at a wavelength longer than the laser output, the elastically scattered laser light can be rejected from OH images with a long-pass filter or band-pass filter. The broadband emission, however, is more difficult to remove. The fluorescence from PAHs is typically very broadband, extending from UV into visible. When excited by a laser beam at 284 nm, most PAH emission is more red-shifted than the OH fluorescence of 308–320 nm (Berlman, 1971). Both high-temperature soot radiation and LII (laser-induced incandescence) signals are characterized by the blackbody radiation. They drop off significantly toward the UV region, being relatively weaker below 320 nm. A short-pass filter can be used to suppress both PAH and soot-related radiation . A combination of long-pass and short-pass filters, or a band-pass filter (308–320 nm). will be effective in minimizing both elastically scattered light and broadband radiation.

Dec and Coy (1996) found that a combination of a 312 ± 16-nm band-pass filter and a 358-nm short-pass filter gave the best compromise between the OH signal strength and low PAH and LII interference. Though the elastic scattering from liquid fuel droplets could be virtually eliminated by adding an additional long-pass filter at 305 nm, it was not excluded from OH images. Because the elastic scattering was found spatially separated from the OH LIF signal, it was used to identify the location of OH distribution with respect to the liquid fuel.

5.3.4 PLIF Measurements of NO

5.3.4.1 Experimental Considerations

The terminology and salient features of NO spectroscopy can be learned from the OH spectroscopy described in Chapter 3. Five electronic states are normally involved in the band spectra of NO in the UV and visible regions: the ground electronic state, $X^2\Pi$, and the first four excited electronic states, $A^2\Sigma^+, B^2\Pi, C^2\Pi,$ and $D^2\Sigma^+$. The corresponding four bands of NO in the UV and visible regions are known as $\beta, \gamma, \delta,$ and ε systems, depending upon the transitions of electronic levels (Pearse and Gaydon, 1965). They are $B^2\Pi \Leftrightarrow X^2\Pi\,(\beta), A^2\Sigma^+ \Leftrightarrow X^2\Pi\,(\gamma), C^2\Pi \Leftrightarrow X^2\Pi\,(\delta),$ and $D^2\Sigma^+ \Leftrightarrow X^2\Pi\,(\varepsilon)$. The δ and ε systems are stronger but lie at shorter wavelengths. The spectral bands are labeled similar to the notation used for the OH radical.

PLIF measurements of NO in IC engines can be realized through one of the three UV laser wavelengths: 193 nm (ε system), 226 nm (γ system), and 248 nm (β system). Because of the shorter excitation and fluorescence wavelengths in the UV region, there are a number of issues that are pertinent to the PLIF measurements of NO.

The first issue arises from the absorption of the excitation laser beam by fuel. Commercial gasoline absorbs at all three wavelengths. Iso-octane, a substitute fuel for gasoline, does not absorb laser beam at 193, 226, and 248 nm. Attenuation of the UV laser beam by partial combustion products is also present. Spectrally resolved broadband transmission measurements revealed strong absorption in the spectral region from 193 to 243 nm (Knapp et al., 1996). At 226 nm, the transmission during the combustion stroke was about 40% in contrast to more than 80% at 248 nm when the in-cylinder pressure was more than 10 bar.

The absorption problem responsible for the laser beam attenuation is also present for fluorescent light at shorter wavelengths, caused by intermediates formed during combustion, including partially oxidized hydrocarbons and PAHs. Additionally, fluorescence trapped by transitions with large absorption rates may occur. This has been observed by Knapp et al., (1997) when fluorescence at 226 nm [$\gamma(0,0)$] and 237 nm [$\gamma(0,1)$] were detected after laser excitation at 248 nm [$\gamma(0,2)$]. To avoid absorption and trapping, the NO emission at 272 nm (0,4) was detected instead.

As the hot-band spectrum of $B^3\Sigma_u - X^3\Sigma_g$ Schumann-Runge system of O_2 (Krupenie, 1972) coincides with that of NO, the fluorescence from O_2 must be accounted for to obtain quantitative LIF measurements of NO. The term *hot-band* here refers to the spectra that appear only at elevated temperatures owing to the Boltzmann redistribution into the upper vibrational levels of the ground electronic state. This may be important for certain combustion situations and excitation-detection schemes. For example, it is difficult to separate oxygen from NO fluorescence when an ArF laser is employed (Sick and Wolfrum, 1993).

5.3.4.2 Application of PLIF to In-Cylinder NO Measurements

In the first excitation scheme, an ArF excimer laser is used, and its output at 193 nm corresponds to the spectral lines of the $D^2\Sigma^+ \Leftrightarrow X^2\Pi$ (0,1) band (Brugman et al., 1993; Tanaka et al., 1997). After excitation to the $D^2\Sigma^+$ state, some NO molecules will jump to the slightly lower level $C^2\Pi$ ($v' = 0$) because of collisional quenching. As a result, the resulting fluorescence displays two clearly distinguishable sequences: $C^2\Pi$ ($v' = 0$) $\leftrightarrow X^2\Pi$ ($v'' = 1$–6), and $D^2\Sigma^+$ ($v' = 0$)$\leftrightarrow X^2\Pi$ ($v'' = 2$–6). The strongest fluorescence originates from the primary excited level. Therefore, the $D^2\Sigma^+ \Leftrightarrow X^2\Pi$ (0,3) transition at 208 nm should be employed for PLIF measurements. A band-pass interference filter consisting of four dielectric-coated mirrors (Schulz et al., 1996) can be used to isolate the NO LIF signal with a maximum transmission of 80% at 208 ± 10 nm. In general, these measurements suffered from severe laser absorption in the presence of combustion. Strong attenuation of the laser beam at the earlier stages of combustion limited their NO measurements to the later stage of combustion. Quantitative measurements of NO using this excitation scheme was further complicated by the difficulty of the spectral interference from oxygen fluorescence excited at 193 nm (Sick and Wolfrum, 1993). Therefore, the NO LIF with 193-nm excitation is more an historical interest than a suitable choice for engines.

A more suitable excitation scheme is achieved by exciting the NO $A^2\Sigma^+ \Leftrightarrow X^2\Pi$ (0,2) band at 247.94 nm using the direct output of a tunable KrF excimer laser (Knapp et al., 1996, Hildenbrand et al., 1998). Two fluorescence peaks can be detected at 226 nm and 237 nm, but the LIF signal at 237 nm experiences less fluorescence trapping than the shorter wavelength. The advantage of NO excitation at 248 nm is the possibility of NO measurements during combustion without severe attenuation. However, probing a vibrationally excited state ($v'' = 2$) results in a lower sensitivity and a signal that is strongly dependent on temperature, because the number density of vibrationally excited NO molecules is a strong function of temperate. The NO excitation and detection at 248 nm is, therefore, particularly suitable for high temperature and high NO concentrations.

The most popular and accurate approach for the PLIF measurements of NO is the excitation of the rotational-vibrational transitions of the $A^2\Sigma^+ \Leftrightarrow X^2\Pi$ (0,0) band near 226 nm. The excitation source is usually an Nd:YAG laser-pumped tunable dye laser. It can also be achieved by the first anti-Stokes output of an H_2 Raman cell excited by a tunable KrF excimer laser (Bräumer et al., 1995)

The most comprehensive PLIF NO measurement was reported by Dec and Canaan (1998) in a single-cylinder diesel engine. A frequency-tripled Nd:YAG laser-pumped optical parametric oscillator (OPO) was employed because of its high pulse energy of 12–15 mJ/pulse at 226 nm. The wavelength at 226.035 was selected because it was at the peak of one of the strongest NO fluorescence excitation lines, but it was located near the middle one of the weakest oxygen absorption region so as to minimize oxygen interference. The subsequent fluorescence from the $A^2\Sigma^+ \Leftrightarrow X^2\Pi$ (0,1) band to (0,4) band at was detected through a band-pass filter of 237–276 nm, to maximize the fluorescence signal while it allowed rejection of elastically scattered light, PAH fluorescence, oxygen interference, and LII. The combined higher laser energy and the selected detection scheme allowed single-shot NO images to be obtained under realistic diesel engine operating conditions without oxygen enrichment of the intake air. In addition to two-dimensional images of NO distribution, total NO formation was examined by integrating the NO PLIF signal over a large probe volume ($15 \times 4.5 \times 140$ mm^3). These total NO results were then corrected for pressure, temperature, and surrounding-species effects on collision broadening and line-shift, Doppler broadening, Boltzmann distribution, and quenching rate using a detailed spectroscopic code developed by Paul et al. (1993; 1994).

5.3.5 Visualization of Formaldehyde by PLIF

Formaldehyde (CH_2O) has long been associated with the autoignition process taking place in the end gas region of a gasoline engine with knocking combustion or in a compression ignition engine. It is known that formaldehyde is formed during the first stage of the autoignition process, and it is responsible for the weak blue light emission of cool flames. As autoignition proceeds into higher temperature region, formaldehyde is consumed rapidly in the second stage of the autoignition process (Benson, 1981). The presence of formaldehyde is, therefore, a good marker of the autoignition process. In particular, it has become one of the most studied topics in the research on CAI (controlled autoignition)/HCCI (homogeneous charge compression ignition) combustion engines.

The principle and experimental setup for PLIF measurements of formaldehyde are similar to that of OH and NO. Because of the greater number of atoms in a CH_2O molecule, there are many more vibrational energy levels in each of the electronic energy state of formaldehyde. As a result, an electronic transition is accompanied by many vibrational-rotational changes spread over a range of frequencies. Figure 5.6 shows the fluorescence excitation spectra of formaldehyde (Miller and Lee, 1975). The fluorescence excitation spectrum has a peak at 353.2 nm corresponding to the 4_0^1 line of the $\tilde{A}^1A_2 \leftarrow X^1A_1$ band. The emission spectrum, slightly shifted to the red, starts at about 360 nm. Most formaldehyde fluorescence is found in the range of 390 to 460 nm. Although the frequency-tripled output (355 nm) of an Nd:YAG laser may be used for the excitation of the $\tilde{A}^1A_2$ state of formaldehyde without the expenses of a dye laser, it induces weak absorption by formaldehyde and hence lower fluorescence intensity. A tunable dye laser is preferred as it can be easily tuned to 353.2 nm for the maximum fluorescence signal.

Formaldehyde distribution in a two-stroke engine was measured by Bäuerle et al. (1994) using PLIF. The excitation at 353.2 nm was achieved by a tunable dye laser with BMQ dye (24 mJ). The fluorescence signal from a 50-mm wide laser sheet was detected by an ICCD camera through the transparent cylinder head. A band-pass filter (400–450 nm) was placed in front of the camera to eliminate spurious laser scattering. PLIF CH_2O images of the complete cylinder charge were obtained from different knocking cycles. This experimental setup was later extended to record a pair of fluorescence images of CH_2O separated by 50 μs so that the temporal progress of the autoignition process could be tracked (Bäuerle et al., 1996). The pair of fluorescence images was obtained by the double-pulse excitation (2.6 and 1.3 mJ, respectively) of the same dye laser sequentially pumped by two excimer lasers.

As part of comprehensive studies on CAI or HCCI combustion in gasoline engines, the effect of charge and temperature stratification as well as the fuel

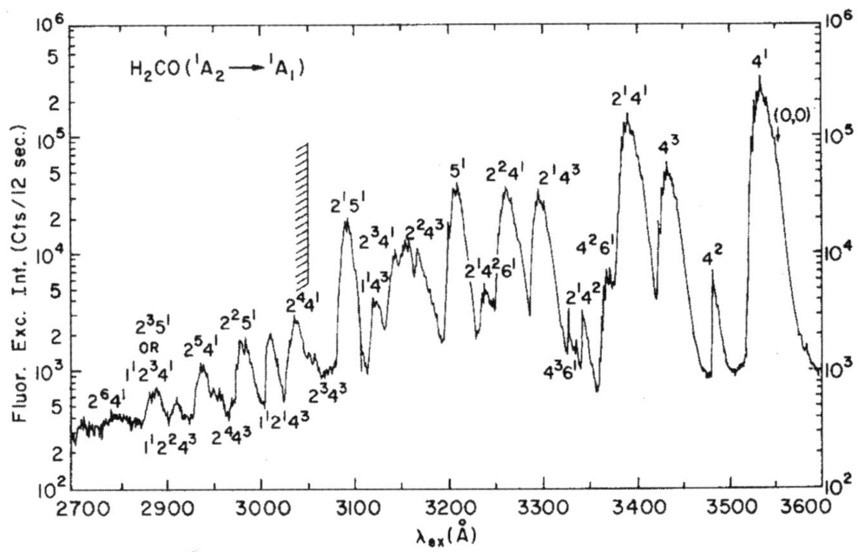

Fig. 5.6 Fluorescence excitation spectra of CH$_2$O, adapted from Miller and Lee (1975).

composition was investigated at Brunel University by PLIF measurements of formaldehyde in a single-cylinder optical engine (Peng et al., 2003).

The output from an Nd:YAG laser (245 mJ at 532 nm) was employed to pump a tunable dye laser. The second harmonic output from the dye laser with pyridine was tuned to the absorption line of the 4_0^1 transition of the $\tilde{A}^1A_2 \leftarrow \tilde{X}^1A_1$. To verify and tune the dye laser output to the 4_0^1 transition line, a premixed formaldehyde/air mixture was placed in a constant volume vessel at 1 bar pressure and room temperature. After it was heated to 100°C, the dye laser output was then tuned to produce the strongest fluorescence signal with a pulse energy of 7 mJ.

The experimental setup was similar to that shown in Fig. 5.1. The optical engine, shown in Fig. 1.1, has a transparent cylinder head and side windows. The output from the dye laser was steered via laser mirrors and then passed through a spherical lens ($f = 1000$ mm) and two cylindrical lenses ($f = -40$ mm, 75 mm) to form a horizontal laser sheet, which passed through entrance and exit windows on the side of cylinder.

Laser-induced fluorescence and radiation from combustion were detected through the top window in the cylinder head. The signal was viewed with a 45° mirror mounted above the cylinder head and imaged onto a gated ICCD camera. As the YAG laser could only operate at 10 Hz, these experiments were carried out at an engine speed of 600 rpm to match the laser-firing

154

frequency. The actual image acquisition rate was limited to less than 1 Hz of the ICCD camera. The laser was triggered to fire at a preset crank angle in the engine cycle via a trigger generator with crank angle and TDC signals from a shaft encoder coupled to the camshaft. The synchronization output from the Nd:YAG laser was then passed to an digital signal-delay generator, from which a TTL (transistor-transistor logic) pulse was sent to operate the intensifier after a short delay. Before the PLIF measurement, the delay of ICCD was increased gradually from the laser firing time until the effect of the laser pulse on the LIF signal disappeared. This was followed by setting the ICCD gate width to a suitable value to obtain a good LIF image. As a result, the gate width was set to 200 ns, and the delay was 885 ns. With such settings, the interference of the laser signal on LIF signal could be removed without any additional optical filter other than the optical lens used.

Figure 5.7 shows the formaldehyde distribution at different crank angles during the autoignition process of CAI combustion operation with n-heptane.

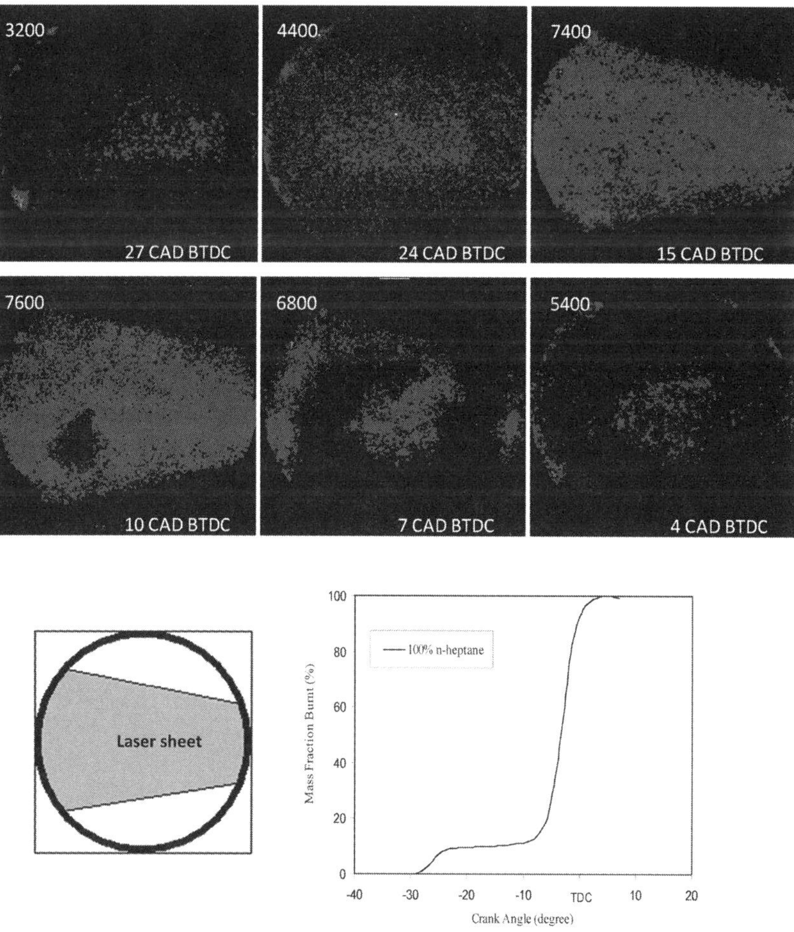

Fig. 5.7 CH$_2$O distributions during the autoignition process in a CAI engine.

As shown in the mass fraction burned curve, the first-stage autoignition started near 30 CA BTDC and ended near 10 CA BTDC. During this period, formaldehyde first appeared at 27 CA BTDC in part and then throughout the laser illuminated area at 15 CA BTDC. The formaldehyde disappeared in a small region at 10 CA BTDC, coinciding with the start of the second stage high-temperature combustion, before it was completely consumed just before TDC. By comparing the formaldehyde distributions for different fuel mixtures, the effect of n-heptane on the two-stage autoignition process was also identified.

5.4 Imaging of Water Vapor and CO by the Two-Photon LIF Technique

Water vapor is present in EGR (exhaust gas recirculation) as a major constituent of burned gases, and it can be used as a marker for the recycled or trapped burned gas. CO is produced due to partial oxidation and incomplete combustion of hydrocarbon fuels. However, CO and water vapor have their absorption lines in the low UV region. The LIF technique discussed so far cannot be applied to the detection of such species. Two-photon excitation, instead, is required to generate detectable fluorescence from these molecules.

Figure 5.8 illustrates schematically the difference between the single and two-photon LIF processes. For LIF measurements discussed in the previous chapters, the molecule under study absorbs one photon. The energy $E_{21} = h\upsilon_{21}$ of this photon of frequency υ_{21} matches the energy difference between levels 1 and 2. The excited species in level 2 may emit fluorescence of frequency υ_f. In the two-photon scheme, the molecule simultaneously absorbs two photons. In this case, the sum of the energies of both photons corresponds to the energy difference between levels 1 and 2, that is, $E_{21} = h(\upsilon_{2\upsilon}+\upsilon_{\upsilon 1})$. In the two-photon process, one can imagine that the first photon excites the molecule to a virtual level and then the second photon lifts the molecule further to the energy level 2. Both frequencies, $\upsilon_{2\upsilon}$ and $\upsilon_{\upsilon 1}$, may be the same, so that a single laser may be used, or they may be different. Similar to one-photon LIF, the excited state emits fluorescence that may be detected as a measure for the species concentration.

In the two-photon fluorescence process, the Einstein coefficient for stimulated absorption B_{12} in Eq. 5.1 would be replaced by $(c/h\upsilon)\cdot\alpha_{12}\cdot I_o$ (Eckbreth, 1996). Hence, the two-photon fluorescence signal can be expressed as

$$P_{flu} \propto \chi_m \, N \, P_o^2 \, \alpha_{12} \frac{A_{21}}{A_{21}+Q_{21}} \tag{5.5}$$

where α_{12} is the two-photon absorption cross section from state 1 to state 2. Since the absorption cross section is very small, the two-photon transitions are difficult to saturate. It shows that the fluorescence signal has a quadratic dependence on the laser power. Fluorescence signal is subject to quenching.

Figure 5.9 shows the energy-level diagram for water vapor including the transitions relevant to combustion diagnostics. If wavelengths near 125 nm are used for excitation from the ground state X^1A_1, the third excited state C^1B_1 can be reached. As there is no laser source at 125 nm, Neij and Aldén (1994) found that the water molecules could be excited through a two-photon excitation process at 248 nm with a tunable excimer laser. The excited water molecules then emit fluorescence light in the range of 400–500 nm. Figure 5.10 shows the excitation and emission spectrum of water vapor. The lines in Fig. 5.10(a) correspond to one or more rotational lines and are numbered according to the notation by Meijer et al. (1986). The emission spectrum in Fig. 5.10(b) exhibits a broad, structureless spectrum that extends from 380 nm to almost 600 nm, with a maximum at approximately 425 nm. The emission spectrum in Fig. 5.10(b) is independent of the rotational absorption lines as shown in Fig. 5.10(a).

Neij and Aldén (1994) estimated that the detection limit at 300 K was 1% water in air. The LIF signal intensity obtained from peak 4 in Fig. 5.10(a) was found to be roughly 20 times stronger than the Raman signal at the atmospheric condition. The fluorescence signal was, however, reduced by about two-thirds as the partial vapor pressure of water was increased from 32 mbar to 320 mbar.

Following the exploratory work by Neij and Aldén (1994), Johansson et al. (1995) applied the two-photon PLIF technique to visualize the water vapor distribution in a spark-ignition engine. The experimental setup of the two-photon LIF measurement is very similar to that used for the PLIF imaging of fuels, which will be discussed in Chapter 7. By tuning the output of an KrF

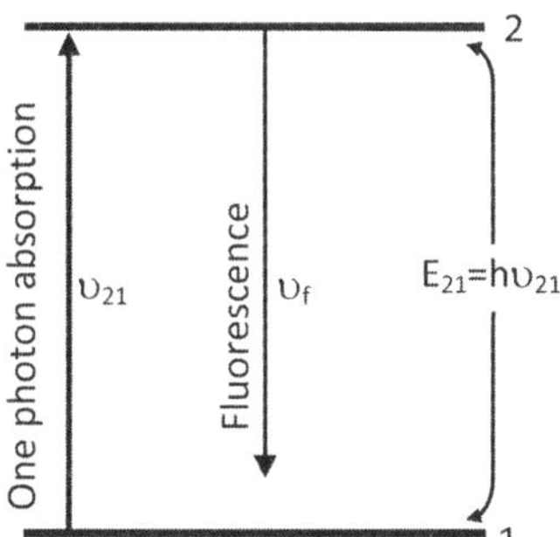

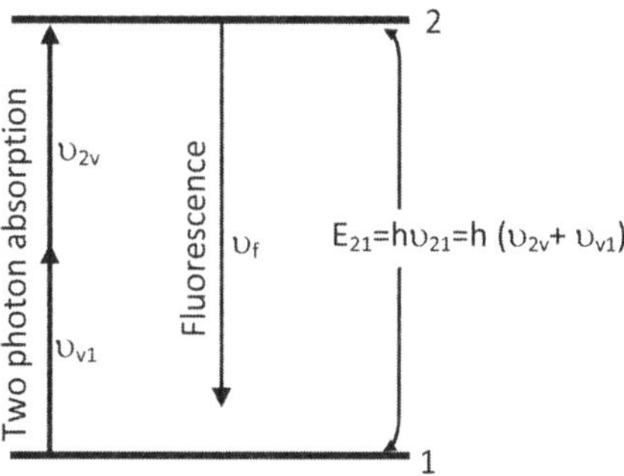

Fig. 5.8 Schematic of the energy-level diagrams of one-photon (left) and two-photon (right) LIF processes.

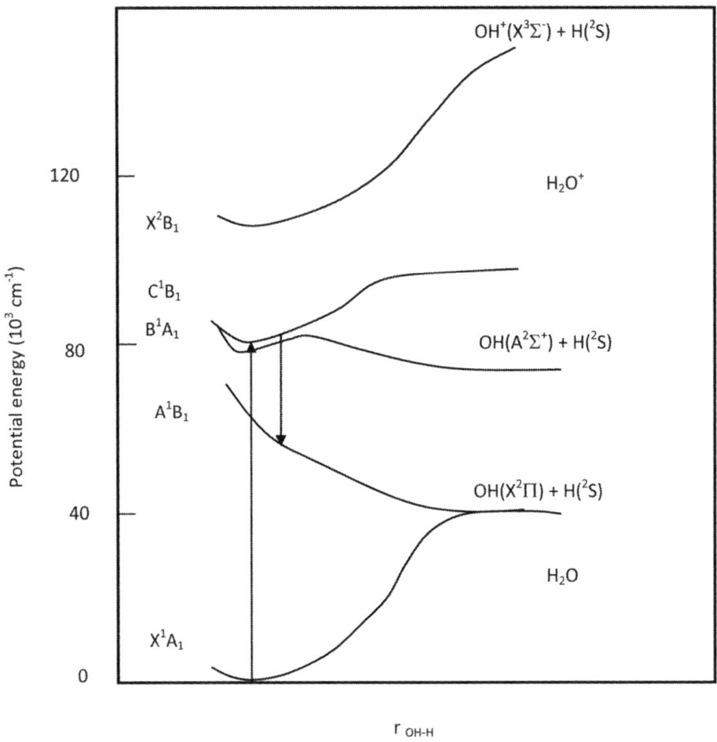

Fig. 5.9 Potential energy-level diagram of water vapor.

excimer laser over the strong peak in the excitation spectrum at 248.285 nm (peak 11) or 248.523 nm (peak 4). Johansson et al. (1995) was able to measure the water vapor (residual gas) distribution near the spark plug in a narrow region of 7×0.1 mm. Because of the second-order laser intensity dependence of the two-photon LIF signal, the laser power output was corrected on a shot-by-shot basis.

To identify the source of CO and UHC (unburned hydrocarbon) emissions from low-temperature diesel combustion, Kim et al. (2008) employed the two-photon PLIF technique to visualize the CO and UHC distributions in an optical diesel engine. The CO LIF measurement is based on the two-photon excitation of the $B^1\Sigma^+ \Leftrightarrow X^1\Sigma^+$ (0,0) transition near 230.1 nm. The excitation laser beam was generated by an OPO pumped by an injection-seeded Nd:YAG laser. The subsequent fluorescence emissions between 440 and 730 nm from the $B^1\Sigma^+ \Leftrightarrow A^1\Pi$ system were detected by an ICCD camera. The precise laser wavelength was found by performing excitation scans in the motored engine with known quantity of CO.

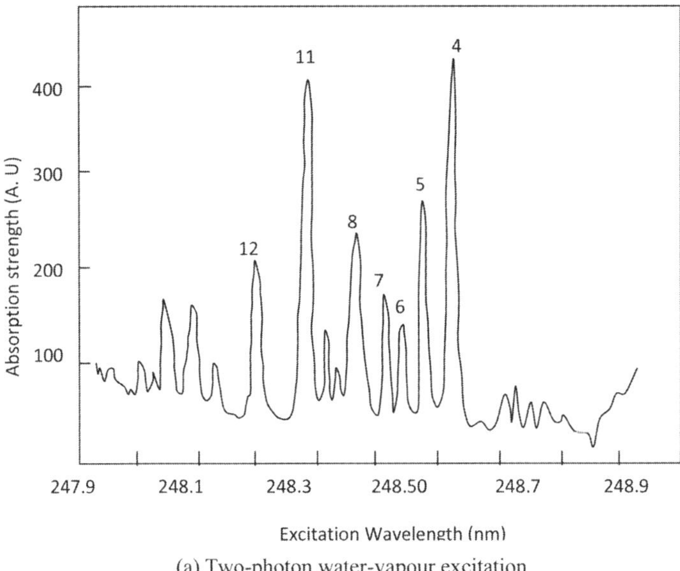

(a) Two-photon water-vapour excitation

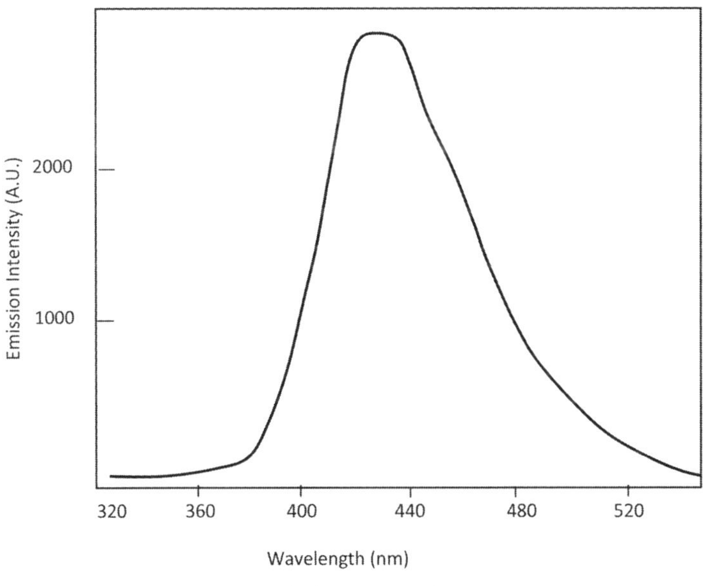

(b) Emission spectrum from water vapour

Fig. 5.10 Two-photon water vapor excitation and emission spectrum.

Compared to the one-photon LIF process, the two-photon LIF is more prone to quenching and its quadratic dependence on the laser intensity. Coupled with all the difficulties with the shorter wavelength laser beam, the two-photon LIF technique is better suited for qualitative in-cylinder visualization than quantitative measurements.

5.5 Summary

Laser-induced fluorescence is the most widely used technique for in-cylinder fuel distribution and combustion species measurements. One of its most attractive features is the ease of acquiring instantaneous, two-dimensional images of in-cylinder fuel and species distributions for both motored and fired operations. The normally red-shifted frequency of fluorescence emission enables the spurious scattered light to be eliminated. As LIF is subject to quenching of different origins, quantitative interpretation of experimental results will require careful and detailed consideration on the experimental setup and data processing. With advance in lasers and cameras, high-speed LIF measurements as well as multispecies PLIF measurements can also be implemented for advanced combustion engines research and development .

References

Andresen, P., Meijer G, Schluter, H. Voges, H., Koch, A., Hentschel, W. Oppermann, W., and Rothe, E. (1990). "Fluorescence Imaging Inside an Internal Combustion Engine Using Tunable Excimer Lasers." *Applied Optics*, Vol. 29, No. 16, pp. 2392–2404.

Baritaud, T. A., and Heinze, T. A. (1992). "Gasoline Distribution Measurements With PLIF in a SI Engine." SAE Paper No. 922355, SAE International, Warrendale, PA.

Bäuerle, B., Hoffmann, F., Behrendt, F., and Warnatz, J. (1994). "Detection of Hot Spots in the End Gas of an Internal Combustion Engine Using Two-dimensional LIF of Formaldehyde." *Proc. 25th Symposium (international) on Combustion,* pp. 135–141. Pittsburgh: The Combustion Institute.

Bäuerle, B., Warnatz, J., and Behrendt, F. (1996). "Time-Resolved Investigation of Hot Spots in the End Gas of an SI Engine by Means of 2D Double-Pulse LIF of Formaldehyde." *Proc. 26th Symposium (international) on Combustion.* Pittsburgh: The Combustion Institute.

Benson, S. W. (1981). "The Kinetics and Thermochemistry of Chemical Oxidation with Application to Combustion and Flames." *Progr. Energy Combust. Sci.* Vol. 7, pp. 125–134.

Berckmüller, M., Tait, N., Lockett, R., and Greenhalgh, D. (1994). "In-Cylinder Crank Angle Resolved Imaging of Fuel Concentration in a Firing Spark Ignition Engine Using Planar Laser-Induced Fluorescence." *Proc. 25th Symposium (International) on Combustion,* pp. 151–156. Pittsburgh: The Combustion Institute.

Berckmüller, M., Tait, N., and Greenhalgh, D. (1996). "The Time History of the Mixture Formation Process in a Lean Burn Stratified Charge Engine." SAE Paper No. 961929, SAE International, Warrendale, PA.

Berckmüller, M., Tait, N., and Greenhalgh, D. (1997). "The Influence of Local Fuel Concentration on Cyclic Variability of a Lean Burn Stratified Charge Engine." SAE Paper No. 970826, SAE International, Warrendale, PA.

Berlman, I. B. (1971). *Handbook of Fluorescence Spectra of Aromatic Molecules*. New York: Academic Press.

Bräumer, A., Sick, V., Wolfrum, J., Drewes, V., Zahn, M., and Maly, R. (1995). "Quantitative Two-Dimensional Measurements of Nitric Oxide and Temperature distributions in a Transparent Square Piston SI Engine." SAE Paper No. 952462, SAE International, Warrendale, PA.

Brugman, M., Klein-Douwel, R., Huigen, G., Walwijk, E., and Meulen, J. (1993). "Laser-Induced Fluorescence Imaging of NO in an n-Heptane and Diesel Fuel Driven Diesel Engine." *App. Physics*, Vol. 57, pp. 405–410.

Dec, J., and Canaan, R. (1998). "PLIF Imaging of NO Formation in a DI Diesel Engine." SAE Paper No. 980147, SAE International, Warrendale, PA.

Dec, J., and Coy, E. B. (1996). "OH Radical Imaging in a DI Diesel Engine and the Structure of the Early diffusion Flame." SAE Paper No. 960831, SAE International, Warrendale, PA.

Deschamps, B., Snyder, R., and Baritaud, T. (1994). "Effect of Flow and Gasoline Stratification on Combustion in a 4-Valve SI Engine." SAE Paper No. 941993, SAE International, Warrendale, PA.

Di Rosa, M. D., and Hanson, R. K. (1994). "Collisional Broadening and Shifting of NO g(0,0) Absorption Lines by O2 and H2O at High Temperatures." *J. Quant. Spectrosc. Radiat. Transfer*. Vol. 52, pp. 515 – 529.

Eckbreth, A. C. (1996). *Laser Diagnostics for Combustion Temperature and Species*, 2nd ed. Amsterdam: Gordon and Breach.

Felton, P., Mantzaras, J., Bomse, D., and Woodin, R. (1988). "Initial Two-dimensional Laser Induced Fluorescence Measurements of OH Radicals in an Internal Combustion Engine." SAE Paper No. 881633, SAE International, Warrendale, PA.

Fujikawa, T., Hattori, Y., and Akihama, K. (1997). "Quantitative 2-D Fuel Distribution Measurements in a SI Engine Using Laser-Induced Fluorescence with Suitable Combination of Fluorescence Tracer and Excitation Wavelength." SAE Paper No. 972944, SAE International, Warrendale, PA.

Ghandhi, J. B., and Bracco, F. V. (1995). "Fuel Distribution Effects on the Combustion of a Direct Injection Stratified Charge Engine." SAE Paper No. 950460, SAE International, Warrendale, PA.

Hildenbrand, F., Schulz, C., Sick, V., Josefsson, G., Magnusson, I., Andersson, O., and Alden, M. (1998). "Laser Spectroscopic Investigation of Flow Fields and NO Formation in a Realistic SI Engine." SAE Paper No. 980148, SAE International, Warrendale, PA.

Itoh, T., Kakuho, A., Hishinuma, H., Urushiahara, T., Takagi, Y., Horie, K., Asano, M., Ogata, E., and Yamasita, T. (1995). "Development of a New Compound Fuel and Fluorescent Tracer Combination for Use With Laser-Induced Fluorescence." SAE Paper No. 952465, SAE International, Warrendale, PA.

Johansson, B., Neij, H., Greger, J., and Aldén, M. (1995). "Residual Gas Visualization with Laser-Induced Fluorescence." SAE Paper No. 952463, SAE International, Warrendale, PA.

Kim, D., Ekoto, I., Colban, W. F., and Miles, P. (2008). "In-Cylinder CO and UHC Imaging in a Light Duty Diesel Engine During PCCI Low Temperature Combustion." SAE Paper No. 2008–01–0–1602.

Knapp, M., Luczak, A., Schluter, H., Beushausen, V., Hentschel, W., and Andresen, P. (1996). "Crank-Angle-Resolved Laser-Induced Fluorescence Imaging of NO in a Spark-Ignition Engine at 248 nm and Correlation to Flame Front Propagation and Pressure Release." *Applied Optics*, Vol. 35, No. 21, pp. 4009–4017.

Knapp, M., Luczak, A., Beushausen, V., Hentschel, W., Manz, P., and Andresen, P. (1997). "Quantitative In-Cylinder NO LIF Measurements with a KrF Excimer Laser Applied to a Mass-Production SI Engine Fueled with Isooctane and Regular Gasoline." SAE Paper No. 970824, SAE International, Warrendale, PA.

Kosaka, H., Nishigaki, T., and Komimoto, T. (1996). "Simultaneous 2-D Imaging of OH Radicals and Soot in a Diesel Flame by Laser Sheet Technique." SAE Paper No. 960834, SAE International, Warrendale, PA.

Krupenie, P. H. (1972). "The Spectrum of Molecular Oxygen." *J. Phys. Chem. Ref. Data.* Vol. 1, pp. 423–534.

Lawrenz, W., Köhler, J., Meier, F., Stolz, W., Wirth, R., Bloss, W., Maly, R., Wagner, E., and Zahr, M. (1992). "Quantitative 2D LIF Measurements of Air/Fuel Ratios During the Intake Stroke in a Transparent SI Engine." SAE Paper No. 922320, SAE International, Warrendale, PA.

Li, Y., Zhao, H., Leach, B., Ma, T., and Ladommatos, N. (2004). "In-Cylinder Measurements of Fuel Stratification in a Three-Valve Twin-Spark SI Engine." SAE Paper No. 2004–01–1354, SAE International, Warrendale, PA.

Meijer, G., Meulen, J., Andresen, P., and Bath, A. (1986). "Sensitive Quantum State Selective Detection of H2O and D2O by (2+1)- Resonance Enhanced Multiphoton Ionization." *J. Chem. Phys.*, Vol. 85, pp. 6914–6922.

Meyer, J., Graul, W., Kiefer, K., Thiemann, J., and Landry, M. (1997). "Study and Visualization of the Fuel Distribution in a Stratified Spark Ignition Engine with EGR Using Laser-Induced Fluorescence." SAE Paper No. 970868, SAE International, Warrendale, PA.

Meyer, J., Haug, M., Schreiber, M., and Unverzagt, S. (1995). "Controlling Combustion in a Spark Ignition Engine by Quantitative Fuel Distribution." SAE Paper No. 950107, SAE International, Warrendale, PA.

Miller, R. G., and Lee, K. C. (1975). "Single Vibronic Level Photochemistry of Formaldehyde: Radiative and Non-Radiative Transitions." *Chemical Physics Letters*, Vol. 33, No. 1, pp. 104–107.

Nakagawa, H., Endo, H. Deguchi, Y., Noda, M., Oikawa, H., and Shimada, H. (1997). "NO Measurement in Diesel Spray Flame Using Laser Induced Fluorescence." SAE Paper No. 970874, SAE International, Warrendale, PA.

Neij, H., and Aldén, M. (1994). "Application of Two-Photon Laser-Induced Fluorescence for Visualization of Water Vapor in Combustion Environments." *Applied Optics*, Vol. 33, No. 27, pp. 6514–6523.

Paul, P. H., Gray, J. A., Durant Jr., J. L., and Thoman Jr., J. W. (1993). "A Model for Temperature-Dependent Collisional Quenching of NO $A^2\Sigma^+$." *Appl. Phys. B*, Vol. 57, pp. 249 – 259.

Paul, P. H., Gray, J. A., Durant Jr., J. L., and Thoman Jr., J. W. (1994). "Collisional Quenching Corrections for Laser-Induced Fluorescence Measurements of NO $A^2\Sigma^+$." *AIAA Journal*, Vol. 32, pp. 1670–1675.

Pearse, R. W. B., and Gaydon, A. G. (1965). *The Identification of Molecular Spectra*, 3rd ed. London: Chapman and Hall.

Peng, Z., Zhao, H., and Ladommatos, N. (2003). "Visualization of HCCI/CAI Combustion Process Using 2-D PLIF Image of Formaldehyde." *Proc. Instn. Mech. Engrs, Part D*. Vol. 217, pp. 1125–1134.

Reboux, J., Puechberty, D., and Dionnet, F. (1994). "A New Approach of Planar Laser Induced Fluorescence Applied to Fuel/Air Ratio Measurement in the Compression Stroke of an Optical SI Engine." SAE Paper No. 941988, SAE International, Warrendale, PA.

Reboux J., Puechberty, D., and Dionnet, F. (1996). "Study of Mixture Inhomogeneities and Combustion Development in a SI Engine Using a New Approach of Laser Induced Fluorescence (FARLIF)." SAE Paper No. 961205, SAE International, Warrendale, PA.

Schipperijn, F., Nagasaka, R, Sawyer, R., and Green, R. (1988). "Imaging of Engine Flow and Combustion Processes." SAE Paper No. 881631, SAE International, Warrendale, PA.

Schulz, C., Sick, V., Wolfrum, J., Drewes, V., Zahn, M., and Maly, R. (1996). "Quantitative 2D Single-Shot Imaging and Mathematical Modeling of NO Concentrations and Temperatures in a Transparent SI Engine." *Proc. of 26th Symposium (International) on Combustion*. Pittsburgh: The Combustion Institute.

Schulz, C., and Sick, V. (2005). "Tracer-LIF Diagnostics: Quantitative Measurement of Fuel Concentration, Temperature and Fuel/Air Ratio in Practical Combustion Systems." *Progress in Energy and Combustion Science*, Vol. 31, No. 1, pp. 75–121.

Serpenguzel, A., Hahn, R., and Acker, W. (1993). "Single Pulse Planar Laser Induced Fluorescence Imaging of Hydroxyl Radicals in a Spark Ignition Engine." SAE Paper No. 932701, SAE International, Warrendale, PA.

Sick, V., and Wolfrum, J. (1993). "Absolute Laser-Spectroscopic Measurements of NO Concentrations to Study the Formation of NO in Laminar Counterflow Diffusion Flames." Joint Meeting of the British/German Sections of the Combustion Institute. Cambridge, U.K.

Stevens, R., Ma, H., Stone, R., Walmsley, H., and Cracknell, R. (2007). "On Planar Laser-Induced Fluorescence with Multicomponent Fuel and Tracer Design for Quantitative Determination of Fuel Concentration in IC Engines." *Proc. IMechE, Part D*, Vol. 221, No. 6.

Suntz, R., Becker, H., Monkhouse, P. and Wolfrum, J. (1988). " Two-dimensional Visualisation of the Flame Front in an Internal Combustion Engine by Laser Induced Fluorescence of OH Radicals." *Appl. Phys.*, B47, pp287-293.

Swindal, J. C., Dragonetti, D. P., Hahn, R., Furman, P. A., and Acker, W. P. (1995). " In-Cylinder Charge Homogeneity During Cold Start Studied with Fluorescent Tracers Simulating Different Fuel Distillation Temperatures." SAE Paper No. 950106, SAE International, Warrendale, PA.

Tanaka, T., and Tabata, M. (1994). "Planar Measurements of OH Radicals in an SI Engine Based on Laser Induced Fluorescence." SAE Paper No. 940477, SAE International, Warrendale, PA.

Tanaka, T., Fujimoto, M., and Tabata, M. (1997). "Planar Measurements of OH Radicals in an SI Engine Based on Laser Induced Fluorescence." SAE Paper No. 970877, SAE International, Warrendale, PA.

Urushihara, T., Nakata, T., Kaduhou, A., and Takagi, Y. (1996). "Effects of Swirl and Tumble Motion on Fuel Vapor Behavior and Mixture Stratification in Lean Burn Engines." *JSAE Review*, Vol. 17, pp. 239–244.

Vannobel, F., Arnold, A., Buschmann, A., Sick. V., Wolfrum, J., Cousyn, B. and Decker, M.(1993). "Simultaneous Imaging of Fuel and Hydroxyl Radicals in an In-Line Four Cylinder SI Engine.", SAE Paper No. 932696, SAE International, Warrendale, PA.

Wolff, D., Beushausen, V., Schlüter, H., and Andresen, P. (1994). "Quantitative 2D-Mixture Fraction Imaging Inside an Internal Combustion Engine Using Acetone Fluorescence." *Proc. COMMODIA 94*. pp. 445–450.

Zhao, F. Q., Taketomi, M., Nishida, K., and Hiroyasu, H. (1994). "PLIF Measurements of the Cyclic Variation of Mixture Concentration in a SI Engine." SAE Paper No. 940988, SAE International, Warrendale, PA.

chapter 6

Fuel and Mixture Composition Measurement by Raman and Rayleigh Scattering Techniques

6.1 Introduction

As discussed in Chapter 3, Rayleigh scattering is characterized by its simple optical setup. It is the strongest of the molecular light-scattering processes, and hence it has the potential for two-dimensional measurements.

Raman scattering occurs with any laser excitation wavelength. It is capable of providing simultaneous multiple-species concentration measurements. Because spontaneous Raman scattering (SRS) does not involve electronic state transitions, stable species (e.g., N_2, O_2, and CO_2) can be measured by SRS in the visible region.

In this chapter, the principle of operation and application of laser Rayleigh scattering and Raman scattering will be presented. A detailed case study will describe the application of SRS to the in-cylinder multispecies measurement in a CAI (controlled autoignition) gasoline engine.

6.2 Fuel Concentration Measurement by Laser Rayleigh Scattering

Laser Rayleigh scattering (LRS) was developed and applied to in-cylinder fuel distribution measurements in the 1980s and early 1990s prior to the rapid advances in the solid-state imaging devices and pulsed lasers of different wavelengths. The fuel concentration measurement by Rayleigh scattering requires the scattering cross section of the fuel to be much greater than air and that it can be represented by one component that has a fixed scattering cross-sectional value. Since what is measured by LRS is the fuel density, information about gas temperature and pressure is required to calculate the fuel concentration.

6.2.1 Principle of Operation

The fundamental principle governing Rayleigh scattering has been discussed in Chapter 3. According to Eqs. 3.23 and 3.24, the mole fraction of fuel vapor in a fuel/air mixture x_f is given by

$$x_f = x_{fo} \frac{P_{fa} \frac{N_o}{N} - P_{ao}}{P_{fo} - P_{ao}} \qquad (6.1a)$$

or

$$x_f = x_{fo} \frac{P_{fa} \frac{p_o}{p} \frac{T}{T_o} - P_{ao}}{P_{fo} - P_{ao}} \qquad (6.1b)$$

where

P_{fa} is the Rayleigh scattering signal of a binary fuel/air mixture

$P_{fo,}$ and P_{ao} are the reference Rayleigh signals for a known fuel mole fraction x_{fo} in the binary mixture and the Rayleigh scattering from air, respectively, at the standard condition p_o and T_o

It is clear from Eq. 6.1b that the measurement of the mole fraction of fuel x_f will require the calibration experiments to be carried out in advance. It is also necessary to know the temperature and pressure in order to derive the mole fraction of fuel from the Rayleigh signal. Generally, the pressure can be obtained by using a pressure transducer. However, measuring the temperature itself is a daunting task, which is one of limiting factors for the application of Rayleigh scattering to IC (internal combustion) engines. In some cases, however, with a judicious design of the diagnostic system and a suitable assumption of the flow under study, Rayleigh scattering can be used to obtain the concentration of a binary mixture system without the requirement of information on temperature, which will be discussed later.

It can be seen from Eqs. 3.21, 3.22, and 6.1 that the values of σ_f and σ_a should be sufficiently different to generate a distinguishable Rayleigh signal from the fuel. Hence, fuel or simulated fuel of a much larger Rayleigh scattering cross section is a prerequisite for the measurement of fuel concentrations. This can be normally achieved by the use of heavy molecules, such as Freon® or hydrocarbon fuels. The ratios of Rayleigh scattering cross sections of some common hydrocarbon fuels to air are 15 for propane, 90 for isooctane, and 305 for diesel.

During the intake stroke, the temperature variation may be assumed to be small, and Eq. 6.1b may be simplified to

$$x_f = x_{fo} \frac{P_{fa} \frac{P_o}{p} - P_{ao}}{P_{fo} - P_{ao}}$$

(6.2)

During the compression stroke, if the port-injected fuel in an SI (spark-ignition) engine is assumed to have evaporated completely by then, the fuel vapor mole fraction is given by (Zhao and Hiroyasu, 1993) the following:

$$x_f = x_{fo} \frac{P_{fa} \frac{P_o}{p} \frac{T}{T_o} \frac{V}{V_c} - P_{ao}}{P_{fo} - P_{ao}}$$

(6.3)

where V_c is the cylinder volume at the start of compression stroke, and V is the cylinder volume at the time of measurement. Therefore, determination of the fuel vapor mole fraction will require measurements of the in-cylinder pressure, gas temperature and scattered as well as the calibration values at standard conditions.

6.2.2 Implementation of LRS

Figure 6.1 shows schematically a typical setup for LRS experiments. The laser output should be vertically polarized in order to maximize the Rayleigh signal. The scattered light is collected at a right angle to enhance the spatial resolution. For a single-point measurement, the laser beam is focused into the test section so that the beam waist of the laser is viewed by a photomultiplier. For two-dimensional imaging, a laser sheet is introduced into the combustion chamber and imaged at a right angle by an intensified CCD (charge-coupled device) camera. To suppress the spuriously scattered light from optical components, the laser beam is made to pass through pinholes or slits. After passing through the combustion chamber, the laser beam is dumped into a light trap. Synchronization among the laser, detector, and the data acquisition system is achieved via a PC computer and a delay generator. Measurements can be made at a particular crank angle by triggering the system from a shaft encoder. Normalization of the scattered light signal to account for variations in laser power output can be effected by monitoring the laser pulse with a photodiode or power meter.

As shown in Eqs. 6.2 and 6.3, if the Rayleigh scattered light intensities from the air and the standard premixed homogeneous air-fuel mixture is determined beforehand under a reference condition, the fuel concentration distribution can be derived from the detected scattered light from the fuel/air mixture, the cylinder volume, and the in-cylinder pressure and temperature. Normally, the calibration can be performed at room temperature and atmospheric pressure.

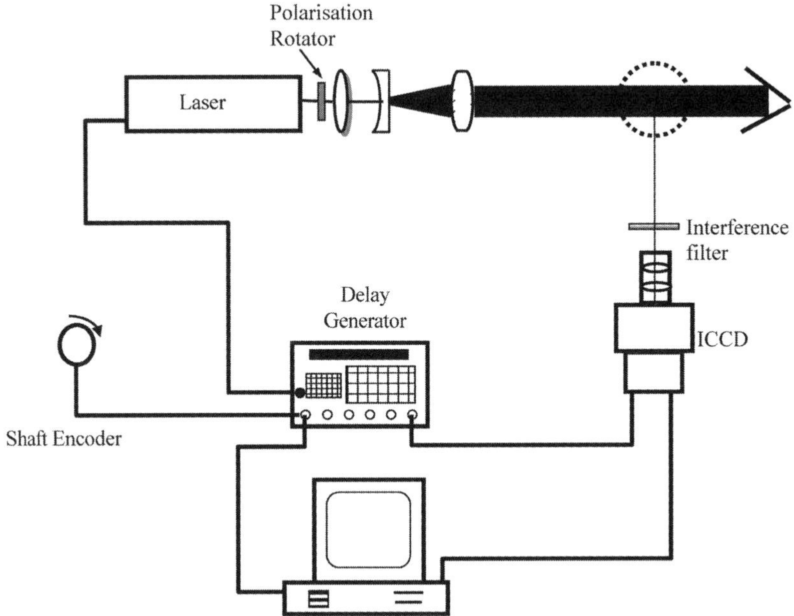

Fig. 6.1 Schematic of the optical setup for laser Rayleigh scattering.

During the intake stroke, the gas temperature may be assumed to be constant, which would not cause any appreciable errors in estimating the mixture concentration. However, the gas temperature will vary during the compression stroke and can be estimated from the in-cylinder pressure measurement by using the equation of state.

6.2.3 Minimization of Background Light in LRS

Since LRS is elastic, it suffers from the interference of background light of the same wavelength or that of continuous spectrum. This includes the Mie scattering from particles, the ambient light, and scattered light from optics, entrance/exit optical windows, other solid surfaces, and flame radiation.

Since Mie scattering from particles is many orders of magnitude stronger than Rayleigh scattering from molecules, it possesses serious interference with Rayleigh scattering measurements. For example, the Mie scattering cross section from a 1-μm particle is about 20 orders of magnitude higher than that of the Rayleigh scattering from air. There are two procedures that can be used to reduce the particle interference, as discussed in detail by Zhao and Hiroyasu (1993). The first is the removal of particles in the measurement volume. This can be realized by supplying bottled air or compressed air with an appropriate filtering system for moderate flow rates. To suppress the lubricant-generated droplets, the engine should be operated with special piston rings. After the data have been collected, spikes of Mie scattering signals can be removed manually or automatically by software. However, submicron particles cannot be removed by the filtering system, and the scattering from them will be more difficult to separate from the Rayleigh scattering from molecules. The post–data processing can become arbitrary, as the effect of the particles is multifaceted, random, and system-dependent.

Flame radiation can be reduced by an interference filter centered at the laser wavelength and a shorter gating time of the detector. Moreover, as Rayleigh scattering from molecules is polarized, the flame luminosity can be halved by a polarization filter. However, the background light yields the most severe limitation of the LRS technique. The magnitude of the background light determines the lower detection limit of the Rayleigh signal that can be detected. The lower the background level, the wider the dynamic range of the LRS system becomes. Furthermore, in IC engines the background light changes both spatially and temporally, which makes data reduction extremely complicated. The key to using the LRS technique is essentially the suppression of undesired light from all sources other than the measured molecules.

To minimize the effect of background light, the optical system should be completely enclosed in a black box. To suppress the spurious reflection of the incident laser beam during its propagation, slits, pinholes, and light traps are all extremely effective. The background light should be measured by imaging the scattered light from the gases with a comparatively smaller scattering cross section, such as helium, whose scattering cross section is about 1.65% that of the air. The signal derived in this way includes the noise in the detection system, stray light due to elastic scattering from the solid surfaces, and the Rayleigh scattering from the helium. As a result, background light intensity can be determined by subtracting the Rayleigh scattering from helium using the known cross sections of helium and air (Fourgette et al., 1986). Note that the background light must be measured with the same optical setup as that used in the actual experiment.

6.2.4 Filtered Rayleigh Scattering

The main difficulty with Rayleigh scattering is the scattering from optics, solid surfaces, particles, and droplets. This can be overcome by filtered Rayleigh scattering. The principle of filtered Rayleigh scattering is illustrated by Fig. 6.2 (Boguszko and Elliott, 2005). Rayleigh scattering from molecules is spread over a narrow spectral range, whereas the scattering from surfaces and particles occurs at a given frequency with little spread in frequency. By passing molecular Rayleigh-scattered light through a filter that blocks the transmission of the scattering at the laser wavelength, part of the Rayleigh scattering can be transmitted in the spectral region between the transmission curve and the wings of the Rayleigh scattering curve.

The filter candidates include mercury at 254 nm, iodine molecules at 532 nm, and rubidium at 780 nm. Mercury is a strong absorbing gas with a large cross section and clean absorption curve but would require an exotic laser (e.g., tripled Alexandrite laser). Iodine is a medium absorption gas with a medium cross section and can be readily used with an Nd:YAG laser, but it suffers from a cluttered absorption curve with fine rotational lines. Rubidium needs to be used with high laser energy at a less common wavelength, and its absorption curve is complicated with fine spectroscopic structures.

Filtered Rayleigh scattering allows measurement to be made close to surfaces and environments with particles or droplets. But it entails a more complex experimental setup than normal Rayleigh scattering (as shown in Fig. 6.3) because, in addition to the atomic (or molecular) gas filters, single-mode tunable lasers are required to provide an excitation source with a very narrow frequency bandwidth. Furthermore, the Rayleigh scattering line shape of hydrocarbon fuels can be complex and changes at high pressure and temperature. For example, more light can pass through the filter as the scattering line broadens. Accurate modeling of the line shape is therefore necessary for quantitative Rayleigh scattering measurement of major species concentration. As a result, it has so far been limited to some preliminary demonstration measurements (Zetterberg et al., 2007)

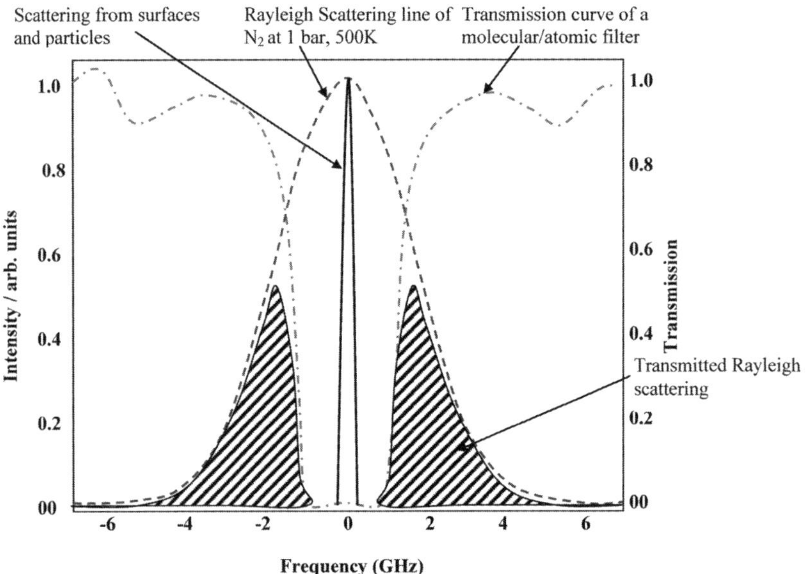

Fig. 6.2 Principle and implementation of filtered laser Rayleigh scattering.

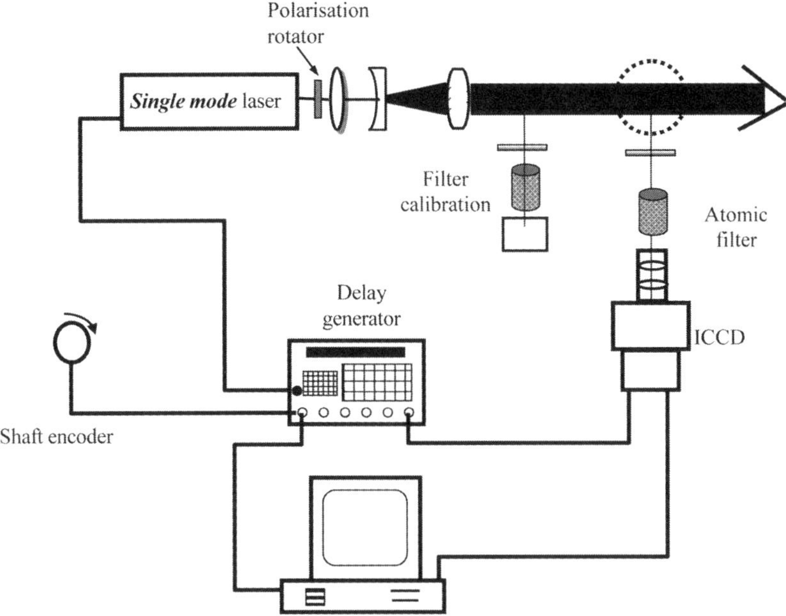

Fig. 6.3 Implementation of the filtered laser Rayleigh scattering technique.

6.2.5 Application of LRS to IC Engines

The most comprehensive development work on the LRS technique and its application to IC engines was carried out in the late 1980s and early 1990s (Zhao et al., 1991). The initial work was done in a steady flow test rig where continuous injection of Freon-12 into the intake port took place during the intake stroke. This was followed by in-cylinder measurements in a motored SI engine. The time histories of both mean and fluctuating fuel concentrations were obtained for both liquid fuels (gasoline, *n*-pentane, *n*-hexane, and Freon-113) and gaseous fuel (Freon-12) injections. The LRS measurements were subsequently extended to two-dimensional fuel vapor concentration in a motored SI engine at 200 rpm by Zhao et al. (1993). An Nd:YAG laser operating at the second harmonic (532 nm) was employed to produce a laser sheet inside the combustion chamber. The scattered light from the measurement plane was detected through the optical window mounted in an extended piston crown by a gated, intensified CCD camera. Freon-12 and propane were used to study the mixing process of fuel and air.

In combination with a spark plug with fiber-optics, LRS was used to measure the fuel concentration in the vicinity of a spark plug gap in a single-cylinder spark ignition engine by Lee and Foster (1995). The pressure, propane concentration, and initial flame development were measured simultaneously by a pressure transducer, LRS, and the fiber-optics in the spark plug. An argon-ion laser beam was focused near the spark plug with an effective beam waist diameter of approximately 75 μm. The scattered beam was collected at 90° to the direction of beam propagation and passed through a 10-nm band-pass filter before it was detected by a photomultiplier. All inner surfaces of the combustion chamber were painted black so as to minimize the spurious scattered light. Any remaining background light from solid surfaces was subtracted from the scattered light. A threshold level was set at each measurement from an estimated maximum Rayleigh signal. All signals above this voltage level were rejected. Such discrimination against Mie scattering was done by visual inspection of signal amplitude and rise time.

Quantitative images of fuel vapor concentrations were obtained in an evaporating and combusting diesel spray using planar LRS by Espey et al. (1994; 1997) in an optical single-cylinder direct-injection diesel engine under motored and firing conditions. Figure 6.2 shows schematically the optical setup used. A frequency-doubled Nd:YAG laser sheet was introduced into the combustion chamber with its polarization perpendicular to the scattering plane to maximize the scattered signal collected in the vertical plane. The scattered images were acquired through a window mounted in place of one of the exhaust valves by a gated intensified CCD camera. A narrow band-pass filter (532 ± 5 nm) was placed in front of the camera to isolate the elastically scattered laser light.

The fuel was supplied from a N_2-pressurized tank to avoid contamination by a normal fuel supply pump system. A mixture of the diesel reference fuels, heptamethylnonane and *n*-hexadecane, was used in the engine. The Rayleigh scattering technique was calibrated to obtain the two-dimensional quantitative fuel concentration measurements in the leading portion of the diesel spray, where the fuel was completely evaporated. Equivalence ratio was then deducted from the fuel vapor concentration using an adiabatic mixing assumption (Espey et al., 1997).

Since the Rayleigh scattering signal was directly related to the input laser power, simultaneous acquisition of the laser energy distribution was monitored through a second, gated, intensified CCD camera by splitting off a small fraction of the laser sheet. By normalization, the shot-to-shot variation of laser energy output, as well as the nonuniform distribution of the laser sheet, was removed. The interference from Mie scattering of particles was minimized by filtering the engine intake air and the fuel. Although background light was recorded and subtracted from the Rayleigh images, it could not be completely eliminated because of temporal variations in the background light. The uncertainties caused by background noise corresponded to as much as 0.3 equivalence ratio units. Other minor uncertainties included the accuracy of the Rayleigh scattering cross section (2.9%), the camera shot noise (0.7%), and the camera calibration curve (1.5%). Taking into account these uncertainties,

Espey et al. (1997) estimated that the total uncertainty in their equivalence ratio images was 21% (0.4 unit of equivalence ratio) for regions of equivalence ratio of 2 and 17% (0.8 unit of equivalence ratio) for regions of equivalence ratio of 4.

Other recent examples of LRS applications include the quantitative analysis of diesel spray evaporation in the high-pressure chamber (Schulz et al., 2004; Idicheria and Pickett, 2007; Pickett et al., 2011) and the in-cylinder dimethyl ether (DME) distribution in a direct-injection heavy-duty engine (Andersson et al., 2001).

6.3 Measurement of Mixture Composition by Spontaneous Raman Scattering

6.3.1 Introduction

Raman scattering is the inelastic interaction between the incident light and the electronic-mediated vibrational-rotational modes of molecules; that is, there is an energy exchange between the photons and scattering molecules. Hence, the scattered photons may gain or lose energy to the scattering molecules, so that the Raman signal has a different wavelength from the incident light. Spectral measurement of Raman scattered light serves to identify the molecular species. The intensity of Raman scattered light depends on the number of molecules that are thermally populated at particular rovibronic (rotational-vibrational) energy levels. Since Raman peaks are spectrally well separated from the incident light source, spurious scattered light can be discriminated against. Raman scattering has the advantage that the interaction time is very short (of the order of femtoseconds), so that quenching effects are not encountered.

With broadband detection, the Raman scattering technique permits the simultaneous measurement of the concentrations of all major species of combustion: H_2O, CO_2, CO, N_2, O_2, and fuel, from which the air/fuel ratio and the residual gas content of the charge can be obtained simultaneously. Since air/fuel ratio is obtained from the ratio of two of the majority species, neither laser power fluctuations nor window fouling affect the measurement precision. Since SRS measures the air/fuel ratio directly, it can be applied to situations where residual gas or EGR (exhaust gas recirculation) gas exists in the cylinder, a situation in which air/fuel ratio cannot be derived directly from the fuel concentration measurement by LIF (laser-induced fluorescence). Because the Raman signal is shifted from the incident laser wavelength, the interference from light scattered off walls can be largely eliminated, and the accuracy of the measurement is typically limited by photon shot noise. The drawback of the technique is the low signal strength, but this is becoming less of a problem with the commercially available high power pulsed UV (ultraviolet) lasers and optical multichannel detection systems with gated, intensified CCD cameras. However, its low-signal strength limits its measurements to a single point in most cases.

173

6.3.2 Principle of Operation

In general, the species concentration is measured from the Raman scattered light intensity in some spectrally integrated portion of the Raman spectrum. The position of the Raman spectrum of a particular species is given by Eq. 3.32 (repeated as Eq. 6.4):

$$\upsilon = \frac{E_v}{hc} = \upsilon_0 \pm \left[\omega_e - 2\omega_e x_e \left(v + 1 \right) - \alpha_e J \left(J + 1 \right) \right] \qquad (6.4)$$

where

υ = the wavenumber (cm^{-1})

υ_0 = the excitation wavenumber (cm^{-1})

ω_e = the equilibrium vibrational frequency of the molecule (cm^{-1})

x_e = the anharmonicity constant of the vibrational mode of the molecule

α_e = vibration-rotation interaction constant

v = the vibrational quantum number of the initial energy state

J = the rotational quantum number of the initial energy state

For diagnostic purposes, Eq. 6.4 can be simplified to:

$$\upsilon = \upsilon_0 \pm \upsilon_{vib} \qquad (6.5)$$

The second term in Eq. 6.4 is replaced by the vibrational frequency, υ_{vib}, in cm^{-1}, which is species specific. Table 3.1 lists vibrational frequencies (cm^{-1}) for molecules of interest to IC engines. The excitation wave number or frequency υ_0 represents the incident light frequency in cm^{-1}. Equation 6.4 or 6.5 defines the spectral positions of the Q-branched vibrational Raman spectrum of a diatomic molecule. The plus and minus signs represent the anti-Stokes lines (frequency upshifted) and Stokes lines (frequency down-shifted), respectively.

The Raman signal P_i from molecular species i, scattered into a solid angle Ω, is given by Eq. 3.34:

$$P_i = \eta_c I_o \Omega V_c N \left(\frac{\partial \sigma}{\partial \Omega} \right) \qquad (6.6)$$

where

η_c = the detection efficiency of the collection system

I_o = the incident laser intensity (W/cm^2)

N = the number density of the Raman scattering gas molecule (cm^{-3})

Ω = the solid angle of collection optics (sr)

V_c = the collection volume (cm^3)

The term $(\partial\sigma/\partial\Omega)$ is the differential Raman scattering cross section of the species. For diatomic molecules, the Stokes Raman scattering cross section can be expressed as

$$\left(\frac{\partial\sigma}{\partial\Omega}\right) = \varepsilon N_i f(T) \frac{\left(\tilde{v}_0 - \Delta\tilde{v}\right)^4}{\Delta\tilde{v}} \tag{6.7}$$

where ε is a constant; $\tilde{v}_0$ is the wave number of incident laser; $\Delta\tilde{v}$ is the wave number of Raman shift; and $f(T)$, the bandwidth factor, describes the temperature dependency of the Raman scattering cross section and is given as

$$f(T) = \frac{1}{1 - \exp\left(-h\tilde{v}_e / 2\pi KT\right)} \tag{6.8}$$

where h is the Planck's constant, K is the Boltzmann's constant, and T is the absolute temperature. The exponential in the Eq. 6.8 indicates that the higher the gas temperature is, the larger the Raman scattering cross section becomes as the higher energy vibrational states become more populated.

For polyatomic molecules, more complex expressions for the scattering cross section arise, but the conclusions drawn from this equation regarding the temperature dependence of the cross section are still valid.

Normally, if the gas temperature is much lower than the characteristic vibrational temperature of the species to be measured, given by $T_V = h\tilde{v}_e / k$, the temperature effect can be ignored. However, when the temperature is higher than or comparable to the characteristic vibrational temperature, the temperature effect needs to be considered.

Table 3.1 lists the Stokes fundamental vibrational Raman cross sections at room temperature for species of interest in IC engines. The cross sections were taken from the tabulation by Inaba and Kobayasi (1972) for a N_2 laser at 337 nm, except those of propane (C_3H_8), which were obtained from Schrotter and Klockner (1979). The values at 248 nm and 308 nm (excimer lasers), 355 nm and 532 nm (Nd:YAG lasers) were calculated from the original values by scaling according to $(\upsilon_0 - \upsilon_{vib})^4$. The scattering cross sections listed in Table 3.1 are obtained for the linear and polarized incident and scattered light perpendicular to the scattering plane. These cross sections are valid and unchanged for any scattering angle between the incident beam and scattering observation axes.

In practice, the species concentration is measured from the Raman scattering signal in some spectral integrated portion of the Raman scattering. According to Eqs. 6.6 to 6.8, the Raman scattering signal strength of species i can be given as

$$P_i = k_i I_0 N_i \tag{6.9}$$

where k_i is a factor dependent on the Raman cross section, the incident laser wavenumber, the geometric optical collection efficiency, the gas temperature and the spectrally region which is selected for the integration. After determining k_i by calibration from Raman scattering in a known sample by independent measurements, the species concentration N_i can be calculated from the measured Raman scattering from Eq. 6.9.

6.3.3 Experimental Setup of SRS

As shown in Fig. 6.4, an SRS system typically comprises a laser source and a multichannel spectroscopic detection unit. In SRS experiments, the selection of an appropriate laser has an important bearing on the signal-to-noise ratio, which is the limiting factor for Raman scattering. As shown in Table 3.1, the Raman cross section has a frequency scaling to the fourth power, that is $(\upsilon_0 - \upsilon_{vib})^4$. Therefore, the laser wavelength greatly affects the signal strength of Raman scattering. However, the strong frequency-scaling effect of Raman cross section is mitigated by the decreasing number of photons at higher frequencies (lower wavelength). Here, we compare the use of second-harmonic output (532 nm) to that of third-harmonic output (355 nm) of an Nd:YAG laser. The Raman cross section of N_2 at 355 nm (2.79×10^{-30} cm^2/sr) is about six times larger than that at 532 nm (0.46×10^{-30} cm^2/sr). Typical conversion efficiency from second to third harmonics is about 30%. As the laser wavelength decreases, the number of photons in each laser pulse is reduced by $(\upsilon_{532} - \upsilon_{vib})/(\upsilon_{355} - \upsilon_{vib})$ or 64%. Therefore, the total number of Raman photons at 355 nm are increased by 15% for 30% of the laser energy input. Therefore, the use of third harmonic output from an Nd:YAG laser can increase the Raman signal by 15%.

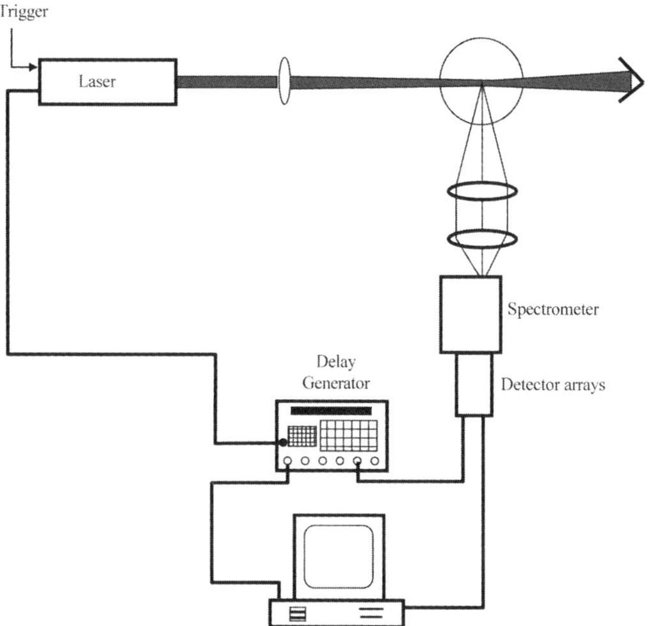

Fig. 6.4 Schematic setup of an SRS system.

In the cases of excimer lasers, a simple calculation shows that by going from 500 mJ at 532 nm of an Nd:YAG laser to 200 mJ at 248 nm of a KrF laser, the total number of Raman photons will be tripled.

From the above discussion, lasers in the UV region seem to be particularly attractive for Raman scattering because of the large increase in the strength of the Raman cross section with increasing frequency. However, LIF interference can become more problematical in the UV and care should be taken to avoid it. KrF lasers (248 nm) normally possess a relatively broad bandwidth on the order of 0.5 nm (80 cm^{-1}), which can excite a number of OH transitions between the ground and third (3,0) vibrational band of the A-X system as well as oxygen transitions in the Schumann-Runge system). In this case, the fluorescence from OH and O_2 is the major interference with the Raman scattering signals in the absence of heavy hydrocarbons. To avoid the interference, the laser needs to be tuned to 248.62 nm and its spectral bandwidth narrowed by injection locking (Grünefeld et al., 1994). For XeCl at 308 nm, OH LIF from the lower vibrational bands (0,0) and (1,0) could pose a problem as well. In the presence of hydrocarbon fuels, extra care needs to be taken to avoid LIF from certain fractions of fuels used. Since the vibrational Raman scattering is highly polarized, the interfering emissions from broadband LIF from aromatic fuels, O_2 and OH can be discriminated by the polarization technique, which requires the use of a vertically polarized laser light and the detection of the vertically polarized Raman scattering. Vertical polarization is achieved by rotating the electric vector of the laser light so that it is perpendicular to the scattering plane defined by the forward direction of incident laser beam and the scattering direction.

Another aspect of Raman scattering must be considered when a laser source is selected. The line separation of the resulting Raman spectra is wider by using a longer excitation wavelength than that by a shorter wavelength laser. This can be seen in Table 3.1, where a comparison is made for several species at 248 nm and 308 nm (excimer lasers) and 355 nm and 532 nm (Nd:YAG lasers). This feature may become important in cases where several species are involved and their measurement and resolution are desired.

The pulsed laser is normally preferred in internal combustion engine applications. The short pulse width of a pulsed laser (<100 ns) enables the instantaneous measurements to be made. More importantly, the use of a short pulse width combined with fast gated detection systems will suppress the background noise from the luminosity of the combustion and substantially increase the signal-to-noise ratio, which is of particular importance for SRS. However, the window's damage threshold and the laser-induced gas breakdown impose a limit on the power density (the ratio of pulse energy and pulse width) of the laser beam that can be used.

Since the Raman shift is comparatively small, a spectrometer is normally required to achieve the desired spectral resolution. Typically, a two-dimensional array detector (an intensified charge-coupled device, or ICCD) and a specially designed imaging spectrometer, as shown in Fig. 6.5, are used to record SRS spectra from a point or multiple point along the focused laser beam.

Standard spectrographs are optimized for the best spectral resolution at the expense of the vertical spatial resolution. In an imaging spectrograph, the optics are specially designed for point-to-point spectrographic imaging and enable many sources to be simultaneously diffracted and resolved as separate spectra on a CCD camera. As shown in Fig. 6.5, each point on the entrance slit produces a spectrum, and hence a series of spectra will be formed on the CCD detector. Therefore, the horizontal axis in the CCD output corresponds to wavelength and contains information of the Raman spectrum from a single point, whereas the vertical direction of the image contains the spatial information along the entrance slit. If the CCD camera is intensified and gateable, the Raman signal-to-noise ratio could further be increased. Scattered light is collected by a pair of achromatic lenses, the second of which should be selected to match the collection solid angle of the spectrograph. Sometimes multiple optical fibers replace the lenses to acquire spectra from different sources. Light at the incident laser wavelength can be minimized by a laser mirror, prior to entering the spectrograph, which reflects light at the laser

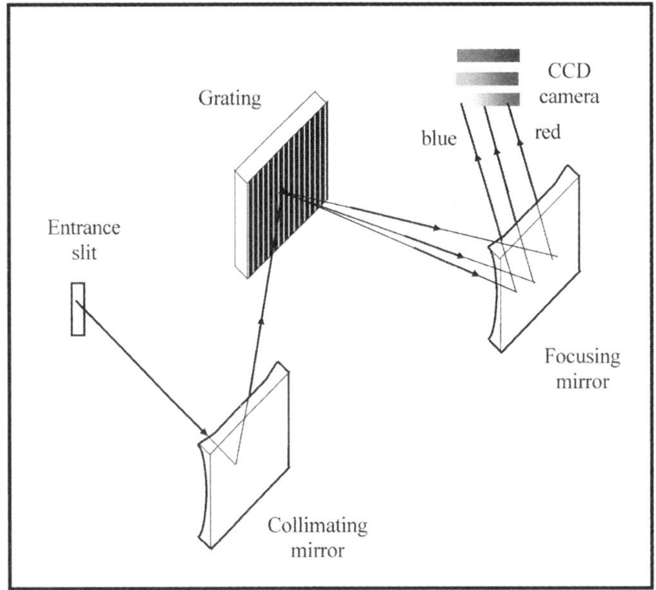

Fig. 6.5 Schematic of an imaging spectrograph.

wavelength but transmits light at other wavelengths. There are basically two types of laser mirrors: holographic notch filter and dielectric notch filter. The holographic notch filter is preferred to the dielectric notch filter, because it has much a higher optical density with a much narrower bandwidth and sharp edges, without the unwanted secondary leakage bands found in some dielectric notch filters.

6.3.4 Background Radiation and Signal-to-Noise Ratio

Because Raman signals are very weak, one must consider the background radiation and other sources of noises, which may interfere with the desired signal and render the Raman scattering useless. There are a number of sources that may contribute to the undesired background radiation. They include Rayleigh scattering, spurious scattering from optical windows and surfaces, Mie scattering from large particles, flame luminosity, soot incandescence, fluorescence from excited molecules, detector dark current, electrical noise, and detector shot noise.

The first three—Rayleigh, spurious, and Mie scattering—are the same frequency as the incident laser beam and thus spectrally separated from the desired signal. Therefore, they can be minimized by using proper filters (e.g., a dielectric laser mirror or holographic notch filter) or spectrographs used for the selection of the desired Raman signals. The interference from the broadband background radiation is particularly serious for SRS. For this reason, most of

179

the applications of SRS measurements are restricted to intake and compression strokes or with premixed combustion. The fluorescence can be a serious problem, particularly if UV Raman scattering is employed. Proper selection of the laser system will eliminate this problem.

Because of its weak nature, SRS can be shot-noise limited. A feasibility study would be useful to establish estimates of signal-to-noise ratios for a particular system and hence its potential for time-resolved measurements. In the following analysis, the signal-to-noise ratio calculations are based on the consideration of photoelectron shot noise and do not include the effects of electronic noise.

For a Poisson distribution of photoelectron arrival times, the shot noise is equal to the square root of the number of signal photoelectrons, which is equal to the total number of photoelectrons minus the background ones. The signal-to-noise ratio can be written as

$$\frac{S}{N} = \frac{N_s}{\sqrt{N_s + N_b}} = \frac{\sqrt{N_s}}{\sqrt{1 + \dfrac{N_b}{N_s}}} \qquad (6.10)$$

where

N_s = the number of signal photoelectrons collected

N_b = the number of background photoelectrons

In the case of multiple pulse operation, the signal-to-noise ratio becomes

$$\frac{S}{N} = \frac{n \cdot N_s}{\sqrt{n \cdot (N_s + N_b)}} = \frac{\sqrt{n \cdot N_s}}{\sqrt{1 + \dfrac{N_b}{N_s}}} \qquad (6.11)$$

where n is the number of pulses used to obtain the averaged signal reading.

The experimental error is given by the reciprocal of S/N, that is,

$$Error(\%) = \frac{\sqrt{1 + \dfrac{N_b}{N_s}}}{\sqrt{n \cdot N_s}} \times 100\% \qquad (6.12)$$

Table 6.1 Estimated Signal Levels for a Mixture of Gases at Different Excitation Wavelengths

Species	Vibrational Frequency (cm^{-1})	Volume Fraction	Raman Signal Photoelectrons at Different Laser Wavelengths				Signal-to-Noise Ratio at Different Laser Wavelengths			
			248 nm	308 nm	355 nm	532 nm	248 nm	308 nm	355 nm	532 nm
N_2	2331	0.755	33104	16445	10169	2512	182	128	101	50
O_2	1556	0.182	10189	5107	3242	856	101	71	57	29
CO_2	1388	0.012	611	306	195	52	25	17	14	7
	1285		450	226	144	39	21	15	12	6
H_2O	3657	0.015	1558	735	445	96	39	27	21	10
C_3H_8	2890	0.036	10697	5172	3180	747	103	72	56	27
	1465		4286	2156	1373	364	65	46	37	19
	857		2525	1281	834	231	50	35	29	15

Eqs. 6.11 and 6.12 indicate that by increasing the number of averaged pulses, one can obtain a larger signal-to-noise ratio or a lower measurement error.

In the following example, we assume that one is able to deliver 100 mJ of 532-nm laser radiation to the probe volume. The Raman signal P is obtained from Eq. 6.6 using the scattering cross sections given in Table 3.1. The number of photons collected is obtained by dividing the scattered light energy P by [$h\,c\,(\upsilon_0 - \upsilon_{vib})$]. Then, the number of photoelectrons is estimated by multiplying the number of collected photons by system collection efficiency and detector quantum efficiency. If one desires a 0.5 mm spatial resolution, and employs an f-number of 5 collection system, the solid angle of the collection optics Ω becomes 3.1×10^{-2} sr. The detector is assumed to have a 20% overall efficiency. The molecular number density 7.34×10^{19} cm^{-3} is used, corresponding to the value in a 10:1 compression ratio engine operating at part load (0.4 bar intake manifold pressure) and a 400 K charge temperature at the bottom dead center before compression. The amount of residual gas fraction used in the calculation is 10%. The calculated signal-to-noise ratios with different excitation wavelengths are presented in Table 6.1.

The values given in Table 6.1 are the shot-noise-limited signal-to-noise ratios. In practice, the signal-to-noise ratios will be lower because of the background radiation and interference.

6.3.5 Data Analysis and Calibration

The Raman spectrum for a gas mixture consists of a number of spectral spikes of the scattering species. Figure 6.6 shows a schematic representation of the

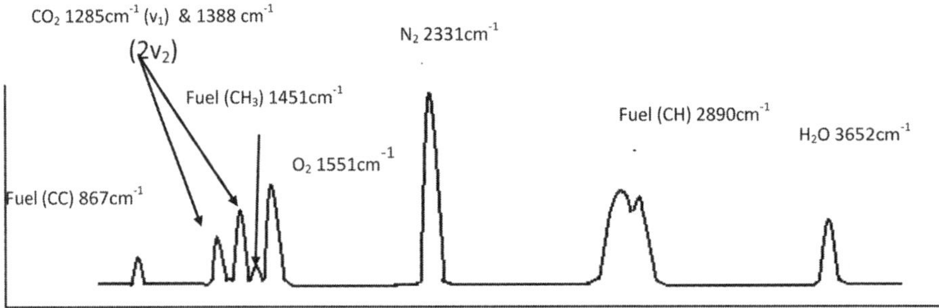

Fig. 6.6 Schematic representation of the Raman spectrum for a mixture of heptanes/octane and air.

measured Raman spectrum of an n-heptane/iso-octane/air mixture with a 532-nm excitation wavelength. The spectral position of Raman scattering from each species is determined by the characteristic vibrational frequency of the scattering molecule υ_{vib} relative to the laser wavelength υ_0. The Raman signal strength is affected by the Raman cross section, the system calibration factor, concentration of the scattering species, the incident laser power, and a spectral function, as given by Eq. 6.8.

The Raman signal of each species is experimentally determined from the Raman spectrograph by integrating over its Raman spectral region. In Fig. 6.6, note that there are three Raman spectral regions associated with the fuel (C_3H_8), which result from its three vibrational modes: C-H, C-C, and CH_3. In this case, the integration over the large C-H region may be used to obtain the Raman signal of propane, as the other two are very small.

The mole fraction ξ_i of each species can be determined from Raman scattering by

$$\chi_i = \frac{S_i/R_i}{\sum_i S_i/R_i} \tag{6.13}$$

where S_i is the Raman signal for species i obtained by integrating the rotational sidebands, which are included in each spectral peak and are not resolved in Fig. 6.6, as well as the central vibrational peak on the spectrum from the experiment. R_i is the relative system response factor obtained from a calibration procedure. From the mole fractions of oxygen/nitrogen and fuel the air/fuel ratio can be determined. Information of the mole fractions of other major species can be used to measure the residual gas or EGR gas concentration.

Calibration may be performed in situ or in a calibration cell for various gases in the calibration mixture. The calibration signal at known conditions is first recorded for a calibration mixture of pure air, where the Raman signals from N_2 and O_2 are obtained by integrating over the rotational sidebands, as well as the central vibration peak. The N_2 system response R_{N2} is assigned a value of unity, from which the relative O_2 system response R_{O2} can be determined. Then a second calibration signal is recorded by adding fuel to the calibration mixture to obtain the relative system response R_{fuel}. Similarly, the relative system response factors are obtained for other species. When there is an overlap between the Raman spectral lines of different species, more elaborate procedure is required (Miles and Dilligan, 1996). For accurate calibration, the effect of temperature on the Raman scattering should be taken into account since temperature affects both the Raman spectral width and its cross section.

6.3.6 Applications of the SRS Scattering Technique to IC Engines

Following the original application of the SRS technique in the late 1970s by Johnston (1979, 1980) to obtain ensemble-averaged measurement of local air/fuel ratios in a motored engine with propane, little progress was made in the 1980s in the development of SRS due to the low signal-to-noise ratio. Advances in lasers and optical multichannel detectors, however, have prompted several researchers to reinvestigate SRS as a useful technique for engine studies. In particular, the availability of intensified cooled CCD cameras, more powerful UV lasers, and the increasing interest in the fuel distribution in the cylinder of IC engines have led to new developments in multipoint multispecies SRS measurements.

Equipped with the imaging spectrograph and ICCD camera, Miles and Dilligan (1996) and Miles and Hinze (1998) were able to carry out in-cylinder measurements of multiple species first in an atypical research engine with side valves and transparent cylinder head and then in a production-type engine. An Nd:YAG laser at 532 nm was used to allow the CO_2 spectral peak to be separated from the O_2 peak. To deliver sufficient laser energy to the measurement volume, optical windows were arranged in a way that the laser beam passed them at Brewster's angle, at which the vertically polarized laser beam experienced little reflectivity. In addition, the laser beam expanded inside the window by the ratio of the reflective indices of window material and the surrounding gases, thereby allowing higher laser energy delivery. More importantly, the laser power was reduced by expanding the laser pulse to 170 ns by means of a pulse stretcher.

When the same SRS apparatus was applied to the production-type single cylinder engine, the flush mounting of the windows precluded the use of Brewster windows. To reduce the risk of breaking windows, an injection seeded Nd:YAG laser was used (Hinze and Miles, 1999). The seeded Nd:YAG laser offered a much better temporal pulse profile, resulting in fewer spikes in instantaneous power. The seeded laser also produced a better spatial profile of the beam, so there were fewer hot spots in the beam that may damage the windows. The absence of Brewster windows called for a larger beam diameter at the windows, which was achieved at the expanse of tighter focusing at the measurement location. The smaller laser beam diameter improved the spatial resolution, but it resulted in occasional laser ignition at the focus as a result of the presence of particles in the charge. The interference of laser ignition on the recorded spectra had to be identified and removed before quantitative results could be obtained. Mole fractions of H_2O, CO_2, O_2, and fuel (C_3H_8) were measured from the start of intake to the end of compression, from which the air/fuel ratio and the mole fraction of residual gases were calculated. Single-shot and cycle-resolved measurements were obtained.

In the mid-1990s, the availability of the tunable excimer laser with ICCD-based imaging spectrograph led to the application of the SRS technique to in-cylinder air/fuel ratio measurements in production engines. A specially designed tunable narrowband KrF excimer laser was tuned to 248.62 nm to avoid LIPF (laser-induced predissociative fluorescence) interference of OH and O_2 around 248 nm, the laser output. A combination of linearly polarized laser beam and a half-wave plate placed in front of the spectrometer further suppressed the remaining O_2 fluorescence, because fluorescence was depolarized and vibrational Raman scattering was highly polarized. To attenuate Rayleigh and Mie scattering in their experiments, a long-pass liquid butyl acetate filter was used, which had a steep cutoff edge between 250 nm and 255 nm, thereby allowing complete recordings of Stokes Raman spectra of major species.

The back-scattered Raman light was collected through the laser entrance window in a production engine (Grünefeld et al., 1995). Since the Raman scattering cross section is independent of scattering angle (Eq. 6.7), the Raman signal is not affected by the direction of detection. One added advantage of the backscattering setup was that the signal was found less affected by oil or any other deposit on the window, because the intense laser beam burned out the deposit and kept a portion of the window clean. This was particularly a useful feature when they carried out experiments during cold start period when large amounts of fuel and combustion products condensed on the window. With the backscattering setup, however, Raman signals were averaged over the whole laser beam. With this setup, Grünefeld et al. (1995) were able to measure the in-cylinder composition for the first 100 consecutive cycles at a fixed crank angle during cold start and warm-up periods.

Knapp et al. (1997) performed both spatially averaged single-cycle and one-dimensional (1-D) spatially resolved ensemble-averaged SRS measurements in a four-cylinder engine. The scattered light was detected at right angle through the third window mounted in a plate between the cylinder head and the engine block. The spatially averaged ($0.5 \times 10 \times 14$ mm^3) measurements were used to detect the gas composition in the residual gas content under different engine operating conditions. For 1-D spatially resolved measurements, a plano-concave lens with 7-mm radius of curvature was used as the detection window to increase the field of view to compensate for the lower signal level encountered. The collection optics imaged a section of 40 mm along the laser beam (4 mm × 1 mm) onto the entrance slit of the spectrograph. The signal-to-noise level was further enhanced by averaging over 1000 cycles for the 1-D spatially resolved SRS measurements.

Other examples of SRS applications include the in-cylinder residual gas concentration measurements (Alger and Wooldridge, 2004; Mori and Shimizu, 2002; Schütte et al., 2000), direct air/fuel ratio measurement in direct-injection spark ignition engines (Beushausen et al., 2003; Schütte et al., 2000) and diesel engine (Aronsson et al., 2009), and the use of Raman scattering as a means of calibration for LIF measurements (Hoffmann et al., 2008; Ipp et al., 2000).

6.3.7 Multispecies Measurements by SRS in a CAI Combustion Engine

6.3.7.1 Experimental Considerations

Figure 6.7 shows the SRS system used for in-cylinder mixture composition measurements in the single-cylinder optical engine with transparent cylinder head, as shown in Fig. 1.1. The polarization plane of the laser beam was set to be perpendicular to the scattering plane. A laser pulse stretcher was employed to extend the laser pulse duration. The laser beam passed the combustion chamber through the side windows. It was focused at the center of the combustion chamber by a focusing lens and delivered into the engine at about 3 mm below the bottom surface of the top window. Scattered light was observed from the top of the engine via a front surface coated mirror. The scattered light was then focused by a pair of achromatic lenses onto the entrance slit of a spectrometer. To block the Raleigh scattering and other spurious light from reaching the detector, a holographic notch filter was placed in the front of the entrance slit of the spectrometer. The spectrometer output was detected by an ICCD camera, in which the Raman spectra of the measured species were recorded.

An UV laser produces much higher signal-to-noise ratio and has been shown to be capable of single-shot air/fuel ratio measurements. However, as shown in Fig. 6.6, the spectral peaks of CO_2, O_2, and one of the fuel spectral mode CH_3 are located very close to each other. To resolve the spectral peaks and detect the Raman spectra of O_2, N_2, H_2O, CO_2, and the fuel mixture simultaneously by the spectrometer, the second harmonic output (532 nm) of an Nd:YAG laser was employed with a 600-mm^{-1} grating, which resulted in a spectral dispersion of 12.6 nm/mm in the plane of the detector. With a 20.8-mm ICCD detection area at the spectrometer exit, this provided approximately 262-nm wavelength coverage. This was sufficient to simultaneously record and resolve Raman spectral peaks of all major species present in the cylinder.

The spectrometer utilizes a Czerny-Turner optical layout with the addition of an astigmatism-correcting mirror. Its output was coupled directly to an ICCD

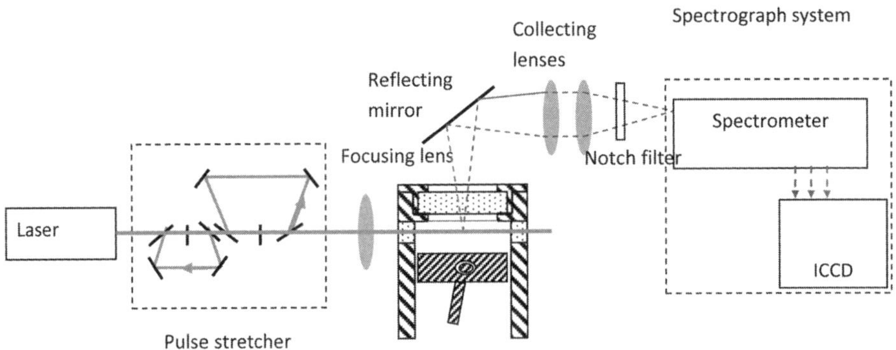

Fig. 6.7 SRS experimental setup for in-cylinder composition measurement in a CAI engine.

camera with a resolution of 1024 × 256 pixels. The entrance slit width was 100 µm, and the magnification of the spectrograph was 1.05.

The need to use a visible excitation source for resolving closely spaced spectral peaks of CO_2 and O_2 molecules significantly reduces the Raman signal strength. In theory, the signal-to-noise ratio can be increased by using a higher power laser. However, the higher power output from a pulsed laser can cause gas breakdown in the measurement volume and damage to the optical window. A pulse stretcher, therefore, was designed and implemented to extend the laser pulse duration, so that the laser pulse peak intensity would be reduced when the total laser energy delivered to the measurement volume was increased. The pulse stretcher operates by splitting off a portion of the beam and passing it through an optical delay loop before recombining it with the original beam. As a result, the duration of laser pulses is extended and intensity peak of each pulse is reduced. Figure 6.8 shows schematically the principle of the optical delay loop used. It comprises two polarizers, two mirrors, and a retarder. The incident laser beam is set to be p-polarized with respect to the surface of the polarizer 1 at the Brewster angle. As the reflectivity at a polarizer for p-polarized light with a Brewster incident angle approaches zero, most of the laser beam will pass through. After passing through the polarizer 1, the laser beam goes through the retarder and its polarization is rotated with an angle θ, $(1 < \theta < 90°)$ by the retarder, before it reaches the polarizer 2 at the Brewster angle. As a result, the laser beam is separated into two beams by the second polarizer, a transmitted beam and a reflected beam. The p-polarized beam leaves the delay loop, whereas the reflected beam is s-polarized and goes around the delay loop, and comes back to the polarizer 2 to produce a second transmitted laser pulse. As a result, the laser beam is perpendicularly polarized with respect to the incident plane for increased Raman signal. This process is repeated a few times until the laser energy in the loop becomes negligible. When considering a pulsed incident laser beam, consecutive laser pulses are produced from the delay loop. If the time delay of the loop is less than the duration of the incident laser pulses, these output laser pulses overlap one and another from the first to the last. As a result, a laser pulse with an extended duration and decreasing intensity profile is produced, as shown by the bottom graph in Fig. 6.8.

If the total reflectance of the loop is R, the intensity profile of the incident laser pulse is $f(t)$, the time delay of laser beam traveling from polarizer 1 to polarizer 2 is τ_0, and the time delay of the laser pulse after passing through delay loop once is τ, then the intensity profile of the first laser pulse output from the delay loop is $\cos^2\theta f(t - \tau_0)$. The intensity profiles of the second, the third, and the nth pulse are $\sin^4\theta f(t - \tau - \tau_0)R$, $\sin^4\theta \cos^2\theta f(t - 2\tau - \tau_0)R^2$, and $\sin^4\theta \cos^{2(n-2)}\theta f[t - (n - 1)\tau - \tau_0]R^{n-1}$, respectively. The total intensity profile of the laser pulses after passing through the delay loop can be written as

$$G(t)=\cos^2\theta f(t - \tau_0) + \sin^4\theta f(t - \tau - \tau_0)R + \sin^4\theta \cos^2\theta f(t - 2\tau - \tau_0)R^2 + \ldots + \sin^4\theta \cos^{2(n-2)}\theta f[t - (n - 1)\tau - \tau_0]R^{n-1}$$

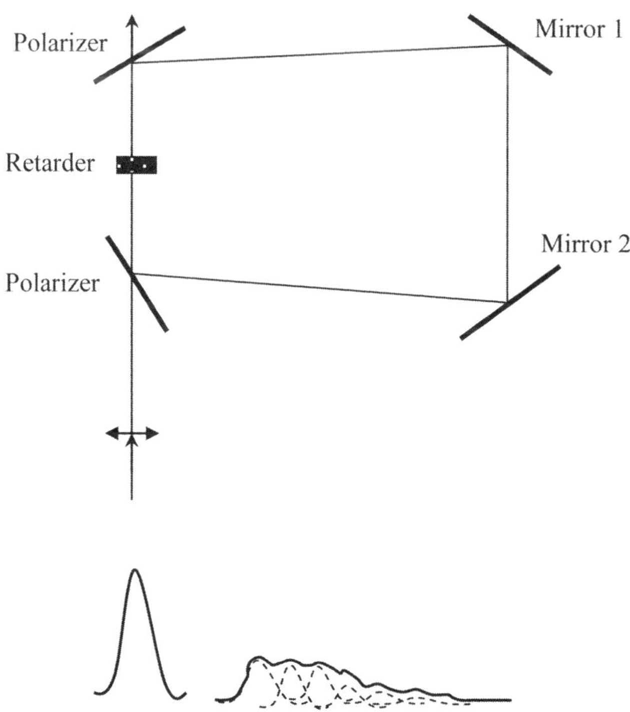

Fig. 6.8 Schematic diagram of a pulse stretcher setup and laser pulses before and after the pulse stretcher.

Note that the peak intensity of either the first or the second pulse is much stronger than the others, and the first two bursts contain most of the laser energy. The stretched pulse will have the lowest peak intensity when the peak intensities of the two pulses are equal, that is, $\cos^2\theta = \sin^4\theta\, R$, which dictates the rotational angel of the retarder, $\theta = 51.9°$. In the current SRS system, two such optical delay loops were included in the pulse stretcher. The optical path lengths of the first and second loops are 2.4 m and 5.2 m. By employing the pulse stretcher, the maximum laser energy that could be delivered to the test section was increased from 25 to 100 mJ without breaking the optical windows or causing the gas breakdown in the measurement volume, and the laser pulse width was extended from 8 to 35 ns.

A spherical lens with 300-mm focal length was employed as the focusing lens, with which the focused beam waist of 102 μm in diameter and distances of 7.6 mm on either side of the waist before the beam diameter doubled was formed in the test section. The scattered light was collected by two identical achromatic lenses with a focal length of 310 mm and clear optical aperture of 78 mm. The collecting lenses gave a f-number of 3.97, which was chosen to be close to the f-number 3.8 of the spectrometer for a high-collection efficiency.

6.3.7.2 Data Interpretation and Calibration

As shown in Fig. 6.6, there are three Raman spectral modes, CH, CC, and CH_3, associated with n-heptane/iso-octane and two modes, v_1 and $2v_2$, associated with CO_2. As fuel CH mode has the strongest signal peak, it was chosen for the measurement of the fuel concentration. As the CO_2 $2v_2$ peak overlaps with fuel CH_3 and the O_2 peaks, CO_2 v_1 peak was selected for the concentration acquisition of CO_2. The signal strength of O_2 was acquired by subtracting the signal strength of the CO_2 $2v_2$ peak and the fuel CH_3 peak from the total signal strength over the three overlapped peaks (CO_2 $2v_2$ peak + fuel CH_3 peak + O_2 peak). According to the principle of Raman scattering, the ratio between the signal strengths of two certain regions of a species should remain constant if the temperature effect can be ignored. Therefore, the signal strengths of CO_2 $2v_2$ peak and the CH_3 peak can be determined from the signal strengths of the CO_2 v_1 peak and the CH peak. The ratios between the signal strengths of the two peaks of CO_2 and the two peaks of fuel had to be determined by calibration.

As shown in Eq. 6.9, the concentration of each species in the mixture can be obtained from the signal strengths by knowing the calibration factor and the delivery laser energy in the measurement volume. However, the result will be affected by pulse to pulse laser variation or altered by the window fouling. In this experiment, using Eq. 6.13, the mole fraction of the species was determined instead as follows:

$$\chi_i = \frac{N_i}{\sum\limits_i N_i} = \frac{S_i/E_0 k_i}{\sum\limits_i S_i/E_0 . K_i} = \frac{\dfrac{S_i}{S_{N_2}} \bigg/ \dfrac{K_i}{K_{N_2}}}{\sum\limits_i \left(\dfrac{S_i}{S_{N_2}} \bigg/ \dfrac{K_i}{K_{N_2}} \right)} \tag{6.14}$$

Which shows that the mole fraction is dependent on the signal strength of a species relative to that of N_2, S_i/S_{N_2}, and a system response factor relative to that of N_2, K_i/K_{N_2}. As both S_i/S_{N_2} and K_i/K_{N_2} are independent on the incident laser energy, errors because of variation in the incident laser energy by the pulse to pulse variation in the laser output and widow fouling are obliterated.

Calibration of K_i/K_{N_2} corresponding species i was performed by putting it with N_2 in a calibration cell at a known temperature. The result is given in Table 6.2, from which the ratios between the signal strength of the two peaks of CO_2 and the two peaks of fuel were determined as 1.2179 and 0.064419, respectively.

Note that the calibration factors given in Table 6.2 could only be used when the in-cylinder temperature was much lower than the characteristic vibrational temperature of the species to be measured. However, the gas temperature in the cylinder could be very high at the later stage (about 1050 K) of the compression stroke, particularly, after autoignition started (more than 1050 K).

Table 6.2 Spectral Peaks and Calibration Factors

Species	Peak Position (nm)	Temperature (K)	Relative Calibration Factor
O_2	579.8	313	1.5411
CO_2	571.0	313	1.2000
	574.4	313	1.4615
Fuel	628.6	313	21.885
	576.5	313	2.9148
H_2O	660.3	453	2.0412

Using Eq. 6.8, the effect of temperature on the relative calibration factor can be estimated for the relative system response as follows:

$$\frac{K_i}{K_{N_2}} = \frac{f_i(T)}{f_{N_2}(T)} \frac{\left(\tilde{v}_0 - \Delta\tilde{v}_i\right)^4}{\left(\tilde{v}_0 - \Delta\tilde{v}_{N2}\right)^4} \frac{\Delta\tilde{v}_{N2}}{\Delta\tilde{v}_i} \frac{\eta_{op,i}}{\eta_{op,N_2}} \tag{6.15}$$

Where $\eta_{op,i}$ and η_{op,N_2} are the optical efficiencies of the system for Raman peaks of species i and N_2, respectively. If the relative calibration factor (K_i/K_{N2}) is known at a reference temperature T_o, its value at any other temperature can be determined by

$$\frac{K_i}{K_{N_2}} = \eta_c \frac{K_{i,T_o}}{K_{N_2,T_o}}, \ and \quad \eta_c = \frac{f_i(T)}{f_{N_2}(T)} \frac{f_{N2}(T_o)}{f_i(T_o)} \tag{6.16}$$

From Eqs. 6.15 and 6.16, the values of temperature correction factor η_c for H_2O, CH peak, O_2, and CO_2 (v_1 peak) were calculated and used to obtain their relative calibration factors (K_i/K_{N2}) as a function of temperature (Fig. 6.9).

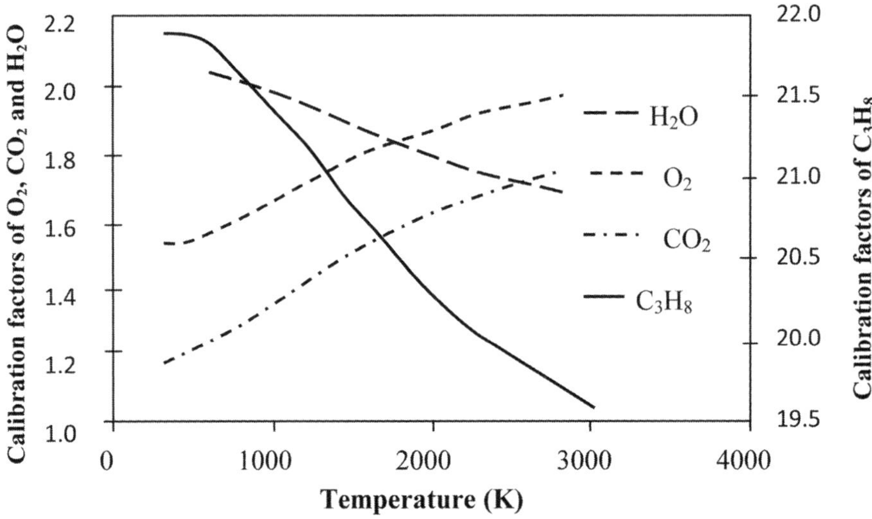

Fig. 6.9 Relative calibration factors as a function of temperature.

In addition, when there is more than one spectral peak in its Raman spectrum of a species, such as CO_2 and hydrocarbon fuel molecules, the ratio of the signal

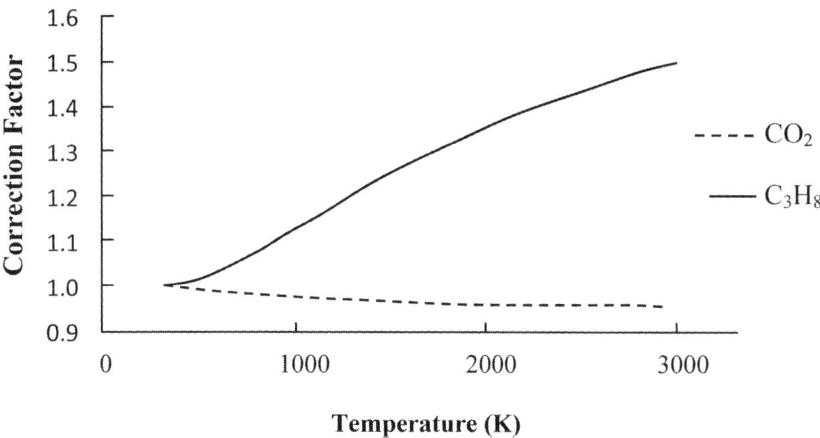

Fig. 6.10 Relative calibration factor between CO$_2$ and fuel peaks.

strength of one spectral peak to another is also affected by temperature and the ratio of the relative calibration factor is given by

$$\frac{K_{i,j}}{K_{i,k}} = \frac{f_{i,j}(T)\left(\tilde{v}_0 - \Delta\tilde{v}_j\right)^4 \Delta\tilde{v}_k}{f_{i,k}(T)\left(\tilde{v}_0 - \Delta\tilde{v}_k\right)^4 \Delta\tilde{v}_j} \tag{6.17}$$

or

$$\frac{K_{i,j}}{K_{i,k}} = \eta'_c \frac{K_{i,j}}{K_{i,k}}, \; and \quad \eta'_c = \frac{f_{i,j}(T)\, f_{i,k}(T_o)}{f_{i,k}(T)\, f_{i,j}(T_o)} \tag{6.18}$$

Figure 6.10 shows the variation of the ratio of the relative calibration factors between the two spectral peaks of CO$_2$, and that between the CH and CH$_3$ peaks. It can be seen that the change of the ratio of the relative calibration factors between the two CO$_2$ peaks is less than 5% up to 2000 K. But the ratio between CH and CH$_3$ peaks increases by 12% as the temperature changes from 400 K to 1000 K. Since the signal strength of O$_2$ is to be determined by subtracting the overlapping CH$_3$ spectral peak, the temperature effect must be included in the determination of CH$_3$ signal strength from the measured CH signal strength when the gas temperature is high.

6.3.7.3 Measurements of In-Cylinder Mixture Composition

During SRS experiments, a trigger pulse at a preset crank angle was generated by the control box and sent to the laser power unit to start the laser firing sequence. The same trigger was also sent to a delay generator, from which a delayed TTL pulse was then produced to operate the ICCD camera. By adjusting the time delay of the delay generator, the gated intensifier was timed to open for 40 ns immediately after laser firing. The laser beam entered the engine through one of the side windows at 3 mm below the top window, about one-half of the engine clearance height. The laser beam was focused at the center of the combustion chamber with a beam waist of 100 µm. The laser beam then exited the other side window before it was terminated by a beam dumper. The scattered light was collected at 90° to the laser beam through a 45° mirror mounted on top of the engine. Based on the focusing and collection optics used, the measurement volume was of cylindrical shape of 100 µm in diameter and 3 mm in length.

The measurements were carried out at an engine speed of 600 rpm. A mixture of iso-octane/n-heptane (40:60 in volume) was used as the fuel. To attain the controlled auto-ignition combustion, the intake air was heated to 100°C and the relative air/fuel ratio was set to 1.6.

Raman spectra over 200 cycles were recorded for each selected crank angle. After the Raman spectra had been recorded and ensemble-averaged, the signal strength, the relative concentration, and the mole fraction of each species were then calculated. The signal strength of each species was obtained by integrating over the appropriate spectral region. Prior to integration, the baseline was found by polynomial fittings and removed by subtracting it from the original spectrum. The random noises were minimized by a five point smoothing method.

Figure 6.11 shows the ensemble-averaged Raman spectra at four different crank angles. At the beginning of the intake stroke shown in Fig. 6.11(a), N_2 (607 nm), CO_2 (571/574 nm), and H_2O (660 nm) spectral peaks were present due to large amount of residual burned gases from the previous cycle. In addition, O_2 (580 nm) spectral peak was also detected because of the lean-burn mixture. The first appearance of fuel (629 nm) was detected at 60° CA ATDC, shown in Fig. 6.11(b), accompanied by noticeable increase in the O_2 spectral peak at 580 nm. As more fuel and air mixture entered the combustion chamber, the signal strength of fuel and O_2 kept on rising, while the signal strength of H_2O and CO_2 dropped significantly, as shown in Fig. 6.11(c). Soon after piston passing the top dead center (TDC), the signal strength of H_2O and CO_2 started rising quickly as fuel and oxygen were burned, seen in Fig. 6.11(d).

Based on the measured Raman spectra, the mole fractions of O_2, fuel, H_2O, and CO_2 as a function of crank angle were then calculated as shown in Fig. 6.12. It was found that without incorporating the temperature correction factors, O_2 and CO_2 would have been overestimated (Zhao and Zhang, 2008).

The relative air/fuel ratio derived from the SRS measurements of fuel and oxygen mole fractions was found to be the same as measured by an exhaust gas analyzer. In addition, it was found that very good agreement was found between the residual gas mole fractions calculated independently from the

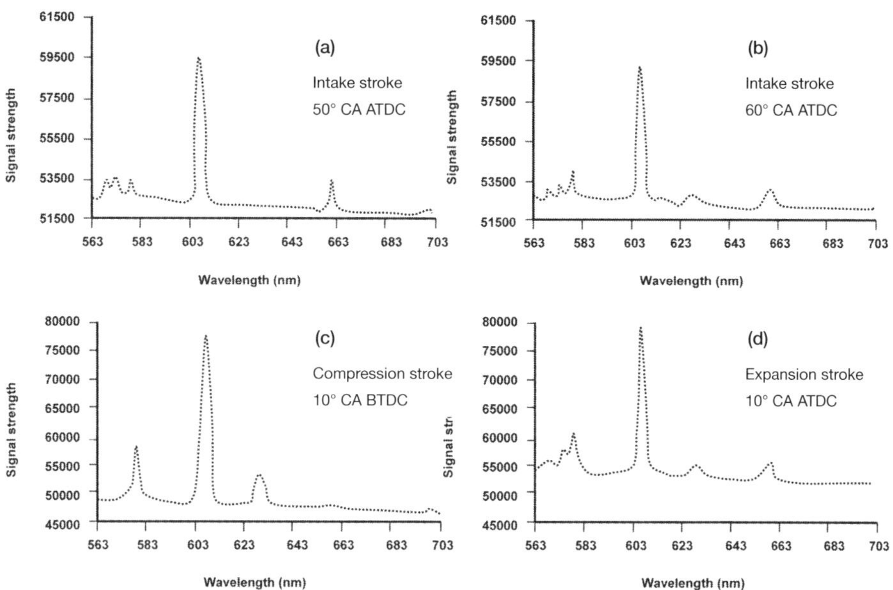

Fig. 6.11 Ensemble-averaged Raman spectra.

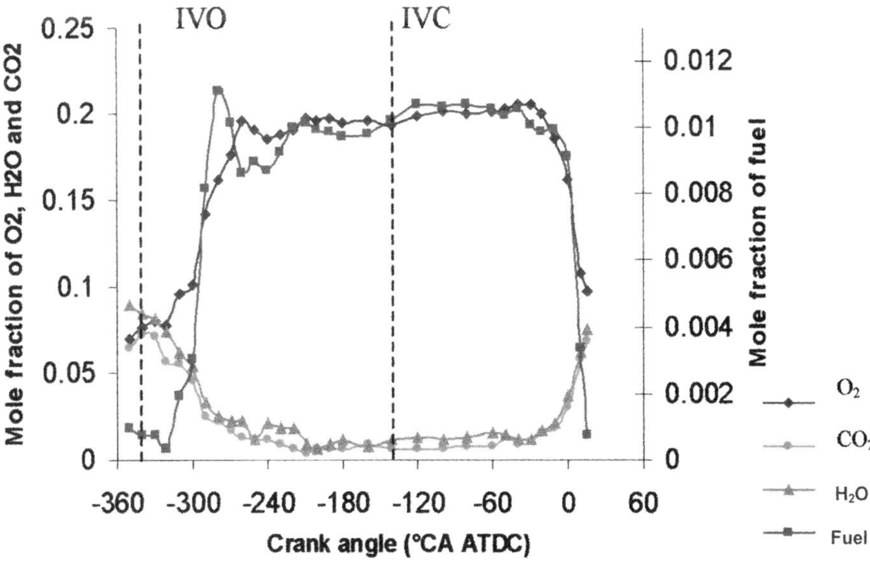

Fig. 6.12 Time history of in-cylinder mixture composition in a CAI engine.

measured mole fraction of each species based on lean-burn combustion reaction for a known relative air/fuel ratio of λ.

6.4 Summary

Raman scattering is a very weak process. In most cases, SRS is limited by the signal-to-noise ratio. High laser intensity and a very sensitive detector are required to obtain SRS measurements. Most major species can be detected by SRS. The high-pressure environment in an IC engine actually facilitates the implementation of SRS, because it increases species concentration and hence the signal-to-noise ratio. The advantage of SRS is its capability of simultaneous detection of multiple major species, from which the fuel/air ratio is directly measured and residual gas concentration can be calculated. When an optical multichannel analyzer, such as an ICCD array coupled to an imaging spectrometer, is used, multipoint and multispecies measurements can be obtained in IC engines. Furthermore, SRS can be employed as a calibration technique for quantitative LIF measurement of fuel vapor concentration.

Both SRS and LRS can be realized at any laser wavelength and their signal strength increase with shorter wavelength. But Rayleigh scattering is a much stronger scattering process than Raman scattering. The optical setup is relatively simple. Because LRS has the same wavelength as the incident laser light, it suffers from the interference from Mie scattering and spurious scattered light from surrounding walls. This has limited most LRS to motored engine operations when extreme caution has been taken to remove particles from the environment. Since it is not species specific, LRS is more suitable to density measurement than concentration measurement. Although it has been superseded by PLIF (planar laser-induced fluorescence) for measuring in-cylinder fuel vapor distribution in most cases, LRS may be better suited to fuel concentration measurement when fuel vapor has a large scattering cross section and lacks an appropriate fluorescence tracer that can match the fuel's evaporation properties.

References

Alger, T., and Wooldridge, S. (2004). "Measurement and Analysis of the Residual Gas Fraction in an SI Engine with Variable Cam Timing." SAE Paper No. 2004–01–1356, SAE International, Warrendale, PA.

Andersson, Ö., Collin, R., Aldén, M., and Egnell, R. (2001). "Laser-Rayleigh Imaging of DME Sprays in an Optically Accessible DI Diesel Truck Engine." SAE Technical Paper No. 2001–01–0915, SAE International, Warrendale, PA.

Aronsson, U., Chartier, C., Andersson, Ö., Egnell, R., et al. (2009). "Analysis of the Correlation Between Engine-Out Particulates and Local Φ in the Lift-Off Region of a Heavy Duty Diesel Engine Using Raman Spectroscopy." *SAE Int. J. Fuels Lubr.* Vol. 2, No. 1, pp. 645–660.

Beushausen, V., Mueller, T., Thiele, O., Wesker, T., et al. (2003). "Spatially Resolved Determination of Fuel/Air Ratio Inside a Direct Injection SI Engine Under Real Fuel Conditions." SAE Paper No. 2003–01–2007, SAE International, Warrendale, PA.

Boguszko, M., and Elliott, G. S. (2005). "On the Use of Filtered Rayleigh Scattering for Measurements in Compressible Flows and Thermal Fields." *Experiments in Fluids*, Vol. 38, No. 1, pp. 33–49.

Espey, C., Dec, J., Litzinger T. A., and Santavicca, A. (1994). "Quantitative 2D Fuel Vapor Concentration Imaging in a Firing Di Diesel Engine Using Planar Laser Induced Rayleigh Scattering." SAE Paper No. 940682, SAE International, Warrendale, PA.

Espey, C., Dec, J., Litzinger, T., and Santavicca, D. (1997). "Planar Laser Rayleigh Scattering For Quantitative Vapor Fuel Imaging In A Diesel Jet." *Combustion and Flames*, Vol. 109, Number 1/2, pp. 65–86.

Fourgette, D. C., Zurn, R. M., and Long, M. B. (1986) "Two-Dimensional Rayleigh Thermometry in a Turbulent Nonpremixed Methane Hydrogen Flame." *Combust. Sci. Technol.* Vol. 44, 307–317.

Grünefeld, G., Beushausen, V., Andresen, P., and Hentschel, W. (1994). "Spatially Resolved Raman Scattering for Multispecies and Temperature Analysis in Technically Applied Combustion Systems: Spray Flame and Four Cylinder in Line Engine." *Appl. Phys. B.*, Vol. 58, pp. 333–342.

Grünefeld, G., Knapp, M., Beushausen, G., Andresen, P., Hentschel, W., and Manz, W. (1995). "In Cylinder Measurements and Analysis on Fundamental Cold Start and Warm Up Phenomena of SI Engines." SAE Paper No. 952394, SAE International, Warrendale, PA.

Hinze, P., and Miles, P. (1999). "Quantitative Measurements of Residual and Fresh Charge Mixing in a Modern SI Engine Using Spontaneous Raman Scattering." SAE Paper No. 9901-01-1106, SAE International, Warrendale, PA.

Hoffmann, T., Hottenbach, P., Koss, H., Pauls, C., et al.(2008). "Investigation of Mixture Formation in Diesel Sprays under Quiescent Conditions using Raman, Mie and LIF Diagnostics." SAE Paper No. 2008-01-0945, SAE International, Warrendale, PA.

Ipp, W., Egermann, J., Wagner, V., and Leipertz, A. (2000). "Visualization of the Qualitative Fuel Distribution and Mixture Formation Inside a Transparent GDI Engine with 2D Mie and LIEF Techniques and Comparison to Quantitative Measurements of the Air/Fuel Ratio with 1D Raman Spectroscopy." SAE Technical Paper No. 2000–01–1793, SAE International, Warrendale, PA.

Idicheria, C., and Pickett, L. (2007). "Quantitative Mixing Measurements in a Vaporizing Diesel Spray by Rayleigh Imaging." SAE Technical No. Paper 2007–01–0647, SAE International, Warrendale, PA.

Inaba, H. and Kobayasi, T. (1972). "Laser Raman Radar–Laser Raman Scattering Methods for Remote Detection and Analysis of Atmospheric Pollution." *Opto-Electronics*, Vol. 4, pp. 101–123.

Johnston, S. C. (1979). "Precombustion Fuel/Air Distribution in a Stratified Charge Engine Using Laser Raman Spectroscopy." SAE Paper No. 790433, SAE International, Warrendale, PA.

Johnston, S. C. (1980). " Raman Spectroscopy and Flow Visualization of Stratified Charge Engine Combustion." SAE Paper No. 800136, SAE International, Warrendale, PA.

Knapp, M., Beushausen, V., Hentschel, W., Manz, P., Grunefeld, G., and Andresen, P. (1997). "In Cylinder Mixture Formation Analysis With Spontaneous Raman Scattering Applied to a Mass Production SI Engine." SAE Paper No. 970827, SAE International, Warrendale, PA.

Lee, K., and Foster, D. (1995). "Cycle-by-Cycle Variations in Combustion and Mixture Concentration in the Vicinity of Spark Plug Gap." SAE Paper No. 950814, SAE International, Warrendale, PA.

Miles, P., and Dilligan, M. (1996). "Quantitative In-Cylinder Fluid Composition Measurements Using Broadband Spontaneous Raman Scattering." SAE Paper No. 960828, SAE International, Warrendale, PA.

Miles, P., and Hinze, P. (1998). "Characterization of the Mixing of Fresh Charge with Combustion Residuals Using Laser Raman Scattering with Broadband Detection." SAE Paper No. 981428, SAE International, Warrendale, PA.

Mori, S., and Shimizu, R. (2002). "Analysis of EGR Cyclic Variations in a Direct Injection Gasoline Engine by Using Raman Scattering Method," SAE Paper No. 2002–01–1646, SAE International, Warrendale, PA.

Pickett, L., Manin, J., Genzale, C., Siebers, D., et al. (2011). "Relationship Between Diesel Fuel Spray Vapor Penetration/Dispersion and Local Fuel Mixture Fraction." *SAE Int. J. Engines*, Vol. 4, No. 1, pp. 764–799.

Schrotter, H. W., and Klockner, H. W. (1979). "Raman Scattering Cross Sections in Gases and Liquids," In *Raman Spectroscopy of Gases and Liquids,* edited by A. Weber, pp. 123–166. Berlin: Springer-Verlag.

Schütte, M., Finke, H., Grünefeld, G., Krüger, S., et al. (2000). "Spatially Resolved Air–Fuel Ratio and Residual Gas Measurements by Spontaneous Raman Scattering in a Firing Direct Injection Gasoline Engine." SAE Paper No. 2000–01–1795, SAE International, Warrendale, PA.

Schulz, C., Gronki, J., and Andersson, S. (2004). "Multi-Species Laser-Based Imaging Measurements in a Diesel Spray." SAE Paper No. 2004–01–1917, SAE International, Warrendale, PA.

Zetterberg, J., Li, Z. S., and Aldén, M. (2007). "Development of Filtered Rayleigh Scattering for Real World Temperature and Fuel/Air Ratio Imaging." Third European Combustion Meeting ECM2007. Crete, Greece.

Zhao, F., and Hiroyasu, H. (1993). "The Applications of Laser Rayleigh Scattering to Combustion Diagnostics." *Prog. Energy Combust. Sci.* Vol. 19, pp. 447–485.

Zhao, F.-Q., Kadota, T., and Takemoto, T. (1991). "Temporal and Cyclic Fluctuation of Fuel Vapor Concentration in a Spark Ignition Engine." SAE Paper No. 912346, SAE International, Warrendale, PA.

Zhao, F.-Q., Taketomi, M., Nishida, K., and Hirorasu, H. (1993). "Quantitative Imaging of the Fuel Concentration in a SI Engine with Laser Rayleigh Scattering." SAE Paper No. 932641, SAE International, Warrendale, PA.

Zhao, H., and Zhang, S. (2008). "Quantitative Measurement of In-cylinder Gas Composition in a Controlled Auto-ignition Combustion Engine." *Measurement Science and Technology*, Vol. 19, pp. 1–10.

chapter 7
Fuel Injection and Spray Characterization

7.1 Introduction

In a diesel or gasoline direct-injection engine, the fuel is directly injected into the combustion chamber in a very short period of time. The liquid fuel atomization and vaporization processes are critical to subsequent fuel and air mixing, combustion, and pollutant formation. To develop more efficient and cleaner engines, significant progress has been made and is continuing in the performance of fuel-injection equipment (FIE), including the fuel-injection pressure and flexibility in the fuel-injection rate and multiple injections. A good understanding of the fuel atomization and evaporation process forms an important part of the research and development process of advanced FIE and high-performance IC (internal combustion) engines.

This chapter begins with a brief overview of qualitative direct-imaging techniques for fuel injection and spray visualization. This is followed by an introduction to quantitative techniques for droplet-sizing measurements by Fraunhofer diffraction and phase Doppler techniques. Finally, temporally and spatially resolved two-dimensional laser imaging techniques, including Mie scattering, laser-induced fluorescence, and exciplex fluorescence techniques, are presented.

7.2 Direct Imaging of Spray and Droplets

The simplest and most used optical technique for fuel spray characterization is the direct imaging of sprays using a single-shot or high-speed camera. Direct imaging of fuel sprays with back illumination has been used extensively to measure spray geometry, penetration depth, wall impingement, and subsequent combustion in the constant volume chamber and inside combustion engines. The direct-imaging technique normally involves a strobe light of short duration that freezes the flow in the fuel spray. Figure 7.1 illustrates a suitable optical geometry for a back-illuminated spray-imaging setup.

Imaging of transient spray requires large quantities of light over a very short time duration. The principal considerations in selecting a light source are its brightness, spectral distribution, and repetition rate. Light sources can be either continuous or pulsed. Pulsed light sources are preferred in the single-shot mode, because a light pulse of very short duration defines the time resolution of the system irrespective of the camera's shutter speed. As energy is delivered

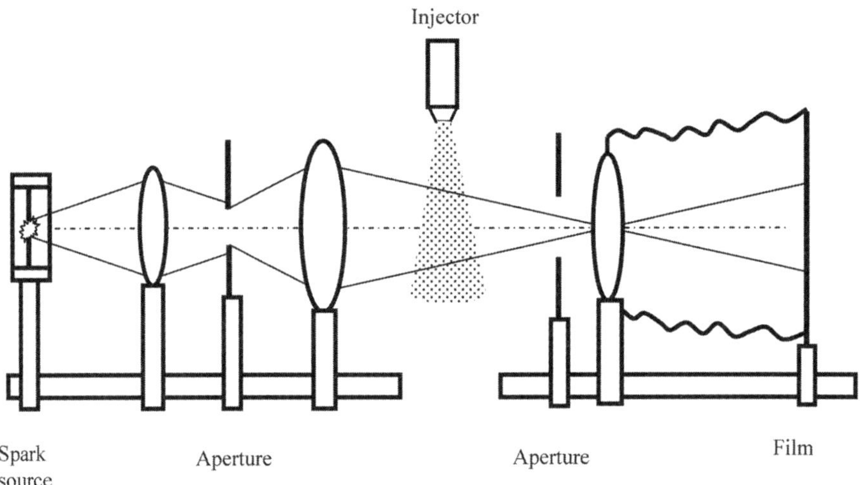

Fig. 7.1 Optical setup for back-illuminated images of sprays.

within a short period of time, a pulsed light source provides more light than an equivalent continuous light source for the high-shutter-speed operation.

There are two types of continuous lamps in the visible spectrum: quartz tungsten halogen (QTH) lamps, and DC arc lamps. QTH lamps have significant advantages over arc lamps: high total visible output, large source area, and a smooth continuum without the sharp peaks of arc lamps. Most pulsed lamps are xenon arc lamps with microsecond pulses. Nanosecond flashlamps are also available with a repetition rate up to 100 kHz. In addition, LED (light-emitting diode) arrays can act as illumination sources for spray visualization.

Lasers offer much more brightness than the conventional light sources mentioned above. Pulsed lasers, such as pulsed Nd:YAG lasers, are better suited for single-shot operations where the time resolution may be determined by the pulse width of the laser output (<20 ns). However, they are of limited use for high-speed continuous recordings because of their low repetition rate (typically, 10 or 20 Hz). When continuous wave lasers, such as HeNe and argon-ion, are used as the illuminating light source for continuous high-speed visualization, the exposure time is determined by the speed of the camera (1/ [number of frames per second]). For example, at 10,000 frames per second, the exposure time is 100 µs. In some applications, this may not be short enough to freeze the fluid motion. A copper vapor laser or a high-repetition-rate Nd:YLF laser, on the other hand, generates laser pulses less than 30 ns and a few millijoules at a repetition rate up to tens of kilohertz. When synchronized with a high-speed continuous recording system, extremely sharp images of fuel spray and combustion can be recorded at very high speeds.

If a fuel vapor field is to be included in the analysis, a back-illuminated imaging setup can be readily upgraded to a shadowgraph or Schlieren system by producing a parallel beam of light using different transmitting optics, as shown in Fig. 7.1. The shadowgraph technique is more suited for studies of fuel injection into ambient or a constant volume chamber when optical access is available for the transmitting and receiving optics. In addition, as the light scattering is proportional to the droplet size, the resulting image will be biased toward larger droplets and dominated by the liquid droplets over fuel vapor. Therefore, subjective selection of a threshold in the image will require a trial-and-error approach to separate liquid and vapor regions. In addition, when a line-of-sight image of the spray and flowfield is generated, the interpretation of the shadowgraph image requires prior knowledge of the spatial geometry of sprays. As will be discussed in Chapter 9, the use of a laser as the light source tends to produce speckle in the shadowgraph images. However, the speckle can be minimized with the introduction of a beam homogenizer or scattering plate in the laser beam delivery.

These images can provide information on not only the general structure of a fuel spray but also estimates of droplet sizes in the well-dispersed region. The droplet size distribution is determined by counting and sizing the diameter of the droplets from the recorded images. The direct-imaging technique can only be applied to relatively dilute regions of the droplet field. The droplet images do not overlap, and the edges of the droplets are not obscured by multiple scattering. Also, as only one plane of the droplet field at a time can be photographed, care must be taken not to include the image of the droplets out of focus into the size analysis.

There is also a trade-off between the minimum droplet size that can be resolved and the detectable area by the direct-imaging technique. Detection of smaller droplets requires larger magnification, which leads to smaller detectable area in the droplet field. In the extreme cases, the detectable droplet field may be restricted to a few millimeters. For quantitative droplet size measurements, the Fraunhofer diffraction method and phase Doppler particle analyzer (PDPA) are required.

7.3 Liquid Droplet Sizing by the Fraunhofer Diffraction Method

7.3.1 Fraunhofer Diffraction

The Fraunhofer diffraction method derives droplet size information from the angular distribution of the elastically scattered radiation. When a relatively large opaque particle ($d \gg \lambda$) is illuminated by a beam of parallel monochromatic light, the Fraunhofer diffraction pattern can be observed on a screen at a great distance from the particle. This pattern is characterized by a series of alternating dark and bright rings that relate to the location of zero and peak intensities, as shown in Fig. 7.2. In practice, the Fraunhofer diffraction pattern is observed at the focal plane using a lens.

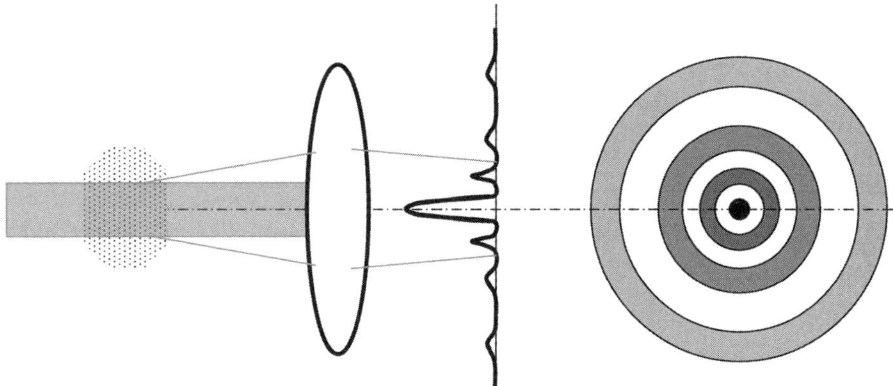

Fig. 7.2 Radial intensity distribution of the Fraunhofer diffraction pattern.

The mathematical description of the Fraunhofer diffraction pattern for a particle of diameter d can be shown to be

$$I = I_0 \left(\frac{2J_1(X)}{X} \right)^2 \tag{7.1}$$

where I_0 is the intensity at the center of the diffraction pattern, $J_1(X)$ is the first-order spherical Bessel function, and X is a dimensionless size parameter given by

$$X = \frac{\pi\, d\, R}{\lambda\, f} \tag{7.2}$$

where f is the focal length of the collection lens, and R is the radial distance in the detection plane (focal plane) as measured from the optical axis.

As shown in Eq. 7.1, the diffraction pattern is not sensitive to refractive index or shape, and it is unique for a particle or a collection of particles of different sizes. It is therefore possible to deduce the size distribution of a collection of particles from their diffraction patterns.

7.3.2 Implementation of the Fraunhofer Diffraction Method

The basic apparatus of the Fraunhofer diffraction method is illustrated in Fig. 7.3. A parallel beam of monochromatic light from a laser (typically a low-power HeNe laser), traverses a cloud of droplets or particles. The transmitted and scattered light is collected by a lens, and its intensity distribution is observed in the focal plane. The intensity distribution is typically measured by a photodetector array.

An apparatus developed by Swithenbank et al. (1977, 1991) has been used extensively and adopted commercially by Malvern Instruments. It collects the light diffracted by a particle using a set of concentric photodetector rings. The light energy within any ring of radii R_1 and R_2 can be shown to be

$$P_{R_1,R_2} = P\left[J_0^2(X_1) + J_1^2(X_1) - J_0^2(X_2) - J_1^2(X_2) \right]$$

(7.3)

where J_0 is the zero-order spherical Bessel function, and P is the energy falling on the particle. For N particles of the same size, the total energy falling on the particles is given by

$$P = NI_0\pi\left(\frac{d}{2}\right)^2$$

(7.4)

If a collection of particles of different sizes is present, then the light energy falling on any ring in the focal plane is the sum of those from individual particles:

$$P_{R_1,R_2} = C\sum_{j=1}^{M} \frac{W_j}{a_j}\left[J_0^2(X_{j,1}) + J_1^2(X_{j,1}) - J_0^2(X_{j,2}) - J_1^2(X_{j,2}) \right]$$

(7.5)

where C is a constant, W_i is the weight fraction of the particles of diameter d_i, and M is the total number of particle size range. Theoretically, the size

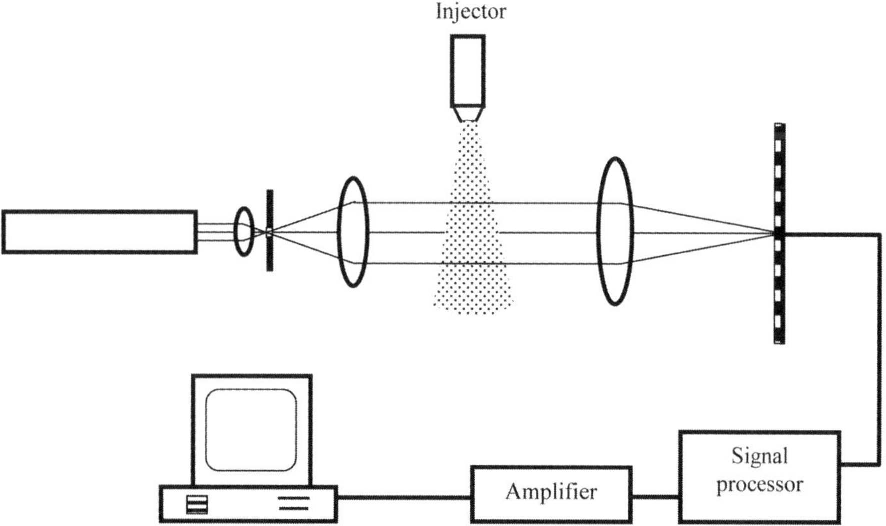

Fig. 7.3 Experimental setup for the Fraunhofer diffraction technique.

distribution may be obtained from the measured data by inverting Eq. 7.5. However, this is difficult to do in practice. Alternative approach to the inversion problem is to solve a discretized version of Eq. 7.5:

$$P(I) = W(J)T(I,J)$$ (7.6)

where $P(I)$ and $W(J)$ are the light energy and weight distributions, respectively, and $T(I,J)$ contains the coefficients that define the light energy distribution for each size range.

Unfortunately, Eq. 7.6 cannot be inverted directly. For this reason, least squares minimization of error solutions are adopted, where a form for $W(J)$ is assumed and iteratively adjusted until the sum of squared errors in $P(I)$ is minimum. This process can be expressed as

$$\sum_{I=1}^{N_d} \left[P_m(I) - W(J)T(I,J) \right]^2 = \min$$ (7.7)

where $P_m(I)$ is the measured light energy distribution, and N_d is the total number of detector rings. Several distributions, which are typically two-parameter functions, have been used, the most common being Rosin-Rammler and log-normal. The two parameters in these functions are varied until a "best fit" to the measured data is found. This approach is rather limited. At best, only an approximation to the true size distribution is achieved unless it fits the assumed form exactly. Further, if the size distribution is multimodal, the assumption of a single-peak will be a very poor representation. To overcome this difficulty, a model-independent algorithm was developed to perform the direct inversion of the diffraction measurements with satisfactory results (Young and Bachalo, 1988).

The ratio of light intensity measured at the center diode, before and after the droplet field is introduced in the field, gives the fraction of the light transmitted by the particles. This ratio is related to the drop concentration by the Beer-Lambert law:

$$\ln\left(\frac{I}{I_0}\right) = -\tau L$$ (7.8)

where L is the optical path length, and τ is given by

$$\tau = 2\sum_{j=1}^{M} N_j \pi \left(\frac{d_j}{2} \right)^2 \qquad (7.9)$$

where N_j is the number of droplets in the size range j per unit volume, and d_j the mean drop diameter of the jth size class. Thus, when the size distribution is known, the light attenuation can be used to determine the droplet concentration. The volume fraction of droplets is given by

$$C_V = -\frac{\ln\left(\frac{I}{I_0} \right)}{3L\sum_{j=1}^{M} \left(\frac{V_j}{d_j} \right)} \qquad (7.10)$$

where V_j is the drop volume distribution that results from the size distribution measurement. For a low-density particle field, Eq. 7.10 cannot be used because of small attenuation. The volume fraction of droplets can be, instead, evaluated from the total scattered light energy measured by the detector:

$$C_V = C'' \sum_{i=1}^{N_d} P(i) \left(L\sum_{i=1}^{N_d}\sum_{j=1}^{M} V(i)T(i,j) \right)^{-1} \qquad (7.11)$$

where $P(i)$ is the total scattered light energy collected from all detector rings other than the central diode, and C'' is a system constant and is determined by calibration.

Using different focal lengths for the receiving lens, droplets in ranging from 0.2 to 2000 mm can be measured. The instrument can be calibrated using particles of known size or calibration reticles that have large numbers of discs of well-determined size photographically etched on to glass.

The method is relative in that it relies on the shape of the scattering pattern rather than the actual intensities. Therefore, it is independent of variation in incident intensity. At a low level of drop concentration, scattering from dust on lenses or windows is a particular problem. To minimize scattering by dust particles, the collection lens and any windows should be kept clean. Practically, the minimum extinction appears to be around 5–10%, as recommended by Malvern Instruments.

If the concentration is too high, multiple scattering occurs and alters the diffraction pattern. Special procedures developed over the years can extend the applicable range the diffraction method to extinction in excess of 90%.

If measurements are carried out in a high-temperature environment, deflections due to density changes will cause the central spot to move. Over a finite time this will have the effect of increasing the spot size and thus reducing the maximum measurable drop size. One way around this difficulty is to gate the readings to that they are only taken when the spot is central. The readings need to be taken in a time that is short compared to the inverse of the fluctuation frequency.

The Fraunhofer diffraction–based instruments are well established as a means for spray characterization. Since the diffracted light is integrated over the measurement volume along the laser beam, it has limited spatial resolution, although an Abel inversion scheme can be used with the assumption of an axisymmetric spray to procure spatially resolved measurements. The method is particularly prone to minor fluctuations in the refractive index, which can cause significant beam deflections. It is very difficult to combine the method with other techniques to produce simultaneous size and velocity measurements of the drops.

7.4 Droplet Sizing and Velocity Measurements by the Phase Doppler Particle Analyzer (PDPA)

7.4.1 The Principle of the PDPA

The Fraunhofer diffraction technique measures a collection of droplets or particles in the expanded laser beam. The results are integrated over the measurement volume, which gives a droplet size distribution in the measurement region. To obtain spatially resolved droplet size distributions, light scattering from an individual particle can be detected and analyzed.

Figure 7.4(a) shows light rays incident upon a liquid droplet. At the first interface, part of the light is reflected from the surface of the sphere, and these rays are referred to as $p = 0$ (reflected rays). The light rays transmitted and refracted by the sphere are designated as $p = 1$ rays (first-order refracted rays). Rays reflected from the internal surface and refracted in the backward direction are $p = 2$ rays (second-order refracted rays).

To compute the phase from ray optics, it is common to reference the actual ray to a hypothetical ray scattered without a phase lag at the center of the sphere, as shown in Fig. 7.4(b). It is apparent that the reflected wave from the first

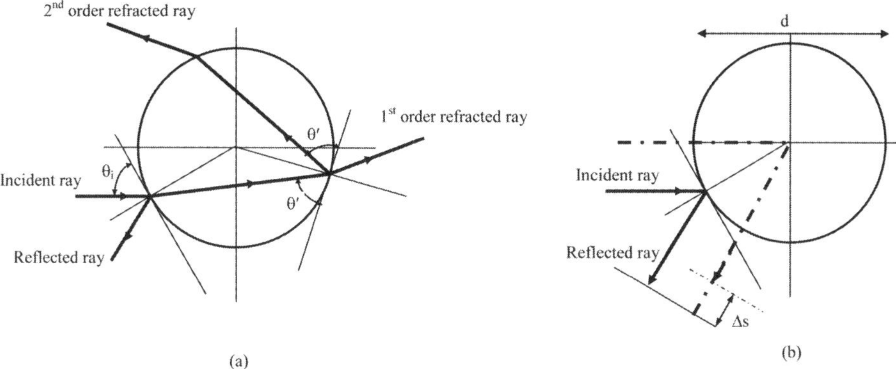

Fig. 7.4 Geometric ray trace of incident ray through a sphere.

surface experiences a shortcut of length Δs relative to the hypothetical wave, and hence a phase difference between the two beams will result:

$$\Delta\sigma_0 = \frac{\Delta s}{\lambda} \tag{7.12}$$

Similarly, the first refracted ray undergoes a change of phase relative to the hypothetical wave.

The general expression for the optical path length of a light ray passing through the sphere relative to the hypothetical wave was derived by Bachalo (1980),

$$\Delta s = 2a\left(\sin\theta_i - p\cdot m\cdot\sin\theta'\right) \tag{7.13}$$

The resulting phase difference is given by

$$\Delta\sigma = \frac{2\pi d}{\lambda}\left(\sin\theta_i - p\cdot m\cdot\sin\theta'\right) \tag{7.14}$$

Where $m = n - ik = m_2/m_1$ is the relative refractive index of the sphere to the surrounding medium, $a = \pi d/\lambda$ is the size parameter, and θ_i and θ' are the angles between the surface tangent and the incident and refracted rays, respectively. The value of the integer p depends on the nature of the ray ($p = 0$, for reflection; $p = 1$, for first refraction; $p = 2$ for second refraction).

Equation 7.14 illustrates that the phase difference is proportional to the diameter of the spherical particle d. The question is now how to measure this phase change. Figure 7.5(a) shows the solution based on a standard dual-beam LDA (laser Doppler anemometry) setup. The rays from each beam are incident upon the particle at different angles and therefore reach the common points on the receiver by different optical paths. The relative phase shift resulting from the differing optical paths is described by

$$\phi = \frac{2\pi d}{\lambda}\left[\left(\sin\theta_{1,i} - \sin\theta_{2,i}\right) - pm\left(\sin\theta_1{}' - \sin\theta_2{}'\right)\right] \tag{7.15}$$

where the subscripts represent beams 1 and 2. Since all the angles are fixed by the transmitting and receiving optics, the phase shift only changes as a result of the particle diameter d.

The phase difference produces an interference fringe pattern on the detector with an intensity distribution as shown in Fig. 7.5(b), which can be described as

$$I \propto I_1 + I_2 + 2\sqrt{I_1 \cdot I_2}\ \cos\left[2\pi v_D t + \phi\right] \tag{7.16}$$

Equations 7.15 and 7.16 illustrate that the spatial frequency (φ) of the fringe pattern is linearly related to the size of the sphere, and the temporal frequency of the fringe pattern is the Doppler difference frequency (v_D) in the LDA technique (see Chapter 4). Therefore, the particle size can be theoretically determined from the phase difference or spatial frequency if it can be measured. Since there is no reference for the phase difference, the position of the particle in the beams has to be known if the absolute value of the phase of the scattered light needs to be determined. The solution is to measure the phase difference between the signals from two different detectors by means of an electronic phase difference processor. The velocity of the sphere is measured simultaneously from the temporal frequency v_D of the scattered signal, as discussed in Chapter 4.

To determine the parameters affecting the phase shift, it is necessary to introduce the coordinates shown in Fig. 7.6. The coordinates are arranged with the z-axis in the direction of the bisector of the two incident beams. The x-axis is in the plane of the beams, and the y-axis completes the coordinate system. The beams are incident in the x-z plane with an angle $\pm\alpha/2$ to the z-axis. The polarization is described by the angle ξ from the y-axis ($\xi = 0°$ perpendicular; $\xi = 90°$ parallel). The scattering direction is described by the off-axis angle ϕ and the elevation angle ψ. The phase difference given by Eq. 7.16, in terms of the geometry of the two laser beams, can then be determined from geometric optics. In the case of reflection ($p = 0$), it is given by

$$\Delta\phi^0 = \frac{\sqrt{2}\pi m_1 d}{\lambda}\left[\sqrt{1 + \sin\varphi\sin\psi\sin\frac{\alpha}{2} - \cos\varphi\cos\frac{\alpha}{2}} - \sqrt{1 - \sin\varphi\sin\psi\sin\frac{\alpha}{2} - \cos\varphi\cos\frac{\alpha}{2}}\right] \tag{7.17}$$

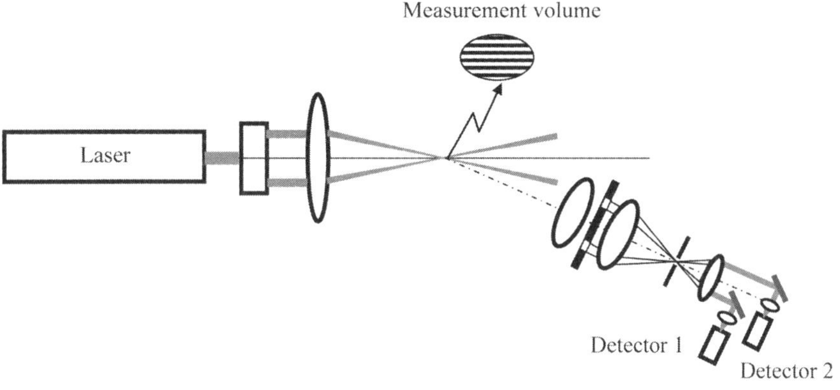

(a) Simple PDF setup

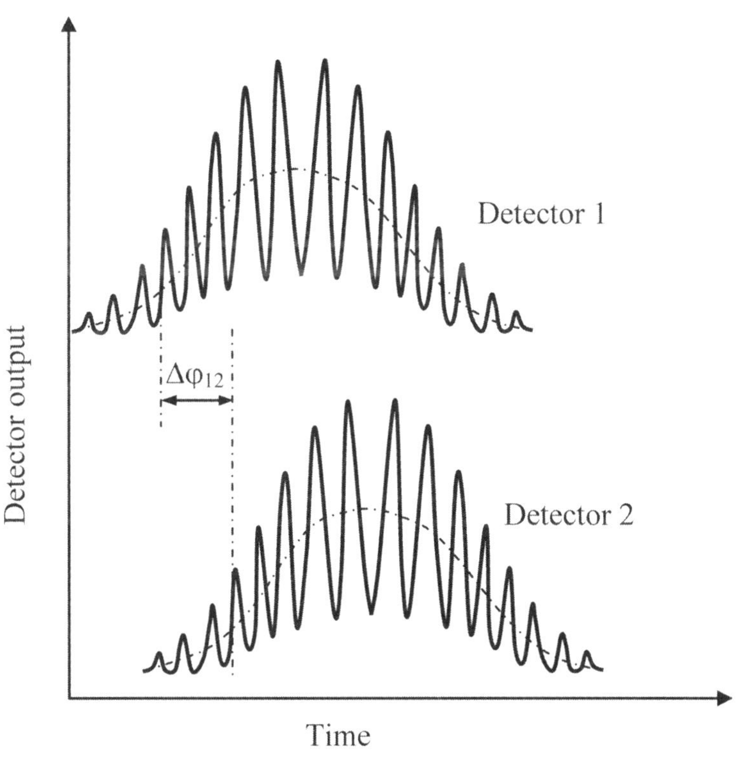

(b) Burst Doppler signals

Fig. 7.5 Optical geometry of a phase Doppler particle analyzer (PDPA) and its burst signal output.

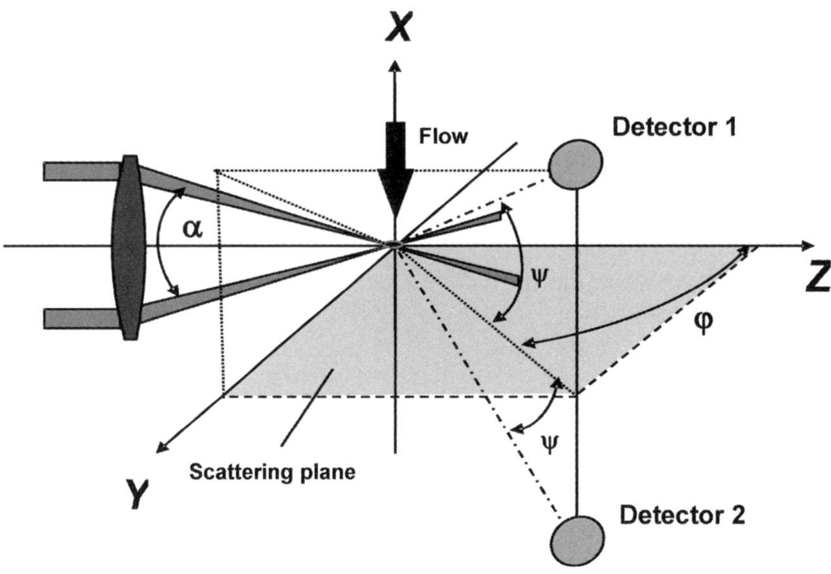

Fig. 7.6 Definition of scattering geometry in a PDPA system.

The phase difference for the first-order refracted beam is given by

$$\Delta\phi^1 = \frac{2\pi d}{\lambda}\left\{\sqrt{1+m^2-\sqrt{2}m\sqrt{1+\sin\varphi\sin\psi\,\sin\frac{\alpha}{2}+\cos\varphi\cos\frac{\alpha}{2}}}\right.$$
$$\left. -\sqrt{1+m^2-\sqrt{2}m\sqrt{1-\sin\varphi\sin\psi\,\sin\frac{\alpha}{2}+\cos\varphi\cos\frac{\alpha}{2}}}\right\} \tag{7.18}$$

The phase difference for higher order refracted rays cannot be expressed in a closed form and must be found by an iterative procedure. In general, both reflected and refracted light will reach the detector, and the phase-diameter relation thus depends on the relative amount of each and can be a highly oscillating function. One of the important considerations in performing phase Doppler-based measurements is to select an observation direction and a detector aperture so that the phase-diameter relationship is a linear relation as given by either Eq. 7.17 or 7.18.

7.4.2 Implementation of the PDPA

The simplest PDPA configuration is shown in Fig. 7.5(a), which is very similar to that of a dual-beam LDA, except that two detectors are required to obtain the information that will allow the determination of the particle size. In this system, the laser beam is split into two beams of equal intensity. The beams are then focused and made to intersect using a transmitting lens. As the particle passes through the beam intersection region, it produces a scattered interference fringe pattern that appears to move past the detectors. Doppler burst signals are produced by each detector, as shown in Fig. 7.5(b). The frequency of each burst signal is used to determine the particle velocity. The phase shift between the two Doppler burst signals, $\Delta\phi_{12}$, can then be determined by measuring the time delay, τ_{12}, between the zero-crossings of the high-pass filtered signals from detectors 1 and 2, and dividing by the measured Doppler period, τ_D:

$$\Delta\phi_{12} = \frac{\tau_{12}}{\tau_D} \cdot 2\pi \tag{7.19}$$

where the measured phase differences are averaged over all the cycles in the Doppler burst signal.

Measurements of the phase shift are then related to the particle size using the relationships given by Eq. 7.17 or 7.18, depending upon if the reflection or the first refraction component of the scattered light is detected. With parallel polarization, the light intensity scattered by refraction from a liquid droplet in air will be approximately 80 times that scattered by reflection when the droplet is subject to a uniform illumination. By careful selection of the light collection angles for the detectors, a linear relationship derived from Eq. 7.18 can be obtained, as shown in Fig. 7.7, assuming the first refraction is the only scattering component received by the detectors. In Fig. 7.7, the slope of the line can be altered by changing the optical parameters, including the laser beam intersection angle (α), collection angle (ϕ and ψ), laser wavelength (λ). In Eq. 7.18, note that the linear relationship is different for droplets of different index of refraction.

In the case of transparent particles (droplets) for which the first refraction is the dominant scattering mode, the preferred light collection position for the receiver is off-axis 30°–70° (ϕ) and in a plane (y-z plane, $\Psi = 0$) that passes through and is orthogonal to the intersection of the two laser beams. Two identical apertures on a single lens are placed symmetrically on either side of the lens center in the y-z plane. When the particle is highly reflective, the light collection position for the receiver should be off-axis between 80° and 110° to detect the reflection.

In practice, three detectors are used in the measurement of phase differences to determine the droplet size, as shown in Fig. 7.8. These three detectors produce two pairs of redundant measurements, which are then compared to estimate the quality of the measurements. More importantly, the introduction of the third detector extends the size range and sensitivity of the PDPA system. In

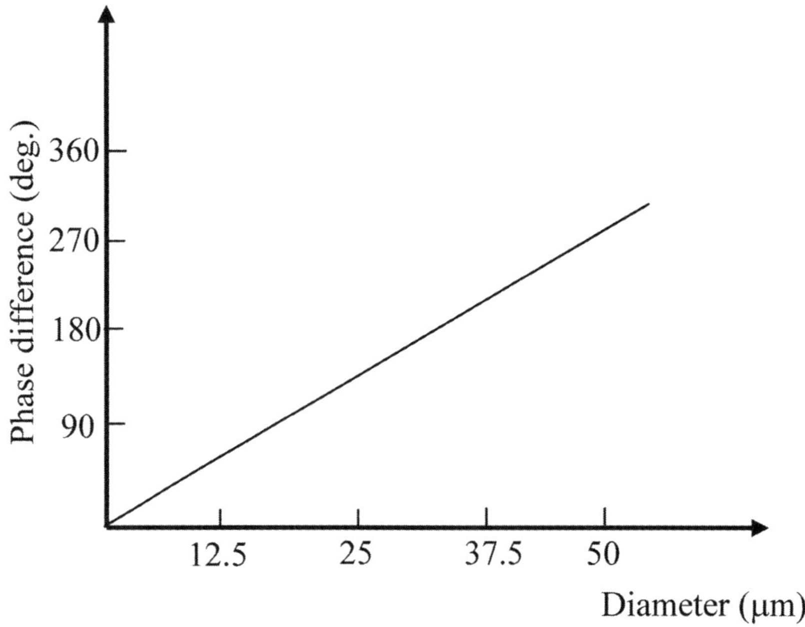

Fig. 7.7 Phase difference and droplet size relationship.

the design of the PDPA system shown in Fig. 7.8, these detectors are spaced in an approximate 3 to 1 spacing. The resulting phase versus diameter curves are plotted in Fig. 7.9. The phase difference between the signals from the closely spaced detectors, D1 and D2, follows the dashed line. The phase between the signals for the detectors with the larger spacing, D1 and D3, follows the solid lines with steeper gradient. The outputs of D1 and D2 produce the phase measurement ϕ_{12}, and those of D1 and D3 yield the phase measurement, ϕ_{13}. With this arrangement, the phase may be measured for detector separations that extend over several fringes (one fringe corresponds to a measured phase shift of 2π). Using the phase measurement of ϕ_{12} allows the determination of which cycle the measurement of ϕ_{13} took place. For example, for a droplet of size d, the value of ϕ_{12} indicates that ϕ_{13} is made on the second cycle of phase, which means that there is an entire interference fringe between the D1 and D3 detectors. And the measurement on detector D3 extends on to the second fringe between the detectors.

The two measurements, ϕ_{12} and ϕ_{13}, are weighted according to their relative sensitivities and averaged to obtain the mean phase value for that particle. Obviously, the measurement of D1 and D3 has the greatest sensitivity, with the change in phase with a change in droplet size being approximately three times the change for detectors D1 and D2.

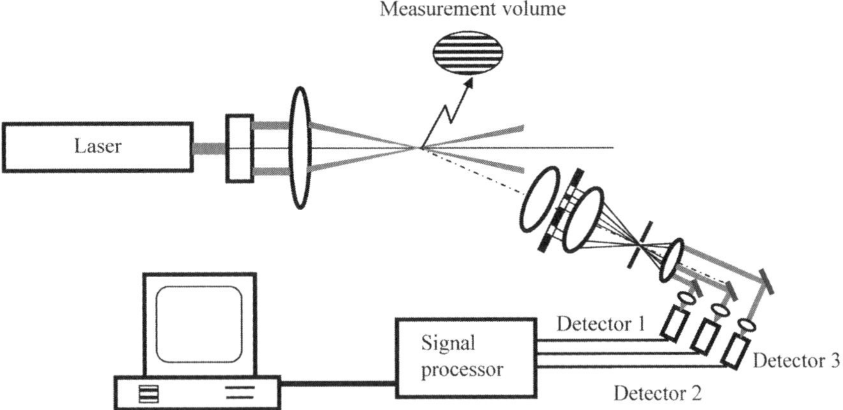

Fig. 7.8 Schematic of a PDPA system with three detectors.

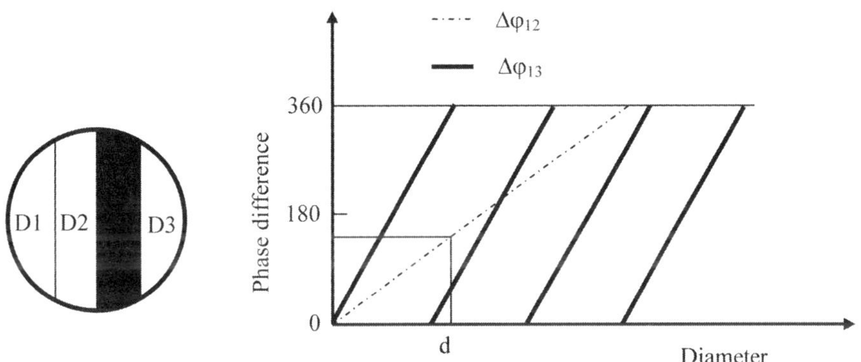

Fig. 7.9 Phase difference and droplet diameter relationship with three detectors.

The Doppler burst signals from the two photodetectors are shown in Fig. 7.5(b), from which measurements must be made to find at least three quantities, namely, the Doppler frequency of a burst for the droplet velocity, the phase difference between the two Doppler bursts for the droplet size, and the number of measurements per second (or the time between bursts) for the droplet flux. The simultaneous measurement of Doppler frequency and phase are typically performed by cross-correlation or fast Fourier transform (FFT) methods by a digital signal processor or computer.

7.4.3 Data Reduction and Presentation

The PDPA technique is a single-particle-counting technique and as such provides the velocity and size measurement for each droplet. The results are normally presented as droplet size distributions in bar charts with the corresponding velocity distribution, as shown in Fig. 7.10. Different mean diameters can be derived from the individual droplet sizes to correlate with the physical process being investigated. The commonly used mean diameters are the Sauter mean diameter (SMD) and the volume mean diameter (VMD), as given by

$$SMD: \quad D_{32} = \frac{\sum N_i d_i^3}{\sum N_i d_i^2} \tag{7.20}$$

$$VMD: \quad D_{30} = \frac{\sum N_i d_i^3}{\sum N_i} \tag{7.21}$$

where N_i is the number of droplets in size class i, and the summation is taken over all the particle size classes.

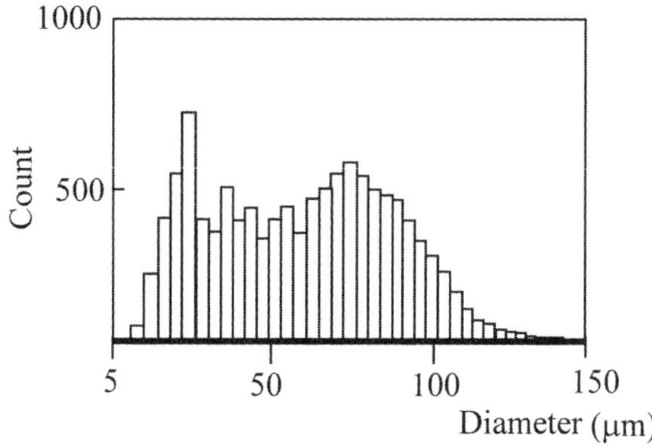

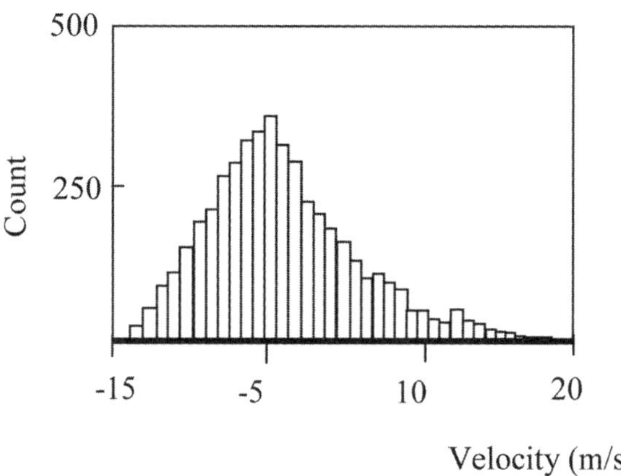

Fig. 7.10 Typical PDPA results showing the droplet size and velocity distribution.

The PDPA technique has the ability to measure the drop number density and volume flux, which are important parameters in the spray atomization process. Volume flux is determined from the measured VMD, D_{30}, sampling cross-sectional area A, and the number of droplets counted N in the sampling time interval τ as given by

$$F = \frac{\pi}{6} D_{30}^3 \frac{N}{A \cdot \tau}$$ (7.22)

where

$$N = \sum_{i=1}^{n} N_i$$

As mentioned previously, size distributions are highly skewed in the larger size classes with a relatively small population, so it is important that a large number of particles be measured to obtain good statistical representation and hence accurate determination of D_{30}. Furthermore, since the volume flux is proportional to the D_{30} cubed, the accurate determination of this quantity is extremely important in realizing good volume flux measurements.

It should be noted that the volume flux measured by the PDPA technique is a temporal distribution. In contrast, the estimation of the volume flux from a photograph would yield a spatial distribution.

The number density N_d is calculated from the measured number of particles N passing the sampling cross section. In practice, the particle number density is determined from the transit time (or residence time) t of each particle as it passes the sample volume relative to the total sampling time τ of each measurement, and the sample volume Ω_s:

$$N_d = \frac{1}{\tau} \sum_i \frac{\sum_j t_{i,j}(d_i)}{\Omega_s(d_i)}$$ (7.23)

Where $t_{i,j}(d_i)$ is the residence time of the jth droplet within the droplet size class d_i. The sample volume $\Omega_s(d_i)$ is written as a function of particle size because of the Gaussian distribution of the laser beam and the droplet trajectory through the cross section of the two beams (Bachalo et al., 1991). For a Gaussian distribution of the illumination beam, smaller particles require a greater illumination to produce a detectable signal. Thus, the effective diameter of the sampling volume is smaller for small particles than the large particles. Therefore, the reliable measurement of the particle number density does not depend on the accurate sizing of the particles but is very dependent on the reliable determination of the measurement volume and the counting of the particles. For one-dimension flow, the effective diameter of the sampling volume can be determined from the transit time and velocity of the particle:

$$D_s\left(d_i\right) = u\left(d_i\right) \cdot t\left(d_i\right) \tag{7.24}$$

Alternatively, it can be measured by counting cycles in the Doppler burst and using fringe spacing. For each size class, an accumulation of fringe counts is collected for each size class. Since particles will pass on all trajectories through the laser beam, the trajectories leading to the maximum number of fringe counts in the signal indicate that these particles passed normally through the diameter of the beam. The product of this fringe counts and the fringe spacing then provides the effective diameter of the sampling volume. This procedure is followed for each particle size class.

For three-dimensional flows, the effective diameter of the sampling volume may be better determined by the signal amplitude method. A maximum signal amplitude is detected when a droplet passes through the center of the Gaussian beam, whereas the minimum detectable amplitude occurs when a droplet travels in a trajectory away from the peak intensity of the Gaussian beam. The ratio of the maximum (I_{max}) and minimum (I_{min}) signal amplitude measured among each droplet class (d_i) can then be used to find the effective diameter (d_{samp}) of the sampling volume for that droplet size class, which is given by

$$\ln\left(\frac{I_{max}\left(d_i\right)}{I_{min}}\right) = \frac{d^2_{samp}}{3r_{laser}} \tag{7.25}$$

where r_{laser} is the laser beam waist diameter at $1/e^2$. Normally the minimum signal amplitude is defined by the threshold signal-to-noise ratio.

7.4.4 Optimization of a PDPA System

The potential source of errors in a PDPA system is the detection of signals produced by the wrong light-scattering component. From Eqs. 7.17 and 7.18, it is noted that different relationships of phase and diameter are applicable to reflection and refraction. As most PDPA systems are designed to make use of the first refraction component of the light scattered, the presence of the reflection component needs to be identified and removed. Otherwise serious errors will occur.

The presence of reflection for a PDPA system optimized for the refraction detection is a result of the problem of trajectory dependent light scattering. As shown in Fig. 7.11, if a droplet, which is relatively large compared to the focused beam waist diameter, passes on certain trajectories, the peak intensity of the beam will fall on a point on the droplet and be reflected to the detector. Although first refraction is a much more efficient process than reflection, on the trajectories shown here, the incident intensities at the points on the droplet scattering light by refraction and reflection may result in the dominance of the refraction being lost. On some trajectories, the scattered light may be dominated by the reflection from large droplets. In these cases, the three-detector arrangement is used to remove the reflection component by determining the direction of the fringe pattern movement. From geometric optics, it is known that reflection produces a pattern that will move in the opposite direction of light scattered by refraction. Additional measures are required to reject the occurrences of unwanted reflection components. For example, the adaptive threshold validation has been developed to minimize this error by setting a minimum detectable intensity or signal amplitude for each drop size class.

Another problem that occurs in the PDPA measurements is that large droplets are more likely to be detected than smaller ones. When using laser beams with Gaussian beam intensity distributions, small particles will need to cross the beams closer to the central peak intensity to produce a detectable signal. Larger particles can pass further out on the edges of the Gaussian beam and still be detected, because particles scatter light in proportion to the diameter squared. This bias toward large droplets must be accounted for in the data reduction.

For an accurate measurement of droplet size, PDPA requires knowledge of the refractive index of the droplet and the surrounding gases. Practical liquid fuels are a mixture of many hydrocarbons, so the density and refractive index, $m = n - ik$, changes according to fuel composition. The imaginary refractive

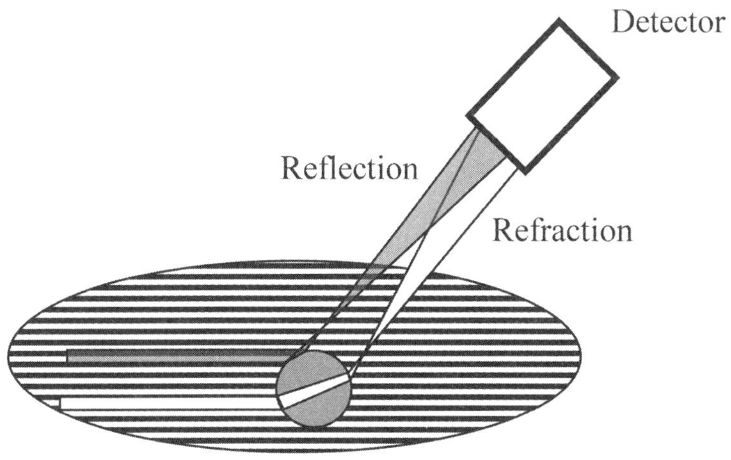

Fig. 7.11 Illustration of reflection interference with PDPA measurements.

index k, which is determined by the light absorption in the liquid, is normally negligible for most hydrocarbons in the visible range. The real refractive indices for some common fuels and fuel components are given in Table 7.1. There are two points to keep in mind about Table 7.1. The first is that diesel or gasoline fuels are multicomponent, with each component having different physical properties. The second is that commercial fuels may contain dyes and therefore exhibit very high absorption and fluorescence, which will drastically modify the angular distribution of the refracted light intensities.

The variation of refractive index with temperature is indirectly obtained from the Eykman relationship, which is derived empirically and specifically for hydrocarbons, as given in Eq. 7.26:

$$\left(\frac{n^2-1}{n+0.4}\right) = \rho \cdot const. \tag{7.26}$$

where fuel density ρ as a function of temperature is readily available.

The theory underlying the PDPA size measurement technique is only applicable to spherical scatters. It is the surface tension that keeps droplets spherical. However, as droplet size increases or its velocity rises relative to the continuous phase, then the droplet will deform and ultimately break up. The Weber number, the ratio of the momentum force on the droplet to the surface tension forces, must be less than 1 for surface tension to dominate and the droplet to remain spherical. In general, nonspherical droplets are rejected in the data reduction stage by means of their departure from sphericity, as described by $\phi_{12} + \phi_{23} + \phi_{31} = 0$.

Therefore, if PDPA were used in the region of a spray where atomization was poor, with liquid jet and ligaments, serious measurement errors could occur.

For a typical commercial PDPA system, the following quantities are claimed:

- Droplet size from 0.5 to 10,000 μm (spherical)
- Dynamic size range of 50:1 at one optical setup
- Size accuracy of 1%
- Droplet velocity up to 300 m/s
- Sample volume determined for each size class
- Particle time-of-arrival 0.6 μs resolution
- Maximum droplet number density 10^6/cc
- Maximum date rate 50,000 samples/s

Table 7.1 Refractive Indices for Fuels and Liquids at Room Temperature in the Visible Range

Liquid	Refractive Index
Water	1.33
Heptane	1.39
Iso-octane	1.39
Benzene	1.50
Toluene	1.50
Diesel	1.45
Ethanol	1.36

7.4.5 Application of PDPA to IC engines

Initial applications of the PDPA technique were used to characterize the spray atomization processes from various gasoline port injectors (Kelly-Zion et al., 1995; Zhao et al., 1995), including the sizes of velocities of the fuel droplets downstream from port fuel injectors in the intake port and in the cylinder under motored conditions. The PDPA techniques was then used to study the liquid fuel behavior in the cylinder of firing spark-ignition (SI) engines technique (Meyer and Heywood, 1997; Posylkin et al., 1994). The effect of open-valve injection and closed-valve injection on the in-cylinder liquid fuel droplets was investigated under warmed-up and cold-start conditions.

Diesel spray and high-pressure, direct-injection (DI) gasoline spray represents a severe challenge to PDPA measurements. It is optically thick, so light transmission through or into the spray is very low. The extremely high drop density requires a very small measurement volume, so that there is only one droplet present. The spray, close to the injector, can be very poorly atomized, with liquid jet and ligaments dominating the flow processes. The presence of liquid jet or ligaments can refract the incident laser light to prevent direct transmission and the formation of a measurement volume. The transient, high-pressure nature of the spray also brings experimental difficulties with very low duty cycles for data arrival and very high velocities. Despite these difficulties, measurements are possible and have been performed. Most of these measurements have been carried out on nonevaporating sprays at atmospheric conditions or pressurized constant volume chambers (Desantes et al., 1998; Nouri et al., 2006; Wigley et al., 2004).

7.5 Two-Dimensional Visualization of Spray Atomization by Mie Scattering and Laser-Induced Fluorescence (LIF)

As discussed previously, the structure of fuel sprays can be characterized by direct imaging with side or back illumination, or the shadowgraph/Schlieren setup. A major limitation of these methods is that they generate information integrated along a line of sight in the measurement region. An alternative approach is to use a laser sheet to illuminate a plane in the spray or a cloud of droplets. These laser-sheet-based methods are capable of giving spatially resolved information with excellent temporal resolution.

The experimental setup for the two-dimensional laser sheet imaging techniques involves the generation of a light sheet from a pulsed laser to illuminate a cross section of fuel spray. The Mie scattering from liquid droplets or the fluorescent emissions from fluorescent species in the fuel is captured at a right angle by an imaging device. The very short laser pulse duration is particularly important in reducing the effective exposure time of the illuminated spray section and obtaining sharp, focused images of fast-moving fuel droplets. In addition, temporal evolution of transient spray can be studied with the aid of a high-speed CMOS (complementary metal-oxide semiconductor) video camera or a multiple-ICCD (intensified charge-coupled device) camera system in conjunction with the high-repetition-rate Cu vapor or solid-state lasers.

7.5.1 Two-Dimensional Spray Imaging Through Mie Scattering or PLIF

The first method for the two-dimensional imaging of sprays and liquid fuel droplets is to use Mie scattering of a laser sheet by the liquid droplets themselves. For a particle of diameter much larger than the wavelength or size parameter $a = \pi d / \lambda \gg 1$, the light-scattering process is termed as Mie scattering. For visible wavelengths, Mie scattering occurs from particles larger than 0.5 μm. The scattering intensity from a group of particles of diameter d is given by

$$I_{sca}(n,\theta,d) = I_0 \cdot C_n \cdot f(n,\theta) \cdot d^2 \tag{7.27}$$

where n is the real part of the refractive index of the particle, θ is the scattering angle, and C_n is the number density of particles. $F(n, \theta)$ is a complicated function of the refractive index, particle shape, and the scattering angle.

The Mie scattering technique is limited to qualitative visualization of the liquid fuel distribution as the scattered light intensity depends on both the drop concentration and size distribution. In addition, as Mie scattering is orders of magnitude stronger than the Rayleigh scattering, it is not possible to detect both liquid and vapor fuel from the scattered laser light.

In contrast to Mie scattering, laser-induced fluorescence (LIF) is generated in both liquid and vapor phases. Therefore, in principle planar laser-induced fluorescence (PLIF) can be used to visualize both liquid and vapor fuel distributions. Furthermore, if PLIF and Mie scattering images can be obtained simultaneously with two separate imaging optics or an imaging doubling optical setup, the fuel vapor distribution can be obtained by subtracting the Mie image of liquid fuel from the PLIF image. However, because of the limited dynamic range of the PLIF images and substantial difference in the liquid and vapor fuel LIF signals, only the most concentrated fuel vapor could be detected in the resulting vapor image.

Further consideration should be given to the scattered light of both Mie and fluorescence from solid surfaces. Nonreflective paint or increased surface roughness help to reduce the flare from solid surfaces. The use of a small f-number lens suppresses the detection of the surface-scattered signals by minimizing the depth of field. In the case of dense spray, additional accuracies can arise from the extinction of the laser sheet by scattering and/or absorption as the beam traverses through the spray and attenuation of the signal light between the laser sheet and the camera by the same effects. Of the different methods, the bidirectional excitation with counterpropagating laser sheets is the most effective to overcome the laser attenuation in the measurement region. The Mie scattering and PLIF techniques have been used for spray characterization, studies of port-injected fuel in the intake manifold and cylinder, in-cylinder spray, and liquid fuel measurements.

Spray characterization of various gasoline injectors was investigated using PLIF (Zhao et al., 1995). The interaction between the flow and fuel spray in the intake port of a port-injection SI engine was investigated by Wensing et al. (1997) using Mie scattering. The geometric structure of the liquid fuel spray entering the cylinder was visualized by Meyer and Heywood (1997) using fluorescence signals of acetone added to the fuel. The effects of fuel volatility on the evaporation process and unburned hydrocarbon emissions were investigated by means of adding fluorescing dopants to the gasoline fuel (Dawson and Hochgreb, 1998). The fluorescing dopants, including acetone, 3-pentanone, and aromatics, were selected to represent the low, middle, and high boiling fractions of fuel, respectively.

To better understand the highly turbulent diesel spray structure and evaporating process, several studies have been conducted using Mie scattering of the liquid fuel (Baritaud et al., 1994; Canaan et al., 1998; Espey and Dec, 1995; Felton et al., 1987; Hodges et al., 1991).

7.5.2 Laser Sheet Droplet Sizing (LSD) by Combined Mie Scattering and LIF

Additional information can also be obtained on the droplet sizes from the combined Mie scattering and PLIF images by means of the so called laser sheet droplet sizing (LSD) technique. When a group of droplets is imaged onto 1 pixel of a solid-state camera, the PLIF signal S_{LIF} is given by

$$S_{LIF} = C_{LIF} \sum_{i=0}^{i=N} d_i^3 \qquad (7.28)$$

Similarly, the Mie signal S_{Mie} is given by

$$S_{Mie} = C_{Mie} \sum_{i=0}^{i=N} d_i^2 \qquad (7.29)$$

where C_{LIF} and C_{Mie} are constants depending on the imaging system and the summation process over all droplets. The ratio of intensity measured by 1 pixel in the PLIF image, to the intensity of the same pixel in the Mie image, is thus given by

$$\frac{S_{LIF}}{S_{Mie}} = \frac{C_{LIF}}{C_{Mie}} \frac{\sum_{i=0}^{N} d_i^3}{\sum_{i=0}^{N} d_i^2} \qquad (7.30)$$

The second term on the right-hand side is the SMD, defined by

$$SMD = \frac{\sum_{i=0}^{N} d_i^3}{\sum_{i=0}^{N} d_i^2} \qquad (7.31)$$

Thus by taking the ratio of the two images and multiplying by a calibration constant that can be determined by referencing a PDPA measurement within a point in the image plane, a map of SMD in the image plane can be generated with high spatial resolution.

Since the LSD technique can use a pulsed laser with a higher power than the continuous wave laser required for a PDPA, signal attenuation by secondary scattering in the spray shall be smaller. The LSD technique measures a ratio of two intensities, which are affected in a similar manner by sources of interference, namely, nonuniformity of the laser sheet profile and secondary scattering by the spray between the object plane and the camera. These interfering effects are expected to some degree to cancel out when the ratios of the Mie and LIF images are taken simultaneously.

However, the dependence of Mie scattering and fluorescence on droplet diameter holds only for transparent and evenly illuminated particles. Because fluorescence is based on light absorption processes, the index n of dependence of the LIF signal on droplet diameter deviates from an ideal value of 3. In addition, the contributions of multiple scattering, optical amplification in the droplet, and droplets at the edge of the light sheet must be considered for quantitative analysis (Greenhalgh & Jermy, 2002). Because of the strong dependence of LIF and Mie scattering on the particle size, their intensities are highly biased toward large droplets. When a range of droplets is present, the intensified CCD camera does not have a large enough dynamic range to account for both the small and large droplets in one measurement. It will either result in the saturation of the camera pixels by the strong signal from the large droplets or the loss of the weak signal from the small droplets as a result of the inability to discriminate it from the background noise. Therefore, the LSD technique is more suited to visualizing the region of uniform droplet sizes. Finally, it is desirable to acquire the instantaneous PLIF and Mie scattering image pairs that are spatially matched either by two cameras or image-doubling optics. Alternatively, the ensemble-averaged Mie and LIF images can be recorded and used to obtain the ensemble-averaged SMD (Jin et al., 2008)

7.6 Simultaneous Visualization of Fuel Vapor and Liquid Fuel by the Laser-Induced Exciplex Fluorescence Technique

7.6.1 Principle of Laser-Induced Exciplex Fluorescence Technique

When the natural fuel fluorescence or a single fluorescence dopant is used to visualize the liquid fuel and the fuel vapor at the same time, it is not possible to separate spectrally the fluorescence from the liquid phase or vapor. This is because the absorption and fluorescence spectra of organic molecules dissolved in nonpolar solvents, such as typical fuels, are virtually identical to the spectra of the same molecules in the vapor phase. Since the liquid fluorescence signal is many orders of magnitude greater than the vapor fluorescence signal, it is difficult to detect fuel vapor in the presence of liquid fuel given the limited dynamic range of photodetectors. To overcome this difficulty, laser-induced exciplex fluorescence (LIEF) was developed by Melton and Verdiect (1984, 1985). In this method, a fluorescent molecule reacts in an excited state, M*, with another molecule, G, to form a second fluorescent species (M–G)*, that is, M*+G $\Leftrightarrow$ (M–G)*.

The newly formed fluorescence species (M–G)* is bound in the excited state rather than in the ground state. The newly formed species (M–G)* is called excited state complex or exciplex. The emission from the exciplex is red-shifted with respect to that of M*. The reaction between M* and G to form (M–G)* is reversible, and in some cases it is possible to adjust the concentration of species

G so that the exciplex (M–G)* is the dominant emitter in the liquid phase and the monomer M* is the dominant emitter in the vapor phase. This is because, for the liquid phase, the equilibrium of the reaction can be driven to the right side (increasing exciplex) with a large concentration of G, but the exciplex is not stable in the vapor phase because of the low density and high temperature.

The peaks of the excited monomer, M*, and the exciplex, (M–G)*, emissions should be sufficiently separated for emitting species M* and (M–G)* to act as markers for the vapor and liquid phases, respectively. If the spray is optically thin and if the species M co-evaporates with the fuel, then the emission from these markers, separated by appropriate filters, will be directly related to the amount of fuel in the vapor and liquid phases. In the absence of quenching effects, the fluorescence intensity is directly proportional to the fuel concentration, allowing qualitative interpretation of results or potentially quantitative measurements with proper calibration.

Quantification of the LIEF signal is limited by several factors. The equilibrium of exciplex formation is temperature dependent, which means that higher temperatures reduce the number of exciplex events. As a result, exciplex fluorescence from a hot liquid phase is weaker. In addition, the emission from the excited monomer increases, leading to a spectral overlap of fluorescence from the liquid phase and vapor. Another problem arises because of the temperature dependence of the fluorescence yield. Furthermore, the exciplex fluorescence is severely quenched by oxygen. Quenching of the vapor phase fluorescence by oxygen can be a serious problem, particularly at high pressures. The quenching of the liquid fluorescence by oxygen is much less of a problem, because the dissolved oxygen can be removed from the fuel prior to the experiment, and any ambient oxygen will not have enough time to diffuse into the liquid droplet during the short droplet lifetime. The quenching of oxygen can be minimized by performing experiments in a nitrogen atmosphere. Quantitative measurements may also be hampered by the necessity to match the boiling temperatures of the particular fuel used and the exciplex generating dopants.

The N, N, N′, N′ tetramethyl-p-phenylene diamine (TMPD)/naphthalene system was the first exciplex system developed for diesel fuels (Melton and Verdiect, 1984). The principal features of LIEF, as discussed earlier, can be illustrated by this exciplex system. A schematic summary of the photo-physics of the TMPD/naphthalene exciplex system is shown in Fig. 7.12. Figure 7.13 shows the fluorescence spectra of the two phases in a mixture of 90% n-decane, 9% naphthalene, and 1% TMPD by mass. The vapor phase fluorescence is dominated by the excited monomer (TMPD*) emission, while the exciplex ([Np + TMPD]*) emission dominates the condensed phase emission. Although the liquid and vapor phase emissions peak at 480 and 390 nm, respectively, an overlap in spectrum between the two fluorescence signals exists over a wavelength range from 380 to 510 nm. This may be attributed to the small

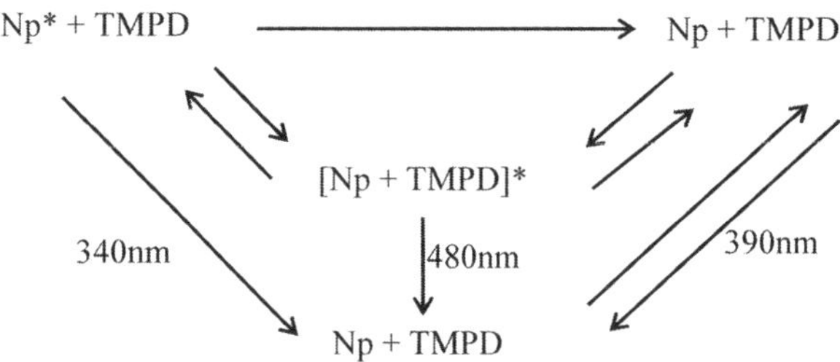

Fig. 7.12 Illustration of the naphthalene and TMPD exciplex system.

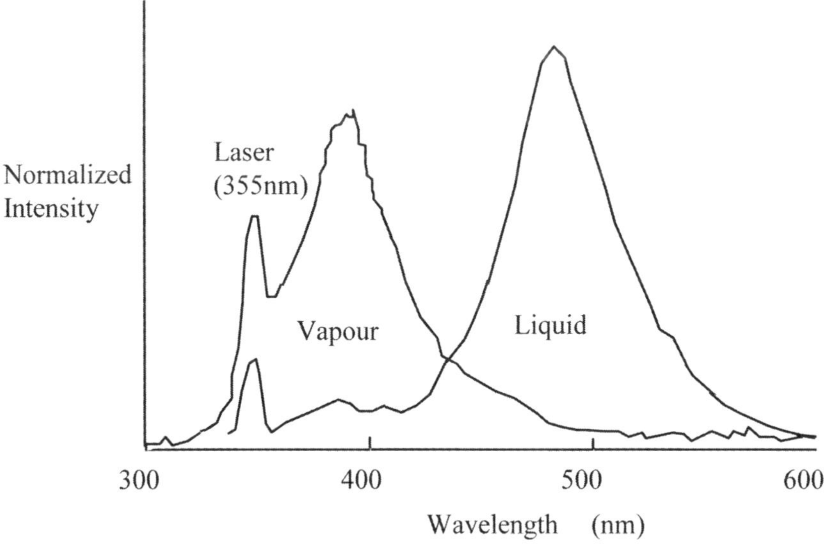

Fig. 7.13 Naphthalene and TMPD fluorescence spectra.

emission of the monomer in the liquid phase and the slight formation of the exciplex in the vapor phase. Normally, the absolute quantity of the fluorescence intensity from the vapor phase is quite small compared with the emission from the liquid phase, because the fluorescence intensity depends on the mass of the excited fluorescent matter. If a long-pass filter is employed to cut off at 520 nm, the cross-talk from the vapor phase could be eliminated. By contrast, the fluorescence signal near 390 nm is always contaminated with liquid phase fluorescence emission. In a region where both liquid droplets and fuel vapor exist in an evaporating spray, the cross-talk from the liquid phase is unavoidable. It will be difficult to obtain quantitative vapor concentration measurements in this region.

As noted previously, the reaction between the excited monomer TMPD* and the naphthalene molecule Np to form the exciplex (Np + TMPD)* is reversible. By adjusting the concentration of naphthalene, the exciplex (Np + TMPD)* can be the dominant emitter in the liquid phase, and the dissociated species TMPD* will be the dominant emitter in the vapor phase.

7.6.2 Experimental Setup of LIEF

Figure 7.14 shows the experimental setup for planar LIEF (PLIEF). To carry out PLIEF measurements, fluorescence signals from both liquid fuel and fuel vapor need be detected simultaneously by two separate cameras, or by one camera with an image doubler in front of the camera. Spectral filters are placed in front of each camera to select the spectral regions of liquid and vapor fluorescence. The selection of the spectral filters depends on the monomer and exciplex fluorescence spectra. For the fluorescence spectra shown in Fig. 7.13,

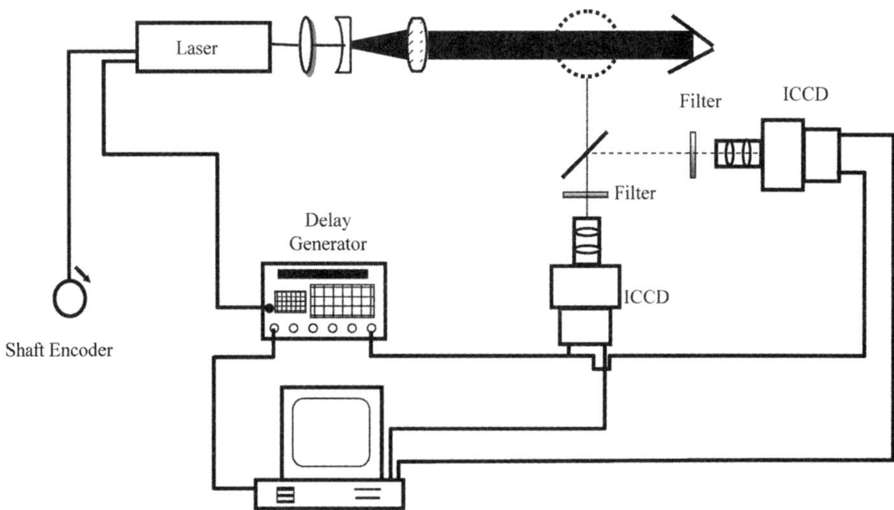

Fig. 7.14 Experimental setup for planar laser-induced exciplex fluorescence (PLIEF).

226

Table 7.2 Summary of Exciplex Dopants and PLIEF Experimental Setups Reported in the Literature

Laser Source	Exciplex Dopants (monomer/exciplex forming dopant)	Detection System Spectral Filters	Application and Reference
Nd:YAG 266 nm, 80 mJ	TMPD/Naphthalene (1.0%, 2.5% mass in cetane)	Camera with colored film Vapor: UG11; Liquid: GG-475	Diesel spray in a confined volume Melton & Verdiect (1984, 1985).
Nd:YAG 355 nm,	TMPD/naphthalene (1.0%, 9.0% mass in decane)	Intensified solid-state camera Vapor: 400 $\pm$12 nm; Liquid: > 450 nm	Diesel spray in a motored two-stroke engine with N_2 Bardsley et al. (1988); Campbell et al. (1995).
Nd:YAG 355 nm, 60 mJ	TMPD/naphthalene (0.5%, 5.0% mass in n-tridecane)	Intensified CCD camera Vapor: 350–390 nm; Liquid: >460 nm	Diesel spray in an RCM filled with N_2 Yeh et al. (1993).
Nd:YAG 355 nm, 100 mJ	TMPD/naphthalene (1.0%, 10% mass in heptane)	Intensified CCD camera Vapor: 400 $\pm$12 nm; Liquid: >435 nm	Diesel spray in a motored four-stroke engine with N_2 Hodges et al. (1991).
Nd:YAG 355 nm, 60 mJ	TMPD/naphthalene (1.0%, 10% mass in n-tridecane)	Intensified CCD camera Vapor: 410 $\pm$10 nm; Liquid: 532$\pm$1 nm	Diesel spray impingement in a constant volume filled with N_2 Senda et al. (1992, 1994, 1997).
Nd:YAG 355 nm, 200 mJ	TMPD/naphthalene (1.0%, 9.0% mass in *n*-octane)	Intensified CCD camera Vapor: 400 $\pm$8 nm; Liquid: 480$\pm$10 nm	Mixture injection SI engine filled with N_2 Fujimoto & Tabata (1993); Tabata et al. (1995).
Nd:YAG, 355 nm, 40 mJ	TMPD/naphthalene (1.0%, 10% mass in decane)	Intensified vidicon camera Vapor: 400 $\pm$ nm; Liquid: 500$\pm$ nm	Fuel spray from an air-assist injector inside a motored two-stroke engine with N_2 Diwakar et al. (1992).
XeCl 308 nm	TMPD/naphthalene (in iso-octane)	Intensified CCD camera Vapor: 2xWG355, UG11 Liquid: WG360, GG455, GG475	Liquid/vapor fuel distribution in an SI engine running with iso-octane Knapp et al. (1996).
Nd:YAG 266 nm	Fluorobenzene/DEMA (0.5%, 9.0% vol. in hexane)	Intensified CID camera Vapor: 295 $\pm$ 11 nm; Liquid: 400$\pm$12 nm	Fuel spray from an air-assist injector inside a motored two-stroke engine with N_2 Ghandhi et al. (1994).
XeCl 308 nm	DMA/naphthalene (5.0%, 5.0% mass in gasoline)	Intensified CCD camera Vapor: 340 $\pm$ 10 nm; Liquid: >380 nm	Quantitative fuel vapor concentration in an SI engine Shimizu et al. (1992).
XeCl 308 nm	DMA/1,4,6-TMN (5%, 7% mass in iso-octane)	Vapor: λ_{peak}=355 nm Liquid: λ_{peak}=405 nm	New exciplex seeding compound Kim et al. (1996).
KrF 248 nm, –	TEA/benzene (2.0%, 2.9% in iso-octane)	Intensified CCD camera Vapor: 307 $\pm$ 12 nm; Liquid: > 325 nm	Fuel distribution in the intake port of an SI engine Münch et al. (1996).

for example, a broad band-pass filter (410 ± 9.5 nm) and a narrow band-pass filter (532 ± 1 nm) have been used to separate the vapor and liquid fluorescence signals (Senda et al., 1992, 1994, 1997). However, because of the overlap in the liquid and vapor emission spectra (Fig. 7.16), some cross-talk will still be present in the exciplex images of liquid fuel and fuel vapor. When a single camera is used, a half-width neutral density filter is also required to attenuate the stronger fluorescence signal from the liquid fuel. The experimental setups and dopants employed for LIEF measurements are summarized in Table 7.2.

7.6.3 Quantitative Analysis of Fuel Vapor Concentration Using PLIEF

As discussed in the previous section, quantification of the LIEF signal is limited by several factors. The fluorescence signals from both exciplex (liquid phase) and excited monomer (vapor phase) are temperature dependent. The fluorescence can be quenched severely by oxygen, which is particularly serious for the vapor phase, which can be minimized by performing experiments in a nitrogen atmosphere. Third, the exciplex fluorescence is affected by the vapor concentration, that is, self-quenched. When the LIEF is applied to a dense spray, the attenuation of laser beam also needs to be taken into account.

In the absence of procedures that relate the measured fluorescence intensities to vapor and liquid concentrations, quantitative interpretation of the images is not possible. Melton (1988) developed procedures based on the calculation of light absorption, quantum yields for emission, refraction within droplets, and the like. The procedures and the photo-physical parameters required are summarized in a "User's Manual" (Melton, 1988). However, the photo-physical function/parameter methods described in the User's Manual are difficult to perform and may not be very accurate. Rotunno et al. (1990) developed a direct calibration method for both the liquid and vapor phases. A known amount of liquid or vapor, located at the plane to be imaged, is irradiated, and the fluorescence intensity is measured. If all excitation and detection parameters are kept constant, the ratio of the measured intensities to the known liquid or vapor concentration can be used as a calibration factor for subsequent spray experiments.

Since an evaporating diesel spray produces a highly dense vapor cloud, the absorption of laser beam by the fuel vapor needs to be considered, as well as the quenching effects by temperature and concentration. Senda et al. (1997) developed a calibration and correction procedure for quantitative measurements of fuel vapor concentration in an evaporating diesel spray using LIEF.

7.6.4 Application of PLIEF for In-Cylinder Fuel Visualization

Given that the boiling temperatures of TMPD (260°C) and naphthalene (218°C) correspond to the middle boiling points of diesel fuels, they are particularly suited to the study of diesel sprays. Initial PLIEF work using TMPD/naphthalene as the exciplex compound was realized using an Nd:YAG laser at its fourth harmonic output (266 nm) as the excitation source (Melton and Verdiect, 1984, 1985). Later on, the third harmonic at 355 nm was adopted (Bardsley et al., 1988; Choi et al., 2001), as a result of much reduced absorption by both TMPD and naphthalene at 355 nm. The fuel used was a mixture of 90% decane, 9% naphthalene, and 1% TMPD by mass. The vapor phase fluorescence was detected at 400 nm, while a long-pass filter (> 450 nm) was employed to isolate the fluorescence signal from the liquid phase. Other details can be found in Table 7.2. Quenching of the vapor phase fluorescence by oxygen was found very strong at high pressures in the engine. Consequently, the experiments were carried out in a motored two-stroke engine with N_2 rather than air.

Yeh et al. (1993) investigated the suitability of TMPD/naphthalene dopants for LIEF in a high temperature and high-pressure environment and found no thermal decomposition during the 10-ms operation period of the rapid compression-expansion machine.

It should be noted that all the above studies were conducted in a nitrogen atmosphere to avoid the vapor phase fluorescence quenching by oxygen. The same exciplex system was used to study in-cylinder fuel vapor distribution of motored SI engines with port fuel injection and mixture injection by Fujimoto and Tabata (1993), Tabata et al. (1995), and Diwakar et al. (1992).

The above studies involved the use of PLIEF in a motored SI engine with N_2. To study the effect of fuel injection timing on mixture distributions and combustion, Knapp et al. (1996) applied PLIEF to a firing SI engine. The quenching of vapor fluorescence by oxygen was partially compensated by the use of a high-power XeCl excimer laser with a shorter excitation wavelength.

Given that the boiling point of TMPD, which dominates the vapor phase fluorescence, is 260°C, its ability to trace the evaporation of gasoline fuel is questionable. To overcome the problem, search for LIEF dopants with matching properties to gasoline fuels has been carried out by several research groups.

Initial work on gasoline-based exciplex compounds focused on mixtures of fluorobenzene, triethylamine (TEA), and iso-octane, following the initial work by Melton (1993). But fluorobenzene was found to evaporate much faster than iso-octane, although the boiling point of fluorobenzene (85°C) is close to that of iso-octane (98°C) (Law, 1989).

Ghandhi et al. (1994) found that in a mixture of fluorobenzene/diethyl-methyl-amine (DEMA) and hexane, co-evaporation was more closely achieved between fluorobenzene and hexane. Since the boiling temperature of hexane (68°C) is low, the results obtained using this mixture will be representative of only the lighter components in gasoline fuels. The LIEF of fluorobenzene/DEMA was excited by an Nd:YAG laser at its fourth harmonic output (266 nm). The emissions of the monomer (fluorobenzene) at 290 nm and exciplex at 380 nm were sufficiently shifted to allow discrimination with interference filters. Ghandhi et al. found that the vapor phase fluorescence from fluorobenzene was severely affected by temperature quenching, which limited their results to the qualitative interpretation of the images. Further study by Lee and Gajdeczko (1996) showed that the cross-talk from Mie scattering onto the vapor phase signal could also be a problem in dense regions of the spray. This was later resolved by adding a WG 305 Schott® glass filter and increasing the concentration of fluorobenzene to 2%. Because the fluorobenzene fluorescence was quenched by oxygen, the studies were limited to motored engines with nitrogen. In a more recent study, Zhang et al. (2011) employed PLIEF with fluorobenzene and DEMA in n-hexane to study flash boiling spray behaviors from a multihole injector.

A mixture of 5% N,N-dimethylaniline (DMA), 5% naphthalene, and 90% gasoline by mass was employed to observe both liquid and vapor phase concentrations of fuel (Shimizu et al., 1992; Leach et al., 2007). DMA was found to account for 85% of the fluorescence at λ_{max} = 355 nm in the vapor phase, when excited at 308 nm by a XeCl excimer laser. The fluorescence of the liquid phase from the exciplex formed by DMA and naphthalene was red-shifted to λ_{max} = 410 nm. As DMA was used to trace the vapor phase of the fuel, the true vapor concentration of fuel may be underestimated in the early stages of evaporation, because its high boiling point (193°C) corresponds to the upper end of the gasoline distillation curve. The vapor phase fluorescence signal was calibrated under motoring conditions for different equivalence ratios at different crank angles. Shimizu et al. (1992) claimed an accuracy of 3% in equivalence ratios from vapor phase fluorescence signals. However, the pulse-to-pulse variations in the laser output were not considered, which would have been an additional source of errors. Only qualitative imaging of liquid fuel in the cylinder was attempted. It is worth mentioning that all the measurements were carried out under firing conditions. Two laser sheets were introduced into the combustion chamber from opposite directions to minimize the effect of light attenuation, which was caused by liquid droplets and affected the quantitative vapor concentration measurement during the early stage of the intake stroke.

Kim et al. (1996) reported a new exciplex mixture suitable for tracking the vaporization of gasoline fuel. The exciplex mixture consists of 5% DMA and 7% 1,4,6-trimethylnaphthalene (TMN) in 88% iso-octane by mass. The excited DMA dominates the vapor phase fluorescence. Because DMA has a boiling temperature of 193°C, it can simulate the heavy components in gasoline. When excited at 308 nm, the vapor phase fluorescence can be detected at a peak of 350 nm, while the liquid phase fluorescence peaks at 405 nm. The exciplex formation was found stable at elevated temperatures (145°C), as was the spectral separation between monomer and exciplex emissions.

Münch et al. (1996) developed another new exciplex mixture, TEA and benzene. The boiling points of benzene and TEA are 81°C and 89°C, respectively, which

230

match closely to that of iso-octane (99°C). When a mixture of 2.9% benzene and 2% TEA in iso-octane was excited by a KrF excimer laser at 248 nm, a red-shifted exciplex fluorescence was observed in the liquid phase with a maximum around 350 nm, while the vapor phase fluorescence peaked at 290 nm.

In addition to the diesel spray characterization (Yamashita et al., 2007), the recent examples of LIEF applications are mainly associated with the in-cylinder measurement of fuel vapor and liquid distributions in the direct-injection spark ignition engines (Gervais and Gastaldi, 2002; Honda et al., 2004; Ipp et al., 2009; Raimann et al., 2003; Skogsberg et al., 2005; Wieske et al., 2006).

7.6.5 Application of LIEF to In-Cylinder Diesel Spray Measurement

To demonstrate the capability of the LIEF technique for full field imaging of the diesel fuel-injection process, a LIEF system was commissioned and applied to a single-cylinder engine with the Bowditch extension (Hertfatmanesh, 2010), as shown in Fig. 7.15.

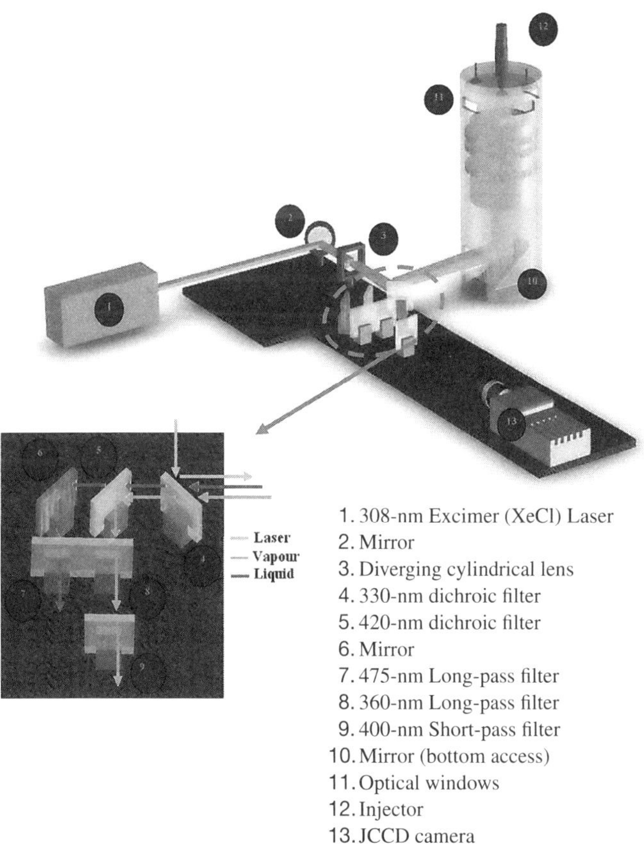

1. 308-nm Excimer (XeCl) Laser
2. Mirror
3. Diverging cylindrical lens
4. 330-nm dichroic filter
5. 420-nm dichroic filter
6. Mirror
7. 475-nm Long-pass filter
8. 360-nm Long-pass filter
9. 400-nm Short-pass filter
10. Mirror (bottom access)
11. Optical windows
12. Injector
13. JCCD camera

Fig. 7.15 Experimental setup of the PLIEF system at Brunel University.

The engine was equipped with a high-pressure common rail fuel injector, and a pneumatic high-pressure pump was used to supply high-pressure fuel to the common rail. The fuel used comprised 88% decane, 10% α-methyl-naphthalene, and 2% TMPD. To minimize the quenching effect by oxygen, the engine was operated with nitrogen supplied from a pressurized nitrogen tank. A pneumatic actuator is installed on the intake manifold which controls the opening and closing of the surge tank.

The output from a XeCl excimer laser was expanded using a cylinder lens with a focal length of 150 mm and subsequently reflected off a 330-nm dichroic filter (DCF). The expanded laser beam illuminated the central part of the combustion chamber through a 45° mirror in the extended cylinder block. The LIEF emissions from liquid and vapor passed the 330 nm dichroic filter and then separated into two paths by the 420-nm dichroic filter. The fluorescence light from the vaporized fuel was reflected off the 420-nm dichroic filter and then transmitted through a 360-nm long-pass filter (LPF), which helped to eliminate any remaining scattered laser and ambient light. An additional 400-nm short-pass filter (SPF) was also placed in front of the camera to minimize the crosstalk with the liquid phase fluorescence, as shown in Fig. 7.16. Meanwhile, after passing through the 420-nm dichroic filter, the liquid signal was transmitted through a 475-nm LPF.

To perform LIEF measurement, it is critical to synchronize the laser, ICCD camera, fuel injection, and nitrogen-filling events. As the camera has an update time of approximately 5 seconds between frames (0.2 Hz), it is necessary to run the engine in the skip-fire mode to prevent window fouling. A trigger box connected to a shaft encoder was set to produce two output signals at a preset

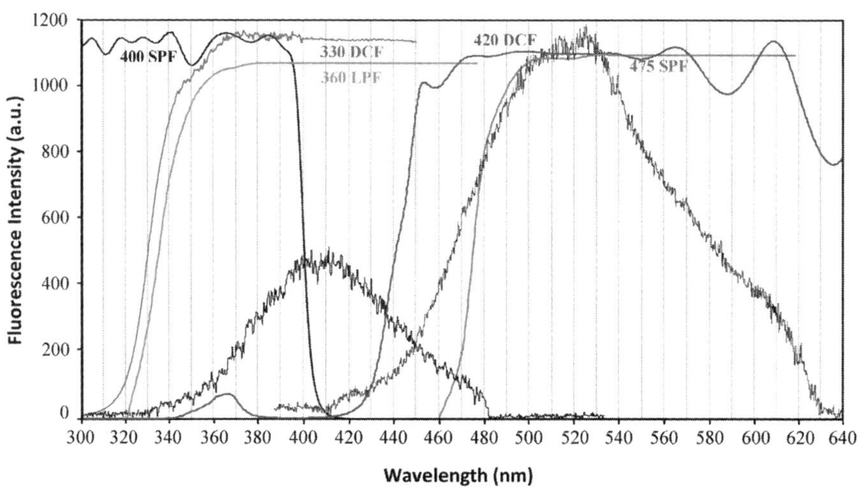

Fig. 7.16 Spectral transmission response of optical filters.

crank angle for every 63 cycles, one to the injector controller (injection signal) and the other to the pneumatic actuator for the nitrogen supply.

Since the injector had a delay of 510 µs and the laser had a delay of 950 ns, the laser was triggered 509 µs after the injection signal, while the camera was triggered 950 ns after the laser by means of a delay generator. The gate width of the ICCD camera was set to 200 ns to minimize the background light.

The other output from the trigger box was set so that the pneumatic actuator was triggered five cycles prior to the injection signal output. Therefore, the fuel injection took place during the sixth motoring cycle with nitrogen at 1.5-bar intake pressure.

During the engine experiments, the reflection from the cylinder head was minimized by thoroughly cleaning the soot deposits from the surface of the cylinder head during each engine rebuild and painting the surface with a black matte paint. Nonfluorescent synthetic oil was used to suppress the fluorescence generated from splashing oil drops. However, the major difficulty encountered was the premature failure of fuel injectors because of the poor lubricity of fuel and dopant mixture as well as the presence of any undissolved TMPD powders.

Figure 7.17 shows the LIEF images of fuel jets at 1200-bar injection pressure. The injection timing was 25° CA BTDC, and the injection quantity was 10 mm³. The spatial distributions of both the vapor and liquid phases can be clearly

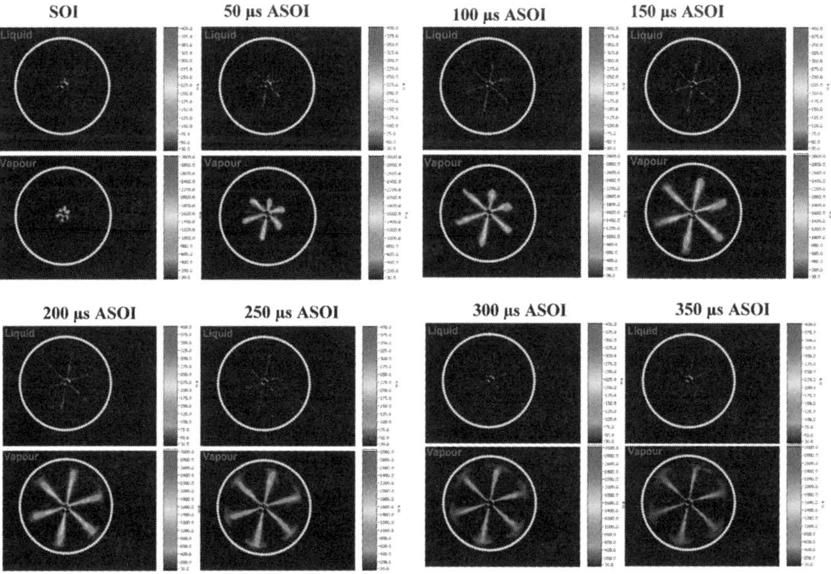

Fig. 7.17 In-cylinder diesel fuel spray and vapor distribution measured by PLIEF.

identified. As expected, the liquid core was thinner and situated in the center of each vapor jet. The fuel impingement occurred with fuel vapor as evidenced by the mushroom-shaped vapor fuel distribution toward the end of the injection. In addition, there was residual liquid detected at the nozzle tip.

7.7 Summary

Spray structure and liquid drops can be directly observed with the aid of a back-illumination source. The sensitivity of the recorded image to the vaporized fuel can be improved if a Schlieren or shadowgraph setup is used.

The laser sheet techniques are particularly suited for spatially resolved measurements of spray and liquid fuel distributions. Images of liquid fuel distribution can be obtained with either Mie scattering or LIF. In addition, liquid and vapor fuel distributions can be identified if both Mie and LIF images of fuel spray can be recorded simultaneously, by subtracting the Mie scattering images (liquid) from the two LIF images (liquid + vapor). Furthermore, the ratio of LIF and Mie intensities can be used to derive spatially averaged droplet SMD distributions in a relatively dense spray region.

Simultaneous measurements of fuel vapor and liquid fuel can be performed by LIEF techniques. LIEF shares many of the features associated with LIF, but the fluorescence from liquid fuel and fuel vapor are separated spectrally by means of an exciplex forming compound. The effect of quenching is more severe in LIEF because of the limited choice of exciplex forming compounds. The quenching of fluorescence from fuel vapor has limited most LIEF applications to nitrogen environment.

There are two approaches to quantitative droplet-sizing measurements: integral methods and particle counting. The integral methods, typically performed by the Fraunhofer diffraction, measure a large number of droplets simultaneously. The resulting intensity comes from integration over space, and a spatial average is obtained. This implies moderate to high droplet concentrations and a significant sample volume. On the other hand, with particle-counting techniques, such as a laser PDPA, droplets are viewed one at a time. This implies low to moderate droplet concentrations and a small sample volume. The higher the concentration, the smaller the volume needs to be. The size distribution by PDPA measurements is built up over time, and a temporal average is obtained. The temporal distribution also provides a measure of mass flux of droplets of different sizes. The spatial and temporal size distributions can be related to each other, if the droplet size–velocity relationship is known.

References

Bachalo, W. D. (1980). "Method for Measuring the Size and Velocity of Spheres by Dual Beam Light Scattering Interferometry." *Applied Optics*, Vol. 19, Part 3, p. 363.

Bachalo, W. D., Rosa, A., and Sankar, S. V. (1991). "Diagnostics for Fuel Spray Characterization." *Combustion Measurement,* edited by N. Chigier, pp. 229–277. Washington, DC: Hemisphere.

Bardsley, M., Felton, P., and Bracco, F. (1988). "2-D Visualization of a Liquid and Vapor Fuel in an I.C. Engine." SAE Paper No. 880521, SAE International, Warrendale, PA.

Baritaud, T. A., Heinze, T. A., and Le Cos J. F. (1994). "Spray and Self-Ignition Visualization in a DI Diesel Engine." SAE Paper No. 940681, SAE International, Warrendale, PA.

Campbell, P., Sinko, K., and Chehroudi, B. (1995). "Liquid and Vapor Phase Distributions in a Piloted Diesel Fuel Spray." SAE Paper No. 950445, SAE International, Warrendale, PA.

Canaan, R. E., Dec, J. E., Robert, M. G., and Daly, D. T. (1998). "The influence of Fuel Volatility on the Liquid-Phase Fuel Penetration in a Heavy-Duty DI Diesel Engine." SAE Paper No. 980510, SAE International, Warrendale, PA

Choi, D., Iwamuro, M., Shima, Y., Senda, J., et al. (2001). "The Effect of Fuel-Vapor Concentration on the Process of Initial Combustion and Soot Formation in a DI Diesel Engine Using LII and LIEF." SAE Paper No. 2001–01–1255, SAE International, Warrendale, PA.

Dawson, M., and Hochgreb, S. (1998). "Liquid Fuel Visualization Using Laser Induced Fluorescence During Cold Start." SAE Paper No. 982466, SAE International, Warrendale, PA.

Desantes, J., Arregle, J., Pastor, J., and Delage, A. (1998). "Influence of the Fuel Characteristics on the Injection Process in a DI Diesel Engine." SAE Paper No. 980802, SAE International, Warrendale, PA.

Diwakar, R., Fansler, T., French, D., Ghandhi, J., Dasch, C., and Hefflinger, D. (1992). "Liquid and Vapor Fuel Distributions from an Air-Assist Injector— An Experimental and Computational Study." SAE Paper No. 920422, SAE International, Warrendale, PA.

Espey, C., and Dec, J. (1995). "The Effect of TDC Temperature and Density on the Liquid-phase Fuel Penetration in a DI Diesel Engine." SAE Paper No. 952456, SAE International, Warrendale, PA.

Felton, P. G., Mantzaras, J., Bardsley, M. E. A., and Bracco, F. V. (1987). "2-D Visualization of Liquid Fuel Injection in an Internal Combustion Engine." SAE Paper No. 872074, SAE International, Warrendale, PA.

Fujimoto, M., and Tabata, M. (1993). "Effect of Swirl Rate on Mixture Formation in a Spark Ignition Engine Based on Laser 2-D Visualization Technique." SAE Paper No. 931905, SAE International, Warrendale, PA.

Gervais, D., and Gastaldi, P. (2002). "A Comparison of Two Quantitative Laser Induced Fluorescence Techniques Applied to a New Air Guided Direct Injection SI Combustion Chamber." SAE Technical Paper No. 2002–01–0750, SAE International, Warrendale, PA.

Ghandhi, J. B., Felton, P. G., Gajdeczko, B. F. and Bracco, F. V. (1994). "Investigation of the Fuel Distribution in a Two Stroke Engine with an Air-Assisted Injector." SAE Paper No. 940394, SAE International, Warrendale, PA.

Greenhalgh, D. A., and Jermy, M. (2002). "Laser Diagnostics for Droplet Measurements for the Study of Fuel Injection and Mixing in Gas Turbines and IC Engines." In *Applied Combustion Diagnostics*, edited by Kohse-Höinghaus, K., and Jeffries, J. B., pp. 408–430. London: Taylor and Francis.

Hertfatmanesh, M. (2010). *Investigation of Single and Split Injection Strategies in a Single Cylinder Optical Diesel Engine.* PhD Dissertation. Brunel University, London.

Hodges, J., Baritaud, T., and Heinze, T. (1991). "Planar Liquid and Gas Fuel and Droplet Size Visualization in a DI Diesel Engine." SAE Paper No. 910726, SAE International, Warrendale, PA.

Honda, T., Kawamoto, M., Katashiba, H., Sumida, M., et al. (2004). "A Study of Mixture Formation and Combustion for Spray Guided DISI," SAE Paper No. 2004–01–0046, SAE International, Warrendale, PA.

Ipp, W., Wagner, V., Krämer, H., Wensing, M., et al. (2009). "Spray Formation of High Pressure Swirl Gasoline Injectors Investigated by Two-Dimensional Mie and LIEF Techniques." SAE Technical Paper No. 1999–01–0498, SAE International, Warrendale, PA.

Jin, S., Brear, M., Watson, H., and Brewster, S. (2008). "An experimental study of the spray from an air-assisted direct fuel injector." *Proc. IMechE Part D*, Vol. 222, No. 10, pp. 883–1894.

Kelly-Zion, P. L., DeYoung, C. A., Peters, J. E., and White, R. A. (1995). "In-cylinder fuel drop size and wall impingement measurements." SAE Paper No. 952480, SAE International, Warrendale, PA.

Kim, J., Golding, B., Schock, J., Keller, P., and Nocera, D. (1996). "Exciplex Fluorescence Visualization Systems for Pre-Combustion diagnosis of an

Automotive Gasoline Engine." SAE Paper No. 960826, SAE International, Warrendale, PA.

Knapp, M., Luczak, A., Beushausen, V., Hentschel, W., and Andresen, P. (1996). "Vapor/Liquid Visualization with Laser Induced Exciplex Fluorescence in an SI Engine for Different Injection Timings." SAE Paper No. 961122, SAE International, Warrendale, PA.

Law, C. K. (1989). "Some Recent Advances in Droplet Combustion." Proc. of the Third International colloquium on Drops and Bubbles. *AIP Conference Proceedings,* Vol. 197, pp. 321–337.

Leach, B., Zhao, H., and Li, Y. (2007). "Two-Phase Fuel Distribution Measurements in a GDI Engine With an Air Assisted Injector Using Advanced Optical Diagnostics." *Proc. Instn. Mech. Engrs Part D: Journal of Automotive Engineering*, Vol. 221, No. 6, pp. 663–673.

Lee, C. F., and Gajdeczko, B. F. (1996). "Quantitative Exciplex Measurements of the Vapor Concentration in Air-Assisted Sprays." Extended abstract in Proceedings of Ninth Annual Conference on Liquid Atomization and Spray Systems. ILASS-Americas 96, pp. 297–301.

Melton, L. A. (1988). "Quantitative Use of Exciplex Based Apour/Liquid Visualization Systems (A User's Manual)." Final Report of Army Research Office Contract DAAL 03–86-k-0082.

Melton, L. A. (1993). "Exciplex-Based Vapor/Liquid Visualization Systems Appropriate for Automotive Gasolines (A Users Manual)." Final Report to Texas Higher Education Coordinating Board. Energy Research Applications Program, Contract 27.

Melton, L., and Verdiect, J. (1984). "Vapor/Liquid Visualization in Fuel Sprays." *Proc. of 20th Symposium (Int.) on Combustion*, pp. 1283–1290.

Melton, L., and Verdiect, J. (1985). "Vapor/Liquid Visualization for Fuel Sprays." *Combust. Sci. and Tech.* Vol. 42, pp. 217–222.

Meyer, R., and Heywood, J. (1997). "Liquid Fuel Transport mechanisms into the Cylinder of a Firing Port Injected SI Engine During Start Up." SAE Paper No. 970865, SAE International, Warrendale, PA.

Münch, K., Kramer, H., and Leipertz, A. (1996). "Investigation of Fuel Evaporating Inside the Intake of a SI Engine Using Laser Induced Exciplex Fluorescence with a New Seed." SAE Paper No. 961930, SAE International, Warrendale, PA.

Nouri, J. M., Hamid, M. A., Yan, Y., and Arcoumanis, C. (2006). "Spray Characterisation of Piezo Pintle-Type Injector for Gasoline Direct Injection Engines." *J. Phys. Conf. Ser. 85 012037*.

Posylkin, M., Taylor, A., Vannobel, F., and Whitlaw, J. H. (1994). "Fuel Droplets Inside a Firing Spark Ignition Engine." SAE Paper No. 941989, SAE International, Warrendale, PA.

Raimann, J., Arndt, S., Grzeszik, R., Ruthenberg, I., et al. (2003). "Optical Investigations in Stratified Gasoline Combustion Systems with Central Injector Position Leading to Optimized Spark Locations for Different Injector Designs." SAE Paper No. 2003–01–3152, SAE International, Warrendale, PA.

Rotunno, A. A., Winter, M., Dobbs, G. M., and Melton, L. A. (1990). "Liquid Visualization of Fuel Sprays." *Combustion Science and Technology* Vol. 71, pp. 247–261.

Shimizu, R., Matumoto, S., Furuno, S., Murayama, M., and Kojima, S. (1992). "Measurement of Air Fuel Mixture distribution in a Gasoline Engine Using LIEF Technique." SAE Paper No. 922356, SAE International, Warrendale, PA.

Senda, J., Fukami, Y., Tanabe, Y., and Fujimoto, H. (1992). "Visualization of Evaporative Diesel Spray Impinging Upon Wall Surface by Exciplex Fluorescence Method." SAE Paper No. 920578, SAE International, Warrendale, PA.

Senda, J., Kobayashi, M., Tanabe, Y., and Fujimoto, H. (1994). "Visualization and Quantitative Analysis of the Fuel Vapor Concentration in Diesel Spray." *JSAE Review,* Vol. 15, pp. 149–156.

Senda, J., Kanda, T., Kobayashi, M., and Fujimoto, H. (1997). "Quantitative Analysis of Fuel Vapor Concentration in Diesel Spray by Exciplex Fluorescence Method." SAE Paper No. 970796, SAE International, Warrendale, PA.

Skogsberg, M., Dahlander, P., Lindgren, R., and Denbratt, I. (2005). "Effects of Injector Parameters on Mixture Formation for Multi-Hole Nozzles in a Spray-Guided Gasoline DI Engine," SAE Technical Paper No. 2005–01–0097, SAE International, Warrendale, PA.

Swithenbank, J., Beer, J. M., Taylor, D. S., Abbott, D., and McGreath, G. C. (1977). "A laser diagnostic technique for the measurement of droplet and particle size distribution." *Prog. Astronaut. Aeronaut.* Vol. 53, pp. 421–427.

Swithenbank, J., Cao, J., and Hamidi, A. (1991). "Spray Diagnostics by Laser Diffraction." *Combustion Measurement*, edited by N. Chigier, pp. 179–227. Washington, DC: Hemisphere.

Tabata, M., Kataoka, M., Fujimoto, M., and Noh, Y. (1995). "In-Cylinder Fuel Distribution, Flowfield, and Combustion Characteristics of a Mixture Injected SI Engine." SAE Paper No. 950104, SAE International, Warrendale, PA.

Wensing, M., Münch, K., and Leipertz, A. (1997). "Characteristics and Application of Gasoline Injectors to SI Engines by Means of Measured Liquid Fuel Distributions." SAE Paper No. 972947, SAE International, Warrendale, PA.

Wigley, G., Goodwin, M., Pitcher, G., and Blondel, D. (2004). "Imaging and PDPA Analysis of a GDI Spray in the Near-Nozzle Region." Experiments in Fluids. Vol. 36, No. 4, pp. 565–574.

Wieske, P., Wissel, S., Grünefeld, G., Graf, M., et al. (2006). "Experimental Investigation of the Origin of Cyclic Fluctuations in a DISI Engine by Means of Advanced Laser Induced Exciplex Fluorescence Measurements," SAE Technical Paper No. 2006–01–3378, SAE International, Warrendale, PA.

Yamashita, H., Suzuki, T., Matsuoka, H., Mashid, M., et al. (2007). "Research of the DI Diesel Spray Characteristics at High Temperature and High Pressure Ambient." SAE Paper No. 2007–01–0665, SAE International, Warrendale, PA.

Yeh, C., Kamimoto, T., Kobori, S., and Kosaka, H. (1993). "2-D Imaging of Fuel Vapor Concentration in a Diesel Spray via Exciplex-Based Fluorescence Technique." SAE Paper No. 932652, SAE International, Warrendale, PA.

Young, B. W., and Bachalo, W. D. (1988). "The Direct Comparison of Three in-Flight Droplet Sizing Techniques for Pesticide Spray Research." *Optical Particle Sizing: Theory and Practice*, edited by G. Gouesbet and G. Grehan, pp. 483–497. New York: Plenum Press.

Zhang, G., Xu, M., Zhang, Y., and Zeng, W. (2011). "Quantitative Measurements of Liquid and Vapor Distributions in Flash Boiling Fuel Sprays using Planar Laser Induced Exciplex Technique." SAE Paper No. 2011–01–1879, SAE International, Warrendale, PA.

Zhao, F.-Q., Yoo, J.-H., and Lai, M-C. (1995). "The Spray Characteristics of Dual-Stream Port Fuel Injectors for Applications to 4-Valve Gasoline Engines." SAE Paper No. 952487, SAE International, Warrendale, PA.

chapter 8

Combustion Spectroscopy and Visualization Techniques

8.1 Introduction

Light emission and absorption from combustion has been used since the early 1920s to obtain a better understanding of the normal and abnormal combustion processes because of its simplicity and easy interpretation. In particular, the full-field visualization of combustion has been extensively employed in the research and development of advanced combustion processes in IC (internal combustion) engines. The rapid advancements in the solid-state CCD/CMOS cameras and high-speed laser systems have made high-speed imaging a highly flexible and user-friendly technique. However, direct combustion visualization is mostly limited to research engines with large optical access. Therefore, multiple probe approaches to combustion visualization have been developed because of the increased need for the techniques that can be readily applied to production engines with minimum engine modifications. Combined with reasonable flame propagation models, these multiple probe techniques can be used effectively in the measurement of both normal and abnormal combustion processes in SI (spark-ignition) engines.

In addition to the detection of light emission, some precombustion events (e.g., autoignition sites) can be studied by shadowgraph and Schlieren techniques, because of their prowess in detecting minute changes in the derivatives of density as a result of charge and thermal inhomogeneity in the flowfield.

In this chapter, following an overview of the origin of light emission in combustion engines, the principle and application of single-point line-of-sight light emission and absorption techniques will be presented. The application of chemiluminescence imaging to autoignition and the combustion process are then given. The implementation of the shadowgraph and Schlieren techniques will be the subject of the final section.

8.2 Combustion Spectroscopy

8.2.1 Light Emission of Combustion

Combustion is often associated with its luminosity, which varies from the faint blue light of the propagating flame in an SI engine to very bright yellow-white diesel combustion. Emission of light during the combustion process can be characterized by the continuum thermal radiation of solid soot particles and/or the banded chemiluminescence emission from gas species. In a diesel or direct-injection gasoline engine, thermal radiation from soot particles can become dominant and be used to measure diesel combustion temperature and soot concentration by means of the two-color technique, which will be discussed in Chapter 10.

In the absence of soot particles, chemiluminescence is responsible for light emission from UV (ultraviolet) to IR (infrared) regions. It occurs as a result of a chemical reaction forming a product in an excited state. Given two species A and B, with an excited intermediate C^+, $[A] + [B] \longrightarrow [C+] \longrightarrow [C] + hv$; where hv represents the emission of light associated with each reaction or chemiluminescence. During the above reactions, the colliding molecules/atoms in their ground electronic states form a new molecule in an excited electronic state, and an electronic transition to its ground state then takes place during the brief time of the collision. The energy is disposed of as light emission hv and the new molecule will be stabilized. The decay of the excited intermediate to a lower energy (either in electronic, vibrational, or rotational level or a combination of them) is hence responsible for the emission of light during the combustion process. Chemiluminescence differs from fluorescence in that the electronic excited state is derived from the product of a chemical reaction rather than the more typical way of creating electronic excited states, namely, absorption.

Free radicals like CH, OH, C_2, CN, and NH are responsible for the strongest band systems from premixed combustion in SI engines, and their emission spectra lie in the visible or near UV. Table 8.1 shows prominent spectral peaks of certain radicals in the visible and near UV region. H_2O and CO_2 emit in the IR spectral region through their vibrational-rotational transitions. Note that the wavelengths given in Table 8.1 are peaks of each emission spectral band, which often spreads out because of different vibrational/rotational levels involved in the transition process, as well as various line-broadening mechanisms. In addition to the high-temperature combustion-generated species, light emission from intermediate species can also be used to investigate the autoignition reactions involved in the CI (compression-ignition) engines, knocking combustion in the SI engine, and in CAI (controlled autoignition)/ HCCI (homogeneous charge compression ignition) type combustion engines. One of the most important species for autoignition study of hydrocarbon fuels is formaldehyde (CH_2O), which has strong emission bands in the visible region as given in Table 8.1. A more complete description of the combustion spectroscopy is given by Gaydon (1974).

In addition, chemiluminescence of the continuous spectrum can also occur during the recombination process of molecules. Of particular

Table 8.1 Strong Emission Peaks of Major Combustion Species in IC Engines

CH (nm)	OH (nm)	CH$_2$O (nm)	CO$_2$ (μm)	C$_2$ (nm)	CHO (nm)
314	281–283	368	2.69–2.77	470–474	320, 330
387–389	302–309	384	4.25–4.3	516	330, 340
431		395		558–563	355, 360
		412–457			380, 385

importance in engine combustion are those due to $CO + O = CO_2 + hv$ and $NO + O = NO_2 + hv$.

The CO + O continuum is mainly responsible for the blue color of flames containing carbon monoxide. The NO + O continuum is responsible for the pale yellowish-green emission from flames containing oxides of nitrogen. Continuous spectra may also be caused by gas-phase processes such as dissociation of molecules, ionization of atoms, and recombination processes.

8.2.2 Light Absorption of Combustion Species

Absorption of light is one of the oldest and still one of the more useful instrumental methods. The wavelength of light that a species will absorb is characteristic of its chemical structure. Specific regions of the electromagnetic spectrum are absorbed by exciting specific types of molecular and atomic motion to higher energy levels. Infrared absorption is associated with vibrational motions of molecules. Absorption of visible and UV radiation is associated with excitation of electrons, in both atoms and molecules, to higher energy states.

Light absorption can occur at any frequency at which light emission can occur, and this allows the absorption lines to be determined from an emission spectrum or vice versa. In the combustion diagnostics, both light absorption and emission can be used to detect the presence of a species and with appropriate calibration to obtain quantitative measurement of species concentration. Absorption measurement is preferred when more sensitive and quantitative measurement is required.

8.2.3 Detection of Emission and Absorption

The intensity of a spectral line in emission is determined by the population of the upper level of the transition N_n and the Einstein coefficient for spontaneous emission A_{nm}, as shown in Eq. 8.1.

$$I_{nm} = \eta_c\, A_{nm}\, N_n\, hv_{nm}$$

(8.1)

Since the population of the jth level N_j is related to the total number density of a particular species N by the Boltzmann's formula,

$$N_j = \frac{N}{Q(T)} g_j \exp\left(-E_j/kT\right) \tag{8.2}$$

Equation 8.1 can be rewritten as

$$I_{nm} = \eta_c N A_{nm} \exp\left[-E_n/kT\right] \tag{8.3}$$

Equation 8.3 states that the emission signal is proportional to the total number density of the emitting species. Since the temperature is often unknown, the measurement of emission is mostly limited to detecting the presence of a species in engine combustion studies.

In the case of the absorption process, the loss of irradiance because of energy absorbed for in an absorbing medium of thickness l, and an incident light intensity $I_0(\lambda)$ is given by:

$$I(\lambda) = I_0(\lambda) \exp\left(-\int_0^L K_{abs}(\lambda)\, dx\right) \tag{8.4}$$

If the absorbing medium is homogeneous, so that the absorption coefficient K_{abs} is independent of the position x, the optical depth of the medium $\int_0^L K_{abs}\, dx$ becomes $K_{abs} L$, and we have

$$I(\lambda) = I_0(\lambda) \exp\left(-K_{abs}(\lambda) L\right) \tag{8.5}$$

from which the absorption coefficient can be experimentally determined by

$$K_{abs}(\lambda) = \frac{1}{l}\ln\left(\frac{I_0(\lambda)}{I(\lambda)}\right) \tag{8.6}$$

which is related to the number density of the absorbing species at the ground state N_1 by

$$\int_{\lambda 1}^{\lambda 2} K_{abs}(\lambda)\, d\lambda = \frac{h\nu_{21}}{c} N_1 B_{12} \tag{8.7}$$

The integration of the absorption coefficient is necessary because the absorption from an absorbing species at a particular spectral line is always spread over a finite frequency in practice by the various broadening mechanisms, as shown in Fig. 8.1(a). As a result, the experimentally determined absorption coefficient has a profile, as shown in Fig. 8.1(b).

Equation 8.7 forms the basis of the absorption technique, where the number density of the absorbing species is determined from the total area under the absorption coefficient profile as a function of wavelength after the absorption coefficient at a particular wavelength has been found from the ratio of the incident light and attenuated light from Eq. 8.6.

Emission or absorption of these band spectra can be easily detected and measured using spectrographic equipment. Fig. 8.2 shows schematically the necessary elements of a spectroscopic system for emission (Fig. 8.1(a)) and for absorption (Fig. 8.1(b)). In emissions, these band spectra are recorded directly by spectrographic equipment, as described in Chapter 2.

In absorption, a background source giving a bright, continuous spectrum over the relevant wavelength range is required. Early work was mostly done with tungsten filament lamps for the visible region, because the brightness temperature of the background source must exceed the flame temperature or, more precisely, the effective excitation temperature of the band system in combustion. The tungsten filament lamp is not adequate for the absorption spectra of engine combustion as they cannot be used above a brightness temperature of about 3000 K which is comparable to that of engine combustion. One of the most favored background sources is the high-pressure xenon arc lamp. The hottest part of the arc may have a brightness temperature exceeding

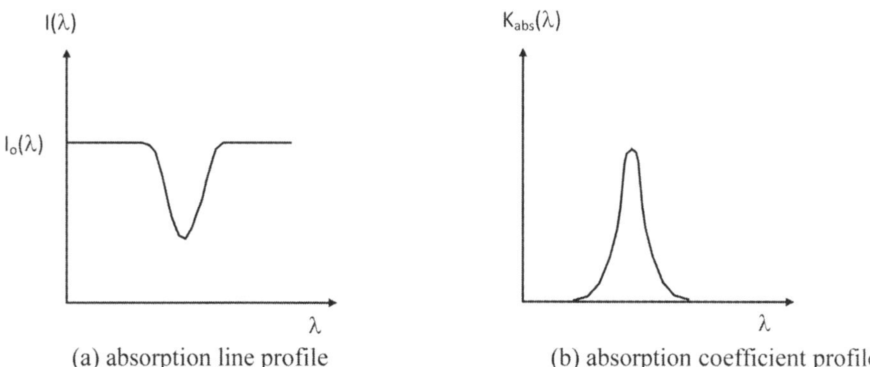

(a) absorption line profile   (b) absorption coefficient profile

Fig. 8.1 Absorption intensity and absorption coefficient profiles.

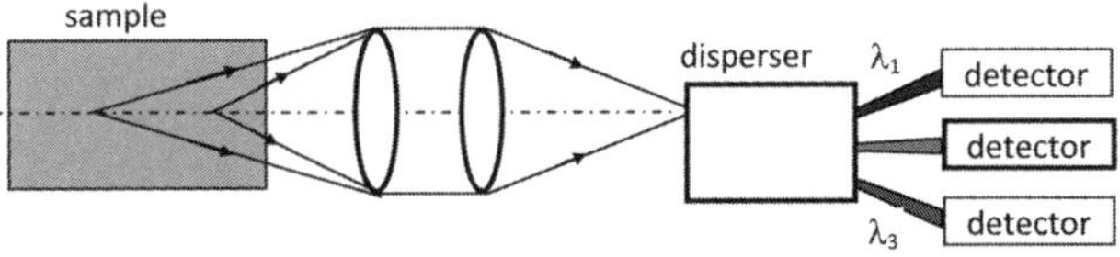

(a) measurement of emission intensity

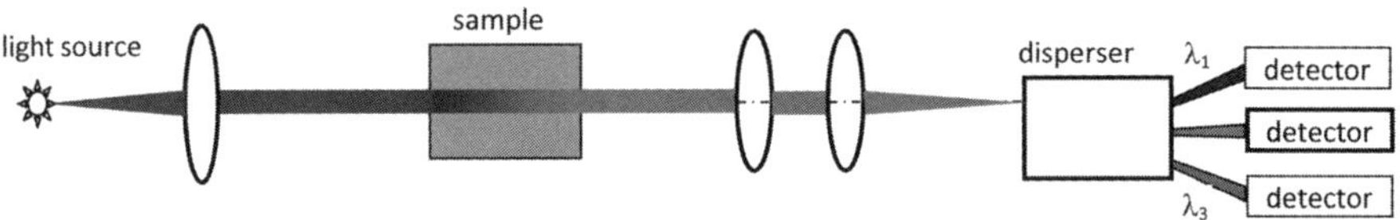

Fig. 8.2 Schematic of emission and absorption systems.

5000 K, especially at shorter wavelengths. Its emission spectrum covers both visible and near UV regions. Either fixed wavelength or tunable lasers over a wide range of wavelengths can be employed for the absorption measurement of specific species in the combustion engine. As mentioned in Chapter 2, the tunable diode laser and QCL (quantum cascade laser) are particularly suited for the detection of some combustion species, such as CO_2 and H_2O, in the IR regions.

Emission and absorption techniques provide measurements averaged over a line of sight. Both techniques have been almost exclusively used to detect the presence of certain species in the cylinder with a few attempts to quantify their concentrations. The absorption technique is preferred when more sensitive and quantitative measurement is required.

8.2.4 Application of Light Emission and Absorption Techniques

Among the earlier investigations are those of Clark and coworkers (Clark, 1928; Clark and Henne, 1927). Light emission and absorption measurements were carried out with respect to knocking combustion. During the 1930s, Withrow and Rassweiler (1931, 1933, 1934) measured the emission/absorption spectrum during both normal combustion and knocking combustion in SI engines. From their extensive measurements, they discovered that afterglow in the burned gas region was caused by the delayed oxidation reaction of CO to CO_2 and that the flame front under normal conditions comprised mainly the band spectra because of CH, C_2, and OH. For knocking combustion, the spectrum sometimes showed a strong continuum in the red as a result of the black body radiation from soot formed by cracking of the lubricating oil at the high temperature of knocking combustion.

From the absorption spectra of end gases in a firing engine, Withrow and Rassweiler (1934) found absorption lines due to formaldehyde appeared prior to the arrival of the flame and were more intense under knocking conditions. This work established that precombustion reactions did occur and that they were oxidative in nature, in that formaldehyde was observed. Measurements of emission spectra of a motored engine by Downs et al. (1953) confirmed that excited formaldehyde was the source of the cool flame radiation.

Spectroscopic measurements in the IR, visible, and UV regions were carried out by Pipenberg and Pahnke (1957) during the oxidation of *n*-heptane in a motored engine. The experimental setup permitted both absorption and emission measurements at discrete times throughout the engine cycle. In addition to the carbonyl group and the various C-C and C-H groups from the fuel, the spectral characteristics related to the formyl radical, formaldehyde, hydrogen, water, carbon monoxide, and carbon dioxide were observed.

A significant step forward was taken by Abata et al. (1978) when they made measurements of the OH radical concentration in the end gas of a firing engine fueled with ethane. This work was the first quantitative OH measurement made in the end gas of a firing engine. The absorption spectroscopy technique was employed using a broad band flash lamp pumped dye laser. The dye laser output was tuned to the $^2\Sigma^+$–$^2\Pi$ electronic transition of OH with a central wavelength of 308 nm and a bandwidth of 1.6 nm. Because of the difficult nature of the measurement technique and the challenges presented in applying it to a firing engine, the results were not extensive. However, the experiments did prove that such measurements were possible.

Noguchi et al. (1979) reported an extensive spectroscopic investigation on the CAI/HCCI combustion process in a two-stroke SI engine. Emission was collected through a sapphire window in the cylinder wall of an opposed-piston two-stroke SI engine. The transmitted light through a slit was equally divided into each of the branched optical passages. The emission from radicals of OH (306.4 nm), CH (431.5 nm), C_2 (473.7 and 516.5 nm), and H (578 nm) was detected separately by a pair of a photomultiplier tube and a narrow band-pass filter via an optical fiber. The emission from CHO, HO_2, and O radicals was observed together at 400 nm. The emission from lithium and sodium, a small quantity of which was added to the fuel, was used to measure flame temperature above 1000°C. The detection of the IR radiation at 4.3 µm from CO_2 gas was employed to determine temperatures below 1000°C. As the optical fiber absorbed both in the infrared and UV regions, a hollow brass pipe was placed in the center of the optical fibers to transmit the emissions from CO_2 (4.3 µm) and OH radicals (306.4 nm).

The time-resolved light emission results showed that, in the case of the conventional spark ignition combustion, all the radicals were observed almost at the same time; in the case of the new combustion process, CHO, HO_2, and O radicals were first detected followed by CH, C_2, and H radicals, and finally, the OH radical. Such observation confirmed that the new combustion process was initiated by the autoignition of premixed mixture due to compression.

Spectroscopic studies of this new combustion process was also reported by Iida (1994,) who investigated the emissions from OH, CH, and C_2 radicals from a two-stroke SI engine running with gasoline and methanol. Shoji et al. (1994, 1996) performed emission and absorption measurements of the preflame reactions of *n*-heptane and iso-octane formaldehyde was detected by emission at 395.2 nm, and the OH radical was measured simultaneously by absorption from a xenon lamp at 306.4 nm. Shoji et al. concluded that formaldehyde and OH radicals were produced at the same time in the end gas region. This is inconsistent with the findings discussed previously, where formaldehyde was found to form earlier than OH radicals. The discrepancy may be ascribed to the fact that the absorption measurement is more sensitive than the emission. Hence, a smaller amount of OH would have been detected by absorption, while emission resulting from formaldehyde would not have been detected until a larger amount had been formed in the end gas region.

The availability of tunable diode lasers had prompted a number of researcher in exploring the tunable diode laser-based absorption techniques for combustion diagnostics and closed-loop control over the past 20 years, most noticeably by at Stanford University (Jeffries and Hanson, 2004). However, most tunable diode laser-based absorption measurements were done in well-controlled, laboratory-scale flames and some on the IR absorption measurement of combustion exhaust pollutants, such as CO, CO2, NO, and so forth (Zahniser et al., 2002).

A more recent example of the tunable diode laser-based absorption techniques for in-cylinder combustion diagnostics is demonstrated by Fitzgerald and Steeper (2010). In-cylinder concentrations of CO during the negative valve overlap period in an optical HCCI engine was measured by a distributed-feedback tunable diode laser near 2.3 μm. High-frequency modulation of the laser output was employed to further enhances the signal-to-noise ratio and detection limits of CO. The system had a sensitivity of 200-ppm for 50-cycle ensemble-average values of CO concentration with 1-ms time resolution. In addition to the limited temporal resolution, the multipasses through the combustion chamber employed to enhance the measurement sensitivity did not produce spatially resolved CO measurement.

8.3 Chemiluminescence Imaging

For a given species, the direct visualization of its chemiluminescence with a high-speed image acquisition system can be used to obtain temporally and spatially resolved images of such species through a particular band-pass filter.

For direct visualization of weak luminous events, an image intensifier can be used with both single-shot and high-speed solid-state cameras. The image intensifier improves a camera's sensitivity by a factor of thousands, and it is capable of extremely short exposure times (nanoseconds) by gating on and off electronically. There is a compromise between the gain and the speed at which the intensifier can be operated. For high-speed continuous recordings, a very rapid decay phosphor is employed with a reduced gain in the hundreds. The use of an intensifier will also compromise the spatial resolution of the imaging system.

Direct imaging of the combustion process is often obtained through large optical accesses, which required substantial modification to the engine and sometimes led to configurations atypical of real engines. To overcome this difficulty, an endoscope-based imaging system requiring very small optical access to the engine can be used, as discussed in Chapter 2. However, the field of view is very restricted with an endoscope, and optical distortion is often present.

Following the pioneering work by Withrow and Rassweiler (1938), flame visualization has been used widely in the studies of flame propagation in spark-ignition engines (e.g., Bates 1989). Dec and Espey (1998) applied the chemiluminescence technique to study the autoignition process in a high-definition (HD) diesel engine under a variety of operating conditions. They used the chemiluminescence from formaldehyde and CH to identify the cool flames during the autoignition process.

More recently, chemiluminescence imaging has been applied to the autoignition and high-temperature heat-release process in the CAI/HCCI engines (Dec et al., 2006; Hultqvist et al., 1999). In the author's laboratory, chemiluminescence imaging was employed to study the effect of injection timing and spark discharge on the CAI combustion with negative valve overlap as detailed below (Yang et al., 2009).

In-cylinder pressure and high-speed chemiluminescence imaging of CAI combustion were recorded simultaneously from a single-cylinder optical gasoline engine with extended cylinder block. Negative valve overlap and an intake heater were employed to achieve stable CAI combustion of the optical engine. The high-speed chemiluminescence images were recorded by a high-speed color video camera with a lens-coupled high-speed image intensifier. In addition, high-speed OH chemiluminescence images through a band-pass filter of 310 nm and CHO chemiluminescence images with a band-pass filter 350 nm were recorded separately.

Figures 8.3 and 8.4 show the chemiluminescence images with and without spark. In the presence of spark ignition, it is noted that the first emission site appeared around the spark plug and expanded as a flame. As the flame

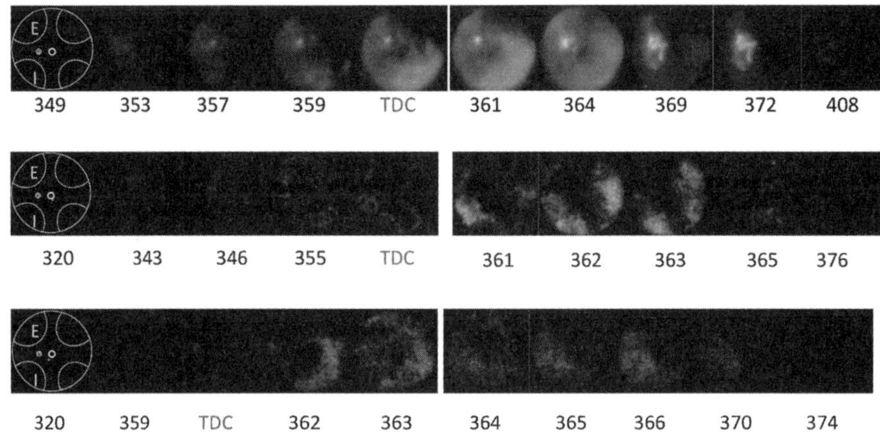

Fig. 8.3 Images of total chemiluminescence (top), CHO (middle), and OH (bottom) of spark-assisted CAI combustion at lambda 1, 1600 rpm, SOI = 80° CA ATDC (with spark).

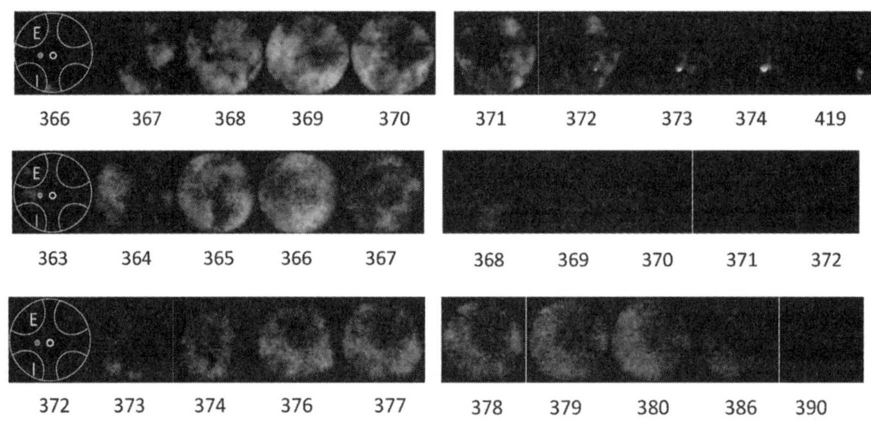

Fig. 8.4 Images of total chemiluminescence (top), CHO (middle), and OH (bottom) of pure CAI combustion at lambda 1, 1600 rpm, SOI = 80° CA ATDC (without spark).

reached its halfway point across the combustion chamber, independent autoignition away from the flame front occurred at 359° CA, as indicated by the chemiluminescence of CHO formed during the low-temperature autoignition reactions. The first formation of CHO radical was captured at an estimated charge temperature of 850 ~ 900 K. Strong CHO emission emerged in front of the flame front at 361° CA and expanded into other regions in the next two images, indicating the start of autoignition sites in such regions.

In the absence of spark discharge, the first combustion emission appeared at 366° CA at sites near the cylinder wall. The autoignited region then expanded rapidly toward the center of combustion chamber before its intensity decreased.

In both cases, the OH images were detected later in the combustion process. The first OH image with spark assistance was detected at 359° CA, and the OH concentration reached its peak at 363° CA when the heat release rate was at its maximum. In the absence of spark discharge, OH radical emission was not detected until 373° CA after the peak heat release rate but coincided with the peak gas temperature. Furthermore, in the absence of spark-ignited flame front, the OH radials distributed more widely in the combustion chamber.

8.4 Schlieren and Shadowgraph Techniques for Nonluminous Flow and Combustion Visualization

Direct visualization of combustion and flame propagation is effective when the flame is luminous. To observe nonluminous events in the cylinder, such as the interaction of flow and combustion, Schlieren/shadowgraph techniques can be used. The terms *shadowgraph* and *Schlieren* have sometimes been used interchangeably, but there are differences between the two. The optical setup of a Schlieren system is similar to that for shadowgraph, with the exception that a knife edge or aperture is used to block part of the light through the test section. Figure 8.5 shows the setup of a typical Schlieren system. Because of the obstruction, the Schlieren method is sensitive to the first derivative of density, whereas the shadowgraph responds to the second derivative of density. Though the Schlieren technique gives greater contrast, the large density gradient in in-cylinder combustion flow and the greater tolerance of the shadowgraph technique to the quality of the optical components can lead to considerable cost savings. More importantly, when a double-pass setup (Fig. 8.7) is used, the rocking of the mirrored piston surface deflects the return image relative to the knife edge. The deflection of the resulting images makes the Schlieren system difficult to use.

Figure 8.6 shows one typical laser shadowgraph system. Traditionally, a high-intensity lamp has been used as the light source. The use of a laser as the light source may be advantageous in some applications. The monochromatic nature of the light avoids the need for expensive achromatic lenses. An additional advantage of monochromatic light is found when studying high-luminosity combustion: Using a narrow band-pass filter, one can eliminate the emission from the combustion, or distinguish combustion emissions from the shadowgraph. Finally, the use of a laser as the light source yields maximum image sharpness. One disadvantage of the laser light is that it enhances background noise because of interference of the light scattered from foreign matters in the test sections. When white light source is used, the interference effect is absent.

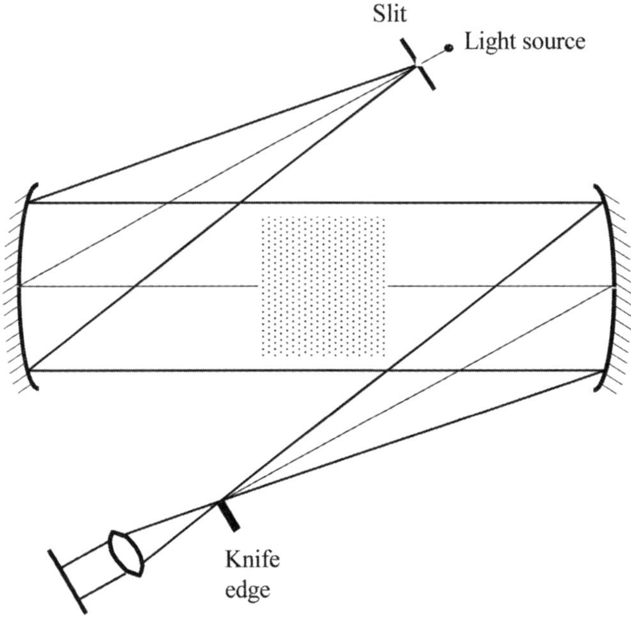

Fig. 8.5 Typical setup of a Schlieren system with concave mirrors.

The laser output normally has a Gaussian distribution of intensity across the beam. As shown in Fig. 8.6, lens L1, usually a microscope objective, focuses the laser beam onto a tiny aperture A. The combination of lens L1 and aperture A acts as a spatial filter, improving the Gaussian intensity distribution of the laser beam, as well as reducing the physical size of the "point" source. The expanding beam exiting the aperture is recollimated by lens L2, which must be larger in diameter than the maximum dimension of the desired field of view. The parallel light is passed through the test section, and then focused by a third lens L3. Lens L4 could be the camera lens or one used to form a real image on a translucent screen.

Such a setup as shown in Fig. 8.6 requires two optical windows mounted at each end of the test section. When only one optical access is available, a double-pass shadowgraph system can be used as shown in Fig. 8.7(a) (e.g., Smith, 1982). Component changes are the elimination of one large lens, since a single lens serves the function of both L2 and L3, and the addition of a beam splitter. The use of a beam splitter permits the incident and return light to be coincident. This is a desirable feature, but it comes at the expense of light intensity at the image plane. Each pass through the beam splitter results in approximately 50% loss in light, such that the efficiency of the system is less than one-quarter that of a single-pass system. Alternatively, the incident beam is tilted slightly such that the reflected beam can be separated from the incident

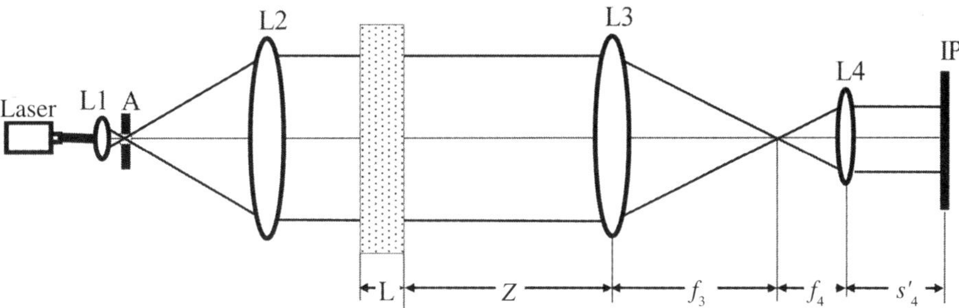

Fig. 8.6 Recollimated laser Shadowgraph setup.

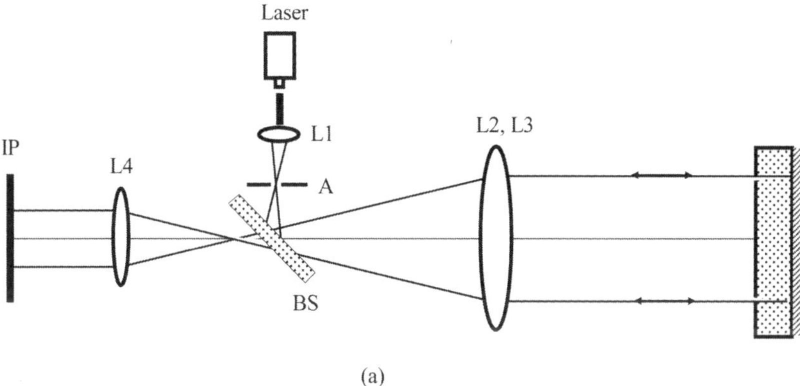

(a)

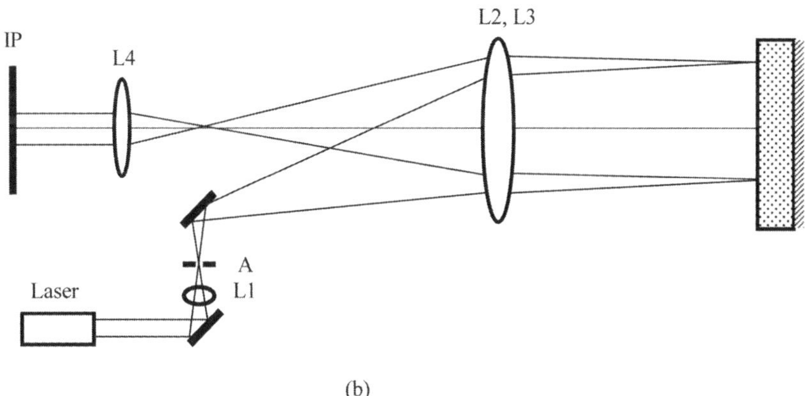

(b)

Fig. 8.7 Double-pass recollimated laser shadowgraph setup.

beam, as shown in Fig. 8.7(b). This system permits a less powerful light source to be used but at the expanse of sensitivity.

For both setups shown in Figs. 8.6 and 8.7, the change in light intensity ΔI for a tiny area, $dxdy$, at the image plane (IP) can be shown to be

$$\frac{\Delta I}{I} = -\frac{\lambda}{n_0} \int_0^L \left(\frac{\partial^2 n}{\partial x^2} + \frac{\partial^2 n}{\partial y^2} \right) dz \tag{8.8}$$

where n is the refractive index, and n_0 is the refractive index at a reference temperature and pressure. The sensitivity of this shadowgraph system λ is given by

$$\lambda = z + (1 - M)\frac{f_3}{M} + \frac{s'_4}{M^2} \tag{8.9}$$

The magnification factor M can be shown to be $M = -\dfrac{f_4}{f_3}$.

Note that the image is inverted. The sensitivity of both systems can be varied by moving only the image plane. This proves to be very convenient for engine experiments where the bulk of the setup is normally fixed. In the double-pass setup, the light passes through the test section twice, and thus sensitivity to density gradients within the test section is doubled. The upper integration limit in Eq. 8.8 becomes $2L$.

Because they are very sensitive to density changes, Schlieren/shadowgraph techniques have been used in the studies of flame initiation and development. Ask et al. (1994) studied the flame kernel development in a single-cylinder SI engine using a Schlieren system similar to that shown in Fig. 8.6.

Using the double-pass shadowgraph system shown in Fig. 8.7(a), stunning images of flame structure were obtained from an optical engine with side valves and transparent cylinder head (Smith, 1982). An argon-ion laser was used as the light source in the shadowgraph system. An acoustic-optic modulator (AOM) was employed to strobe the laser light synchronously with the engine. The expanded laser beam passed through the engine combustion chamber and reflected off a mirror bonded to the piston top. Single-shot photography and high-speed photography was used to record shadowgraphs of flow and flame. In the initial experiment, shadowgraphs were directly imaged onto the camera. Because of the rocking of the piston mirror, shadowgraph images moved around. In the worst case, this motion was so large as to deflect the image partially off the film. This motion was maximum at the focal plane of the source image. A Mylar® screen was therefore placed behind lens L4 in Fig. 8.7 to form a real image. The image on the Mylar screen was in turn focused by the camera lens onto the film. By recording shadowgraph images indirectly through the Mylar screen, the image movement on the film was minimized.

In addition to observing the initial flame development in an SI engine, Schlieren/shadowgraph techniques have been used extensively in autoignition and knock studies. Because of their sensitivity to density variations, these techniques are effective in detecting autoignition sites in the end gas region and the pressure waves generated during knocking combustion (e.g., Ball, 1955; Hayashi et al., 1984; Konig and Sheppard, 1990; Miller, 1947; Male, 1949).

8.5 Summary

Visualization of the in-cylinder process can be readily realized through chemiluminescence imaging over a broad spectral band. With the aid of a narrow band-pass filter, chemiluminescence imaging can be used to detect the presence of a known species, if the continuum thermal radiation from soot particles is absent. The rapid advancement in the high-speed digital imaging system has significantly simplified the experiments and image postprocessing process. As discussed in Chapter 5, PLIF (planar laser-induced fluorescence) is another very useful technique to provide spatially and temporally resolved two-dimensional images of a species, if the species can be excited by a laser and emits fluorescence in the UV and visible region.

In the case of a nonluminous process, the shadowgraph and Schlieren techniques are excellent choices to detect minute changes in the species concentration and thermal conditions. However, both full-field imaging and shadowgraphy are limited to research type of engines with large optical access. The endoscopic imaging system can be better suited to production type engines as long as it is capable of capturing sufficient light with minimal image distortion.

The tunable laser absorption is more suited for measurement of species at very low concentration from a volume of gases along the laser beam. The lack of spatial resolution of the technique has limited its application to in-cylinder measurements.

References

Abata, D. L., Myers, P. S., and Uyehara, O. A. (1978). "Spectroscopic Investigation of Hydroxyl Radical Formation in the End Gases of a Spark Ignited Engine Utilizing a Dye Laser." SAE Paper No. 780970, SAE International, Warrendale, PA.

Ask, T. O., Almas, T., Valland, H., and Paulsen, H. (1994). "Study of the Initial Flame Growth in a Spark Ignition Engine." *COMMODIA* 94, pp. 69–83. Japan.

Bates, S. (1989). "Flame Imaging Studies of Cycle-by-Cycle Combustion Variation in a SI Four-Stroke Engine." SAE Paper No. 892086, SAE International, Warrendale, PA.

Ball, G. A. (1955). "Photographic Studies of Cool Flames and Knock in an Engine." *Proc. 5th symposium (International) on Combustion.* Pittsburgh: The Combustion Institute.

Clark, G. L., and Henne, A. L. (1927). "Ultraviolet Spectroscopy of Engine-Fuel flames." *SAE International Trans.*, Vol. 22, No. 1, pp. 15–23.

Clark, G. L. (1928). "Spectroscopic Study of Fuels and Analysis of Detonation Theories." *SAE International Trans.*, Vol. 23, No. 2, pp. 351–357.

Dec, J. E., and Espey, C. (1998). "Chemiluminescence Imaging of Autoignition in a DI Diesel Engine." SAE Paper No. 982685, SAE International, Warrendale, PA.

Dec, J. E., Hwang, W., and Sjoberg, M. (2006). "An Investigation of Thermal Stratification in HCCI Engines Using Chemiluminescence Imaging," SAE Paper No. 2006–01–1518, SAE International, Warrendale, PA.

Downs, D., Street, J. C., and Wheeler. R. W. (1953). "Cool Flames in a Motored Engine." *Fuel*, Vol. 32, p. 279.

Fitzgerald, R., and Steeper, R. (2010). "Application of a Tunable-Diode-Laser Absorption Diagnostic for CO Measurements in an Automotive HCCI Engine." *SAE Int. J. Engines*, Vol. 3, No. 2, pp. 396–407.

Gaydon, A. G. (1974). *The Spectroscopy of Flames,* 2nd ed. London: Chapman and Hall Ltd.

Hayashi, T., Taki, M., Kojima, S., and Kondo, T. (1984). "Photographic Observation of Knock with a Rapid Compression and Expansion Machine." SAE Paper No. 841336, SAE International, Warrendale, PA.

Hultqvist, A., Christensen, M., Johansson, B., Franke, A., Richter, M., and Alden, M. (1999). "A Study of the Homogeneous Charge Compression Ignition Combustion Process by Chemiluminescence Imaging." SAE Paper No. 1999–01–3680, SAE International, Warrendale, PA.

Iida, N. (1994). "Combustion Analysis of Methanol Fueled Active Thermo Atmosphere (ATAC) Combustion Engine Using a Spectroscopic Observation." SAE Paper No. 940684, SAE International, Warrendale, PA.

Jeffries, J. B., and Hanson, R. K. (2004). "Tunable Diode Laser Sensors for Practical Combustion Applications." Optical Society of America, Conference on Lasers and Electro-Optics, San Francisco, CA, May, 2004.

Konig, G., and Sheppard, C. G. W. (1990). "End Gas Autoignition and Knock in a Spark Ignition Engine." SAE Paper No. 902135, SAE International, Warrendale, PA.

Male, T. (1949). "Photographs at 500,000 Frames per Second of Combustion and Detonation in a Reciprocating Engine." *Proc. 3rd Symposium on Combustion*, pp. 721–726.

Miller, C. D. (1947). "Roles of Detonation Waves and Autoignition in Spark Ignition Engine—Knock as Shown by Photographs Taken at 40,000 and 200,000 Frames per Sec." *SAE Trans.* Vol. 1, pp. 98–143.

Noguchi, M., Tanaka, Y., Tanaka, T., and Takeuchi, Y. (1979). "A Study on Gasoline Engine Combustion by Observation of Intermediate Reactive Products During Combustion." SAE Paper No. 790840, SAE International, Warrendale, PA.

Pipenberg, K. J., and Pahnke, A. J. (1957). "Spectrometric Investigations of n-Heptane Preflame Reactions in a Motored Engine." *Ind. Engng Chem.*, Vol. 49, p. 2067.

Shoji, H., Tosaka, Y., Yoshida, K., and Saima, A. (1994). "Radical Behavior in Preflame Reactions Under Knocking Operating in a Spark Ignition Engine." SAE Paper No. 942061, SAE International, Warrendale, PA.

Shoji, H., Shimizu, T., Nishizawa, T., Yoshida, K., and Saima, A. (1996). "Spectroscopic Measurement of Radical Behavior Under Knocking Operation." SAE Paper No. 962104, SAE International, Warrendale, PA.

Smith, J. R. (1982). "The Influence of Turbulence on Flame Structure in an Engine." *Flows in Internal Combustion Engines*, edited by T. Uzban, pp. 67–72. New York: ASME.

Withrow, L., and Rassweiler, G. M. (1931). "Spectroscopic Studies of Engine Combustion." *Ind. Eng. Chem.*, Vol. 23, pp. 769–776.

Withrow, L., and Rassweiler, G. M. (1933). "Absorption Spectra of Gaseous Charges in a Gasoline Engine." *Ind. Eng. Chem.*, Vol. 25, pp. 923–931.

Withrow, L., and Rassweiler, G. M. (1934). "Formaldehyde Formation in Preflame Reactions in an Engine." *Ind. Eng. Chem.*, Vol. 26, pp. 1256–1262.

Withrow, L., and Rassweiler, G. M. (1938). "Motion Pictures of Engine Flames Correlated with pressure Cards." *SAE Trans.*, Vol. 43, pp. 185–204.

Yang, C., Zhao, H., and Megaritis, T.(2009). "In-Cylinder Studies of CAI Combustion with Negative Valve Overlap and Simultaneous Chemiluminescence Analysis." SAE Paper No. 2009–01–1103, SAE International, Warrendale, PA.

Zahniser, M., Nelson, D., and Kolb, C. (2002). "Tunable Infrared Laser Differential Absorption Spectroscopy (TIDAS) Sensors for Combustion Exhaust Pollutant Quantification." Chapter 26 in *Applied Combustion Diagnostics*, edited by Kohse-Hoinghaus, K., and Jeffries, J. London: Taylor and Francis.

chapter 9
Gas Temperature Measurement

9.1 Introduction

Since physical probes, such as thermocouples or thermistors, typically lack the necessary spatial resolution, time response, and durability, in-cylinder gas temperature measurements have been dominated by optical techniques. Before the advent of laser-based techniques, all temperature measurements by optical techniques were carried out using line-of-sight techniques based on radiation thermometry. Thus, the classification of temperature measurement techniques according to their spatial resolution seems to be a natural choice from both historic and technical points of view. The measurement of combustion temperature in a diesel engine by means of the two-color method will be discussed in Chapter 10, because it is often used to measure in-cylinder soot loading.

The common feature of the laser-based techniques is their capability for nonintrusive, in situ measurements with high spatial and temporal resolution. Therefore, these techniques offer the unique opportunity to measure accurately in-cylinder gas temperature distributions and their effect on subsequent combustion and pollutant formation processes. More recently, two-dimensional temperature measurements have been done with Rayleigh scattering or PLIF (planar laser-induced fluorescence), with latter offering a greater potential to be applied to more realistic engine conditions.

9.2 Radiation Thermometry
9.2.1 Principles of Radiation Thermometry

We are all familiar with the dull red glow of coals in a fire and the bright white glow of incandescent lamps. We know that the whiter and brighter an object appears, the hotter it is. This is in fact the very basic principle on which radiation thermometers or pyrometers are based.

Because our eyes are only sensitive to a narrow range of radiation spectrum (0.4–0.7 µm), the visible spectral region in Fig. 9.1, most of us associate thermal radiation only with objects that are hot (>500°C). But as their temperature reaches above 1500°C, objects become so bright that our eyes have difficulty observing them comfortably.

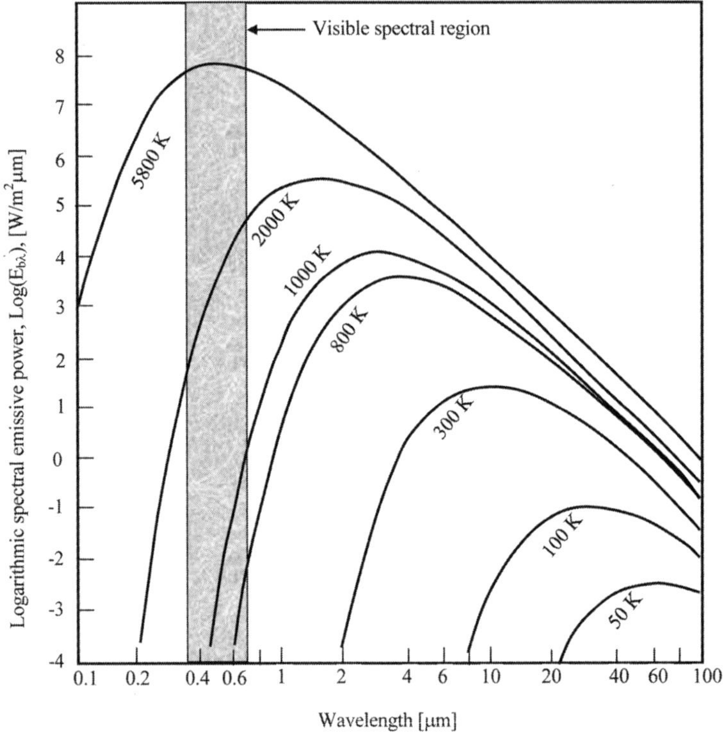

Fig. 9.1 Monochromatic blackbody emissive power distribution as a function of wavelength.

The relationship between the intensity of radiation and the temperature shown in Fig. 9.1 is given by the Planck's law for a perfect emitter and absorber of radiation, known as blackbody:

$$E_{b,\lambda}(T) = \frac{C_1}{\lambda^5 [e^{(C_2/\lambda T)} - 1]} \tag{9.1}$$

where

$E_{b,\lambda}$ = monochromatic emissive power of a blackbody at temperature T (W m^{-3})

λ = wavelength (μm)

T = temperature (K)

C_1 = the first Planck's constant = 3.7418×10^{-16} W m^2

C_2 = the second Planck's constant = 1.4388×10^{-2} m K

For several practical reasons, most radiation thermometry is carried out at wavelengths in the visible or near infrared (IR) region. In this portion of the spectrum, $(C_2/\lambda T) > 4.6$, and the Planck's law is approximated to 1% or better by Wien's law:

$$E_{b,\lambda}(T) = \frac{C_1}{\lambda^5 \exp\left(C_2/\lambda T\right)} \tag{9.2}$$

However, all practical objects emit less radiation than a blackbody at the same temperature. The fraction of full radiation that is actually emitted is called the monochromatic emissivity ε_λ of the object. Thus, for a nonblackbody its radiation is given by

$$E_\lambda(T) = \varepsilon_\lambda \frac{C_1}{\lambda^5 \left[\exp\left(C_2/\lambda T\right) - 1\right]} \tag{9.3a}$$

or

$$E_\lambda(T) = \varepsilon_\lambda \frac{C_1}{\lambda^5 \exp\left(C_2/\lambda T\right)} \tag{9.3b}$$

The monochromatic emissivity ε_λ is normally dependent on temperature, wavelength, and concentration, and it is the most important and yet the most limiting factor for accurate determination of gas temperature using the radiation thermometry. If the emissivity ε_λ is independent of wavelength, the emitting object is considered to be a gray body.

Another important concept in radiation thermometry is the apparent temperature or brightness temperature T_a, which is the temperature of a blackbody that will emit the same radiation intensity as a nonblackbody at temperature T, that is $E_{b,\lambda}(T_a) = E_\lambda(T)$. Combining Eqs. 9.2 and 9.3b leads to

$$\frac{1}{T} = \frac{1}{T_a} + \frac{\lambda}{C_2} \ln \varepsilon_\lambda \tag{9.4}$$

In practice, the amount of radiation detected by a thermometer is given by

$$P = \int_{\lambda_1}^{\lambda_2} \eta_A \tau_\lambda S_\lambda E_\lambda(T) d\lambda \tag{9.5}$$

where

η_A = the geometrical factor of the optical system

τ_λ = the spectral transmission of the optical system

S_λ = the spectral response of the detector

Hence, the gas temperature can be determined from the detector output for a given optical setup with known spectral transmission and spectral response of the detection system, if the emissivity is known.

For gases that are not opaque, according to Kirchhoff's law, the monochromatic emissivity ε_λ is equal to the monochromatic absorptivity α_λ:

$$\alpha_\lambda = 1 - I/I_o = 1 - e^{-K_{abs}(\lambda)L} \tag{9.6}$$

where

$K_{abs}(\lambda)$ = the absorption coefficient (cm^{-1}) at the wavelength λ

L = the optical path length (cm)

I_o = the light intensity without absorption

I = the light intensity after absorption

The absorptivity, and hence the emissivity of a gas, depends on the gas composition, concentration, and temperature as well as the gas thickness.

The application of radiation thermometry to in-cylinder gas temperature measurements relies on the detection of thermal radiation of gases present in the cylinder. Elementary gases such as O_2, N_2, and dry air have a symmetrical molecular structure, and they neither absorb nor emit radiation unless they are heated to such extremely high temperatures that they become ionized plasmas and electronic energy transformations take place. On the other hand, gases like CO_2, CO, H_2O, and hydrocarbons, are asymmetric in one or more of their modes of vibration so that they absorb and emit infrared radiation in bands. The most important gases for thermal radiation detection are CO_2 and water vapor (H_2O).

As discussed in Chapter 3, changes in energy levels resulting from changes in vibrational frequency manifest themselves in a strong peak at the wavelength corresponding to the vibrational transformation, accompanied by multiple rotational energy changes slightly above or below the peak. This process results in absorption or emission bands. The shape and width of these bands depend on the temperature and pressure of the gas, while the magnitude of the monochromatic absorptivity is primarily a function of the thickness of the gas layer. The most important absorption/emission bands for water vapor lie between 1.7 and 2.0 µm, 2.2 and 3.0 µm, and 4.8 and 8.5 µm. The absorption/emission bands of CO_2 are between 2.6 and 2.8 µm, and 4.2 and 4.5 µm. One or two of these bands are used for the gas temperature measurements by radiation thermometry.

When both CO_2 and H_2O exist in a mixture and the radiation is detected from an overlapping spectral band (e.g., around 4.5 µm), the emissivity of the gases will not simply be the sum of the two but will have a lower value because of absorption/emission in overlapping wavelength bands. The precise method of calculating the absorptivity or emissivity of a gas mixture is quite complex. A comprehensive treatment of this subject is beyond the scope of this text, and the reader should consult Hottel and Egbert (1942), Tien (1968), and Siegel and Howell (1993) for details of the subject. In engine applications, the value of emissivity is normally determined experimentally because of the number of unknown parameters involved.

9.2.2 Implementation of Radiation Thermometry

The implementation of radiation thermometry involves the setting up of a reference source, optics, and a detector (Fig. 9.2).

Silicon photodiodes are widely used in the visible and near IR region because of their low cost and ease of operation. Photomultipliers are used if greater sensitivity and very fast response are required. PbS and PbSe detectors can be used in the IR region of 2 µm to 5 µm. For better performance in the region and beyond, the HgCdZnT family of detectors can be employed. These IR detectors are normally cooled to improve their performance.

The optics serve two functions. First, they define the spectral region of the detection by means of spectral filters or spectrometers. Second, they define the angular field of view of the pyrometer, θ. The field of view θ determines the spatial resolution of the system and is defined by the use of a field stop and an aperture stop of diameter a and b, respectively, as shown in Fig. 9.2(b). The use of optical lenses ensures that only certain part of the gases or flame is imaged on to the detector. The focal length and the position of the optical lens determine the size of the image D of the emitting object d.

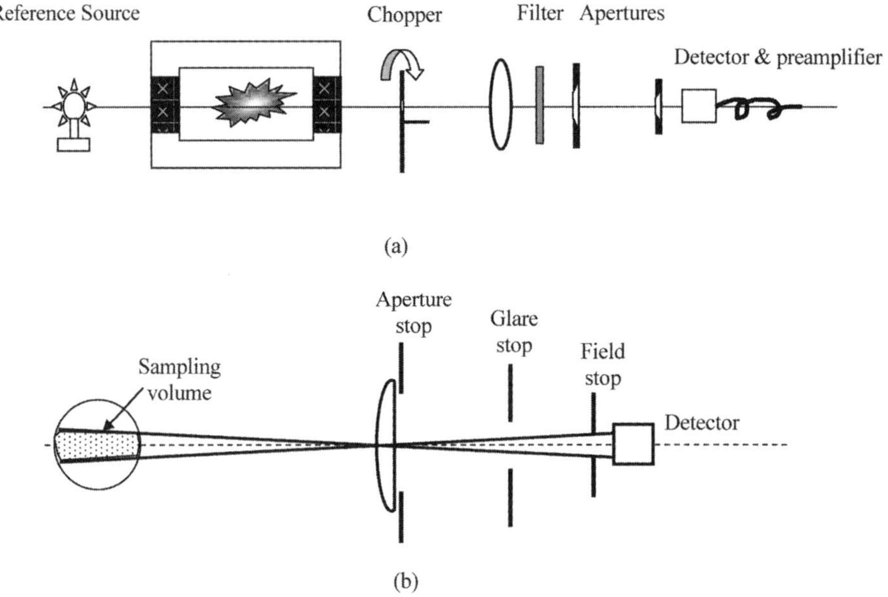

Fig. 9.2 Schematic of a radiation thermometry system.

The optical windows and lenses should be selected to be transparent in the spectral region of operation, as discussed in Chapter 1. For many applications, infrared radiation is detected. Calcium fluoride has a high transmission (>90%) over the spectral range from 0.2 to 8.0 microns. It is slightly soluble in water and susceptible to thermal shock and hence not suitable for high-temperature and high-pressure applications. However, calcium fluoride lenses can be used as an alternative to the more expensive sapphire ones. Germanium and zinc selenide are transparent up to 16.0 microns. Their transmission without optical coating is much lower. Sapphire offers excellent mechanical and optical properties as the window material. It has a transmission band that extends from the UV (ultraviolet) throughout the visible and near IR regions to 5.5 microns.

Finally, a reference source is normally required for performing the required temperature measurements. A reference source of blackbody radiation should be used, though a calibrated tungsten filament lamp may be used as a low-cost alternative. The temperature of the tungsten filament can be adjusted easily by the electrical current flow through the filament. A calibration curve at a particular wavelength can be obtained for the filament's apparent temperatures as a function of the supplying current.

The radiation laws that provide the basis for radiation thermometry apply to radiation of thermal origin only. Chemiluminescence, which is the light emission from an electronically excited state resulting from a chemical reaction, should be avoided by proper selection of the detection wavelength range when gas temperature is measured by a radiation thermometer.

9.2.2.1 The Emission-Absorption Method

The temperature measurement based on the direct measurement of monochromatic radiation using Eq. 9.5 offers the advantage that only one window is required. However, the measurement of emissivity is mandatory and it is often itself proves to be very difficult. During their study, Takagi et al. (1990) performed measurements of the absorptivity of CO_2 gas at three different bands and found that its value remained constant at 0.88 ± 0.02 at wavelength 4.3 µm and air/fuel ratio of 15:1 when the CO_2 concentration exceeded 3%. The absorptivity of CO_2 gas as a function of crank angles was

estimated from $\alpha_{CO2} = 1 - \dfrac{P_2 - P_3}{P_1}$, where P_1, P_2, and P_3 are the radiation

signal from the blackbody radiation source, total radiation from the blackbody and the gas volume, and total radiation from the gas volume alone, respectively.

One solution is to measure both the absorption and emission from the same medium so that the emissivity factor may be ignored if the reference source radiation can be detected without the gas absorption, as shown below.

Consider a reference source, gas volume, optics, and detector arranged as shown in Fig. 9.2. In the absence of gas, radiation received by the detector from the reference source through a narrow band-pass filter centered at λ is given by

$$P_1 = \eta_c \, \tau_{w1} \tau_{w2} \, \varepsilon_b E_{b,\lambda}\left(T_b\right) \tag{9.7}$$

where η_c includes the geometric factor and spectral response of the detection system; τ_{w1} and τ_{w2} are the window transmission of windows 1 and 2; and ε_b is the emissivity of the reference source.

With the gas present at a given temperature between the reference source and the detector, the radiation received by the detector includes radiation from the reference and the gas volume is given by

$$P_2 = \eta_c \tau_{w2} \left\{ \tau_{w1} \varepsilon_b \, E_{b,\lambda}\left(T_b\right)\left[1 - \alpha_g\left(\lambda, p, T_g\right)\right] + E_{g,\lambda}\left(T_g\right)\varepsilon_g\left(\lambda, p, T_g\right) \right\} \tag{9.8}$$

If a chopper is placed between the reference source and gas, the radiation from the gas alone will be detected:

$$P_3 = \eta_c\,\tau_{w2}\,\varepsilon_g\left(\lambda, p, T_g\right) E_{g,\lambda}\left(T_g\right)$$

(9.9)

Combining Eqs. 9.7, 9.8, and 9.9 with Kirchhoff's law ($\alpha_g = \varepsilon_g$) leads to

$$\ln\left[\frac{P_1 + P_3 - P_2}{\varepsilon_b\,\tau_{w1}\,P_3}\right] = \frac{C_2}{\lambda}\left(\frac{1}{T_g} - \frac{1}{T_b}\right)$$

(9.10)

The application of the emission-absorption method (also known as the chopped method) to an IC (internal combustion) engine can be accomplished by detection of the reference source radiation P_1 during the intake stroke when the emission of gas is negligible. If this radiation is absorbed by more than a negligible amount by the gas during intake, then the absorptivity $\alpha_g(\lambda, p_1, T_{g1})$ needs to be considered, so that Eq. 9.7 becomes

$$P_1 = \eta_c\,\tau_{w1}\tau_{w2}\,\varepsilon_b\left[1 - \alpha_g\left(\lambda., p_1, T_{g1}\right)\right] E_{b,\lambda}\left(T_b\right)$$

(9.11)

Equations 9.8, 9.9, and 9.11 can then be combined as follows:

$$\ln\left[\frac{P_1 + P_3 - P_2}{\varepsilon_b\,\tau_{w1}\,P_3}\frac{1}{\left(1 - \dfrac{\alpha_g\left(\lambda, p_1, T_{g1}\right)}{\alpha_g\left(\lambda, p_2, T_{g2}\right)}\right)}\right] = \frac{C_2}{\lambda}\left(\frac{1}{T_{g2}} - \frac{1}{T_b}\right)$$

(9.12)

where T_{g2} is the gas temperature measured during compression or combustion.

This method is capable of producing gas temperature history as a function of crank angle on an individual cycle basis. As it is difficult to obtain accurately the absorptivity of the gas under measurement conditions, the second term on the left-hand side of Eq. 9.12 should be ideally zero or negligible. This can be achieved by performing the calibration at a point during the intake stroke when the absorption is minimal. The most comprehensive evaluation of the emission-absorption method and its application to measure the end-gas temperature in an SI (spark-ignition) engine was carried out by Burrows et al. (1961).

A similar setup was also used to measure the burned gas temperature in the main combustion chamber of a divided-chamber spark ignition engine (Ryu and Asanuma, 1984; Sakurauchi et al., 1987; Takagi et al., 1990). The infrared radiation from CO_2 gas was selected by a band-pass filter centered around 4.3 μm and focused on to a cooled InSb detector by calcium fluoride lenses. The initial measurements were carried out using the chopped method where both the absorption and emission of CO_2 gas were measured.

9.2.2.2 Null Method

The null method is considering a simplified setup of Fig. 9.2 where the windows are absent. The radiation from a blackbody source received by the detector without (Eq. 9.13) and with (Eq. 9.14) the presence of gas is given as follows:

$$P_1 = \eta_c \, E_{b,\lambda}\left(T_b\right) \tag{9.13}$$

$$P_2 = \eta_c \left\{ E_{b,\lambda}\left(T_b\right)\left[1 - \alpha_g\left(\lambda, p, T_g\right)\right] + E_{g,\lambda}\left(T_g\right)\varepsilon_g\left(\lambda, p, T_g\right)\right\} \tag{9.14}$$

When the radiation from reference source P_1 and that from the gas and source P_2 become equal, then Eqs. 9.13 and 9.14 can be equated and written in the following form:

$$E_{b,\lambda}\left(T_b\right)\alpha_g\left(\lambda, p, T_g\right) = E_{g,\lambda}\left(T_g\right)\varepsilon_g\left(\lambda, p, T_g\right) \tag{9.15}$$

According to Kirchhoff's law, the emissivity of the gas is the same as its absorptivity. Hence, Eq. 9.15 becomes, $E_{b,\lambda}\left(T_b\right) = E_{g,\lambda}\left(T_g\right)$ or $T_b = T_g$.

This equality establishes the conditions for the determination of the gas temperature at the null point: the radiation from the reference source P_1 is the same as the radiation received at the detector from the gas media and the reference source P_2.

In deriving the null point, Eqs. 9.14 and 9.15 do not include the effect of the windows. The radiation from both windows can be minimized by cooling. In addition, the transmission of the window between the reference source and the gas does affect the null point and should be considered. Furthermore, when the null method is applied to an IC engine, the reference source radiation is detected during the intake stroke. The effect of the absorptivity of the gas during the intake system should be considered. The effect of both window's transmission and gas absorptivity can be analyzed by referring to Eq. 9.12. At the null point, Eq. 9.12 becomes

$$\ln\left[\frac{1}{\varepsilon_b\,\tau_{w1}\,P_3}\frac{1}{\left(1-\dfrac{\alpha_g\left(\lambda,p_1,T_{g1}\right)}{\alpha_g\left(\lambda,p_2,T_{g2}\right)}\right)}\right]=\frac{C_2}{\lambda}\left(\frac{1}{T_{g2}}-\frac{1}{T_b}\right) \tag{9.16}$$

which gives the corrected gas temperature at the null point.

The null method is a point-by-point technique; that is, one gas temperature is measured at a particular time or crank angle, and a certain amount of time is required to build up a temperature-time history. Unlike the emission-absorption method, by which the gas temperature can be measured as a function of time for the entire engine cycle with a given source temperature, the null method generates the gas temperature time history from different cycles as it requires the adjustment of source temperature for different crank angles.

The sodium line-reversal technique is one of practical implementation of the null method. The spectral wavelengths employed in this case are the D lines of sodium, as detailed Withrow and Rassweiler (1932). The sodium line-reversal method produces spatial and time averaged temperature measurements of burnt gases. Because of the tedious procedure involved, it is more suited to stationary situations.

9.2.2.3 Two-Color Gas Thermometer/Pyrometer

According to Eq. 9.9, if the emissivity ε is known and the detector output can be compared with that of a blackbody at the same wavelength, then the gas temperature T can be determined at once. Unfortunately, the emissivity is not generally known. It is possible, however, to cancel the need for the parameter or reduce the influence of it if measurements are made at two wavelengths rather than at one, from the ratio of outputs at the two wavelengths:

$$\frac{P_{\lambda1}}{P_{\lambda2}}=\frac{\eta_{c1}\varepsilon_{\lambda1}}{\eta_{c2}\varepsilon_{\lambda2}}\left(\frac{\lambda_2}{\lambda_1}\right)^5\frac{\left(e^{\left(C_2/\lambda_2 T\right)}\right)-1}{\left(e^{\left(C_1/\lambda_1 T\right)}\right)-1} \tag{9.17}$$

where the spectral response factors η_{c1} and η_{c2} can be readily determined from calibration with a blackbody radiation source. If the radiation source behaves like a gray body, the emissivity ratio will be 1 and the radiation body temperature can then be obtained from Eq. 9.17. Otherwise, the emissivity ratio must be determined or some law for the dependence of emissivity on other parameters needs to be assumed.

At General Motors, Agnew (1960) developed a two-color system for the measurement of the end-gas temperature in an SI engine. Radiation from the end-gas region in the cylinder head was focused on to the entrance slit of a monochromator and detected by a lead sulfide detector. Radiation was detected from water vapor bands at $\lambda_1 = 1.89$ and $\lambda_2 = 2.55$ microns from different cycles, though at the same crank angle. An optical chopper was placed in front of the monochromator to provide a zero reference when the radiation was blocked and to prevent the extremely intense hot flame emission from reaching the detector by blocking radiation. Self-absorption by water vapor present between the window and detector was minimized by continuously flushing the spectrometer with nitrogen.

The crucial part of this technique lies in the accuracy of the emissivity ratio. In this study, the emissivity ratios were determined from measured radiation-intensity ratios and temperatures calculated from measured gas pressure, airflow, and exhaust gas measurements, using the ideal gas law under motored conditions with air. The emissivity ratios were determined by a function of the optical path. The results showed a rather large effect of water vapor concentration on the measured emissivity ratios.

The major difficulty associated with this two-color method is the accurate determination of the emissivity ratios. It is a recognized fact that the radiation temperature corresponds more closely to the maximum gas temperature, whereas the temperature calculated from pressure and the perfect gas law should correspond more closely to the mass-averaged temperature. As the difference between the two could be as much as 10%, the emissivity ratio used may be better determined by means of the emission-absorption measurements at the two wavelengths.

Another difficulty with this method of temperature measurements is related to the fact the emissivity ratio is critically dependent on the optical path of the gas or water vapor concentration. As water vapor was produced in the end-gas region during the autoignition process, an iterative correction technique was devised by Agnew (1960) to account for the variation of emissivity ratio caused by extra water vapor.

Compared with the monochromatic absorption-emission technique (the chopped method), the two-wavelength method requires only one window instead of two. The effect of window fouling on the accuracy of temperature measurements is minimized using the two-color method as the intensity ratio of two lines is measured.

9.3 Laser Rayleigh Scattering (LRS) Thermometry

9.3.1 Principle of Laser Rayleigh Thermometry

As discussed in Chapter 6, Rayleigh scattering is proportional to the total number of scattering molecules. If gases are assumed to be ideal, Rayleigh scattering intensity is proportional to the pressure and the mixture-scattering cross section and is inversely affected by the gas temperature, as given by

$$I = I_0 \frac{pA_0}{RT} \sum_{i=1}^{j} \chi_i \sigma_i \tag{9.18}$$

If the Rayleigh signal from the same gas mixture at a reference temperature T_o and pressure p_o is I_{cal}, then the temperature T can be obtained from the Rayleigh scattering intensity I:

$$\frac{T}{T_o} = \frac{\sum_{i=1}^{j} \chi_i \sigma_i}{\sum_{i=1}^{j} \chi_{i,o} \sigma_{i,o}} \frac{p}{p_o} \frac{I_{cal}}{I} \tag{9.19}$$

Calibration is usually done at room temperature. Normalization with respect to the instantaneous laser signal is employed to reduce uncertainties caused by variations in laser power. As indicated by Eq. 9.19, an important attribute of Rayleigh thermometry is its inherently large dynamic range. The simplicity and relatively large scattering cross sections of Rayleigh scattering render it suitable for two-dimensional temperature measurements.

In Eq. 9.19, the gas pressure can be measured accurately by a pressure transducer. The main difficulty for the accurate interpretation of temperature lies in the precise determination of the mixture's scattering cross section. In a homogeneous nonreacting flow of a known mixture composition, the mixture's scattering cross section remains constant. The gas temperature can be determined accurately using this equation. However, in a reacting flowfield, since both the mixture number density and the gas composition vary simultaneously, the application of the LRS technique for quantitative temperature measurement becomes difficult.

9.3.2 Implementation and Application of Rayleigh Thermometry

The experimental setup for Rayleigh thermometry is identical to that used for density measurement, as shown in Fig. 6.1. A vertically polarized laser output should be used as the excitation source to maximize the Rayleigh signal. Since Rayleigh thermometry relies on the absolute level of signal, it is necessary to normalize the scattered signal to account for variations in laser output by monitoring the laser output on a pulse-to-pulse basis. The biggest problem with Rayleigh thermometry is the interference of spurious scattered light from airborne particles and solid walls, which significantly reduce the signal-to-noise ratio and the dynamic range of temperature measurements. This limits Rayleigh thermometry to very clean environment and nonreflective surroundings. Procedures for minimizing background noise as discussed in Chapter 6 should be followed, including the use of filtered Rayleigh scattering for measurement in near-surface or particle-laden environments.

At the University of Heidelberg (Bräumer et al., 1995; Orth et al., 1994; Schulz et al., 1996), LRS was used to obtain two-dimensional temperature distributions in a transparent square piston SI engine. A tunable KrF excimer laser was used because of its high energy output and short wavelength, to boost the Rayleigh signal, which increases with $1/\lambda^4$. Scattered light was imaged by a gated ICCD (intensified charge-coupled device) camera at a right angle. Narrow band-pass filters were used to stop fluorescence light reaching the detector.

The temperature at each point (x,y) in the laser sheet was calculated from the measured Rayleigh intensity using Eq. 9.19. The calibration signal was obtained from the fuel/air mixture in a motored engine at the crank angle when the measurement was to be made. This eliminated the difference in scattering cross sections between air and the fuel/air mixture if calibration had been done with air only. The calibration temperature T_0 was calculated from the polytropic compression relationship and engine geometry. The pressure ratio was evaluated from a pressure transducer.

Additional measures were taken to obtain quantitative temperature measurements. Since LRS is an elastic scattering process, the spurious scattered light from surfaces and airborne particles cannot be directly removed by an optical filter. To overcome this problem, images were first taken of helium in the engine. Because the Rayleigh cross section of helium was negligible, the output was assumed to be from the cylinder walls. The subtraction of the helium image (background image) from the measured Rayleigh images, therefore, minimized the effect of spurious light. Background correction was also used to exclude offsets of the digitizers and camera noise. All the images were subject to normalization to a reference image to correct for laser intensity distribution and nonuniform response of the camera. The reference image was obtained from a homogeneous medium, such as air, for normalization of all other images.

Other variables that had to be considered were the scattering cross sections of different mixtures. For the unburned mixture, accurate temperature measurements can be achieved if the scattering cross section of the mixture is known and mixture is homogeneously distributed in the cylinder. Otherwise, Rayleigh thermometry will suffer from errors associated with temporal and spatial variations of the scattering cross section as a result of mixture inhomogeneity. In a reacting flow, the scattering cross section of burned gas may be very different from that of the unburned mixture. To overcome this difficulty, Orth et al. (1994) used methane in their experiments. Because the scattering cross section of methane and the air mixture varied less than 2% before and after combustion, accurate two-dimensional temperature measurement was achieved. When the burned gas temperature distribution in the engine was measured for propane fuel, the uncertainty due to variation in the mixture-scattering cross section was estimated to be 15% (Bräumer et al., 1995; Schulz et al., 1996). Apparently, LRS is not suitable for measuring burned gas temperatures if higher hydrocarbons or gasoline fuel are used, because of the larger variations in scattering cross-sections between burned and unburned mixtures.

As discussed in Chapter 6, filtered Rayleigh scattering can be used to remove the elastic scattering from surfaces and particles so that the temperature measurement can be done in a dirty environment such as a sooting flame (Hofmann and Leipertz, 1996).

9.4 Temperature Measurement by Spontaneous Raman Scattering (SRS)

9.4.1 Principle of Raman Thermometry

9.4.1.1 Raman Spectrum Synthesis

The theoretical treatment of spontaneous Raman scattering (SRS) in Chapter 3 provides the fundamental formulation upon which SRS temperature measurements are based. Species concentration derives from the strength of the Raman scattering as discussed in Chapter 6, whereas temperature is from the shape of the Raman spectrum.

In Chapter 3, the vibrational Raman scattering signal integrated over the vibrational band spectrum was derived. However, Raman thermometry requires the knowledge of the shape of a Raman spectrum, as well as its intensity, as a function of temperature. The process of calculating a spectrum is known as spectral synthesis. The process involves the construction of a spectrum by examining the spectral position of each transition, the line strength of that particular transition, and the number of molecules in the initial quantum state. Then convolutions are performed over the spectrum to account for the excitation laser line width, and the detection bandwidth, and shape of the measuring instrument. The shape of the Raman spectrum is a convolution of the spectral power and the slit function of the spectral disperse device, such as a monometer.

Here we concentrate on the Raman spectrum of diatomic molecules. The individual line strength of vibrational-rotational Raman scattering spectrum for diatomic molecules is given by (Long, 1977) Eq. 9.20:

$$I_{rot,vib} = \frac{\pi^2}{\varepsilon_0^2}\left(\nu_{ex} \pm \nu_{rot,vib}\right)^4 N_{ini}\,\varphi\left(\alpha'^2,\gamma'^2,{}^i\rho_s\right)I_o \tag{9.20}$$

where ${}^i\rho_s$ indicates the polarization state, for example, ${}^{\parallel}\rho_{\perp}$ shows that the incident light is plane-polarized parallel ($\parallel$) to the scattering plane and the light-scattering component perpendicular($\perp$) to the scattering plane is detected. The scattering plane is defined in Fig. 3.3. N_{ini} is the number of molecules in the initial vibrational-rotational state, given by

$$N_{J,ini} = \frac{N_{v,ini}}{Q_{rot}}\,g_{ini}\,(2J+1)\exp\left(\frac{-B_v hcJ(J+1)}{kT}\right) \tag{9.21}$$

where $Q_{rot} = kT/hcB_v$, g_{ini} is the statistical weight of the initial vibrational-rotational state, and for N_2 molecules, it has values of 3 and 6 for odd- and even-numbered rotational energy levels, respectively. $N_{v,ini}$ is the number of molecules in the initial vibrational state, as prescribed by the Boltzmann distribution:

$$N_{v,ini} = N\exp\left(\frac{-hc\,\mathrm{v}\,\nu_{vib}}{kT}\right)\left[1-\exp\left(\frac{-hc\,\nu_{vib}}{kT}\right)\right] \tag{9.22}$$

The grouping hc/k has a value of 1.44 K deg. cm^{-1}.

Values of the function $\phi(\alpha'^2,\gamma'^2,{}^i\rho_s)$ for calculating the individual transition line strength from Eq. 9.20 are dependent on the polarization state of incident light and scattering light, as well as the change in vibrational-rotational states (Long, 1977),

$$\varphi\left(\alpha'^2,\gamma'^2,{}^{\perp}\rho_{\perp}\right)=\kappa(v)\left[(a')^2 + \frac{4}{45}b_{J,J}\,(\gamma')^2\right]$$

or $\hspace{4cm} \Delta v = \pm 1,\ \Delta J = 0;\ Q-branch \hspace{1cm} (9.23)$

$$\varphi\left(\alpha'^2,\gamma'^2,{}^{\perp}\rho_{\parallel}\right)=\kappa(v)\left[\frac{1}{15}b_{J,J}\,(\gamma')^2\right]$$

$$\varphi\left(\alpha'^{2},\gamma'^{2},{}^{\perp}\rho_{\perp}\right)=\kappa(v)\left[\frac{4}{45}b_{J\pm2,J}\left(\gamma'\right)^{2}\right]$$

or $\qquad\qquad\qquad\qquad\Delta v=\pm1,\ \Delta J=\pm2;\ O-\ and\ S-branch$ (9.24)

$$\varphi\left(\alpha'^{2},\gamma'^{2},{}^{\perp}\rho_{\text{II}}\right)=k(v)\left[\frac{1}{15}b_{J\pm2,J}\left(\gamma'\right)^{2}\right]$$

where $k(v)$ depends on whether it is a Stokes and anti-Stokes line,

$$k_{s}(v)=\frac{h}{4\pi\omega_{e}}(v+1)\qquad\qquad\text{Stokes Line}$$

$$k_{as}(v)=\frac{h}{4\pi\omega_{e}}v\qquad\qquad\text{Anti-Stokes Line}$$

where v is the vibrational number of the initial energy level. In the case of N_{2}, $\omega_{e}=\tilde{v}/c=2360$ cm^{-1}.

The Planck-Teller coefficients $b_{J',J''}$ are given by

$$b_{J+2,J}=\frac{3(J+1)(J+2)}{2(2J+1)(2J+3)} \tag{9.25}$$

$$b_{J-2,J}=\frac{3J(J-1)}{2(2J+1)(2J-1)} \tag{9.26}$$

$$b_{J,J}=\frac{J(J+1)}{(2J-1)(2J+3)} \tag{9.27}$$

Note that from Eqs. 9.20 to 9.24 the O- and S-branch transitions arise solely from the anisotropy invariant γ' of the derived polarizability tensor, while the Q-branch transitions arise from both the mean α' and the anisotropy invariant γ' of the derived polarizability tensor. For linear molecules, the value of α' greatly exceeds that of the anisotropy invariant γ'. This means that in vibrational-rotational Raman scattering, Q-branch transitions are much stronger than the O- and S-branch transitions. Furthermore, the O and S branches are spectrally quite spread out and can be neglected in the limited spectral region about the Q branch. The anisotropy contribution can also be ignored because of its weak contribution. Within a given vibrational band, then, the individual Q-branch vibrational-rotational line strengths will mainly scale with the initial rotational state population.

Equations 9.20–9.24 also show that the relative strength of the various vibrational bands will scale as $(v + 1)$ for Stokes and v for anti-Stokes, in addition to weighting by the relative populations. The effect of relative populations is that anti-Stokes bands are much weaker than Stokes bands because of less populated upper vibrational energy states.

Here we will use N_2 as the example because of its importance in combustion thermometry. The spectra of other diatomic molecules are similar. Because of their relative strong signal strength and hence wider use in Raman thermometry, we will consider the fundamental Q branches. The fundamental Q-branch vibrational-rotational line locations are given by

$$\nu = E_v / hc = \nu_{ex} \pm \left[\omega_e - 2\omega_e x_e \left(v + 1\right) - \alpha_e J\left(J + 1\right) \right] = \nu_{ex} \pm \nu_{vib} \tag{9.28}$$

The plus and minus signs represent the anti-Stokes and Stokes vibrational bands, respectively. From Eq. 9.28, it is apparent that without significant vibrational-rotational interaction, that is, $\alpha_e = 0$, all Q-branch transitions would reside upon one another. In N_2, with $\alpha_e = 0.017$, a given vibrational band is shaded toward smaller Raman shift with increasing J. The effect of anharmonicity, $\omega_e x_e$, is to shift and thus separate the various vibrational sequences from each other. Without this, all the vibrational bandheads would

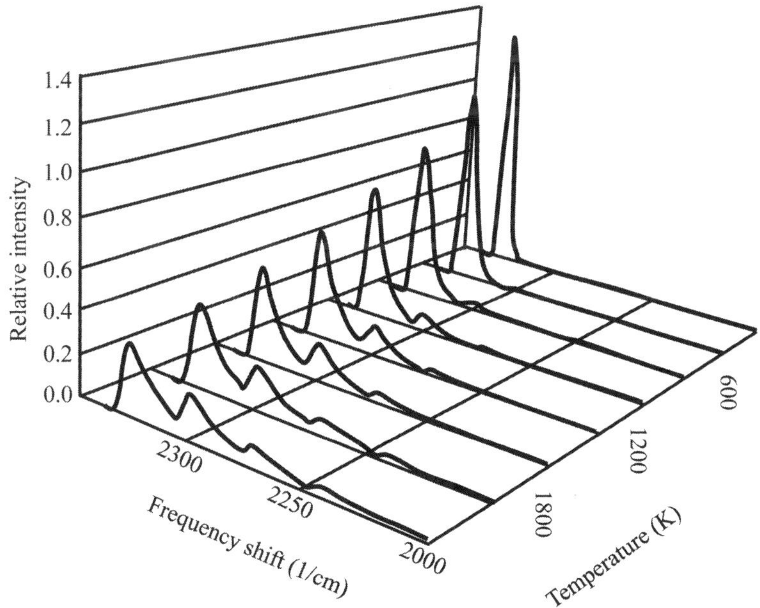

Fig. 9.3 N$_2$ vibrational Q-branch Raman spectrum as a function of temperature at constant concentration.

coincide with each other as well. These features are illustrated in the calculated Stokes vibrational-rotational Raman spectrum as a function of temperature as shown in Fig. 9.3. The shifting of the vibrational-rotational state populations with temperature is reflected in the increasing appearance of wings in the vibrational-rotational Raman spectrum.

9.4.1.2 Raman Thermometry

Figure 9.4 illustrates the four common methods of determining temperature from N_2 vibrational Raman scattering as given by Lapp (1974) and Drake et al. (1982). All the techniques rely on the increased population in the upper N_2 vibrational levels with rising temperature. In the band area method shown in Fig. 9.4(a), the temperature is determined from the relative intensities of the upper and ground level vibrational bands within the Stokes Q branch. For the contour fit method in Fig. 9.4(b), the temperature is determined by theoretical replication of the relative Stokes spectral intensity profile. In the band peak intensity method shown in Fig. 9.4(c), the temperature is obtained by using the intensity ratio for two spectral bands at the peaks of the contour.

The Stokes/anti-Stokes method in Fig. 9.4(d) involves a wider spectral band-pass and hence provides the maximum collected signal. The ratio of Stokes/anti-Stokes fundamental vibrational Raman scattering for N_2 is related to temperature by

$$\frac{P_s}{P_{as}} = \frac{\left(\nu_o - \nu_{vib}\right)^4}{\left(\nu_o + \nu_{vib}\right)^4} \exp\left(-hc\nu_{vib}/kT\right) \frac{f^s(T)}{f^{as}(T)} \qquad (9.29)$$

The Raman shift ν_{vib} corresponds to the energy difference between the ground and first excited vibrational levels. The exponential term in Eq. 9.29 indicates that the higher the gas temperature is, the smaller the ratio becomes as the anti-Stokes signal increases. The bandwidth factors $f(T)$ are included to account for the fraction of the total Raman scattering actually "seen" by the detector, as discussed in Chapter 7. In Fig. 9.5, the anti-Stokes/Stokes bandwidth factor ratio in N_2 is displayed for the laser excitation at 532 nm and various interference filter bandwidths.

The Stokes/anti-Stokes method for N_2 cannot be used at temperatures below about 800 K because of an insufficient number of photons in the anti-Stokes signal (Drake et al., 1982). The other three methods provide reasonably accurate measurements down to 300 K, since these techniques employ only the Stokes signal.

Pure rotational SRS is not widely employed for combustion diagnostics. The Raman shifts are small, that is, less than a few hundred wave numbers,

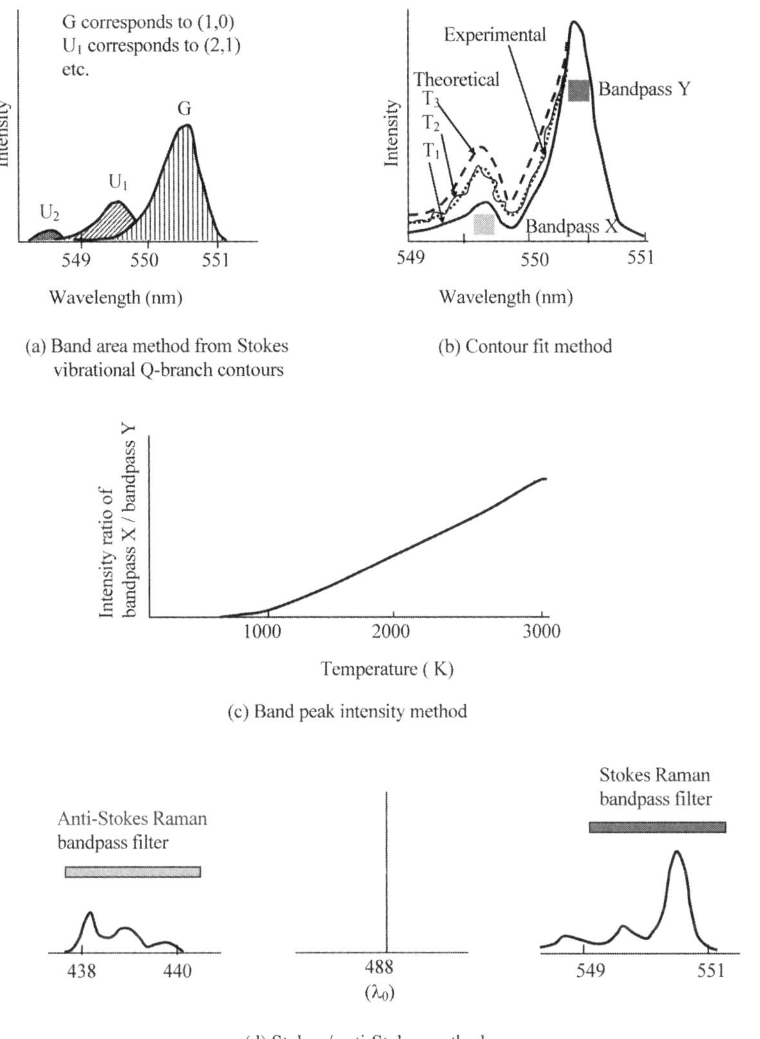

(a) Band area method from Stokes vibrational Q-branch contours

(b) Contour fit method

(c) Band peak intensity method

(d) Stokes / anti-Stokes method

Fig. 9.4 Schematic representation of Raman thermometry for N_2 vibrational Raman scattering, adapted from Drake et al. (1982).

and in a multicomponent mixture, spectra interference and overlap making interpretation difficult. Discrimination against the incident laser light is also difficult to achieve because of the small spectral separations. However, pure rotational Raman scattering cross section for a single transition is generally an order of magnitude stronger than that for an entire vibrational branch. Advances in both atomic/molecular and thin film filters are helping to mitigate this problem. However, pure rotational Raman scattering will likely remain to be limited to simple mixtures or for situations where one constituent is dominant.

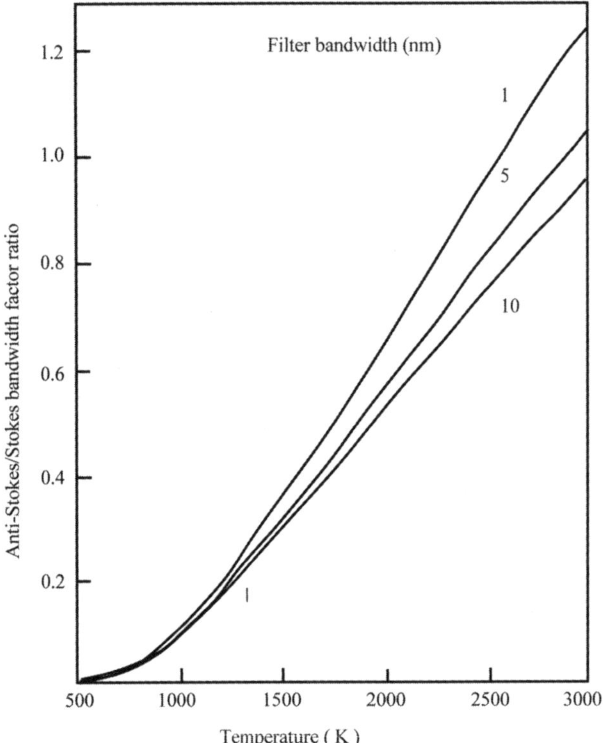

Fig. 9.5 Temperature dependence of the anti-Stokes/Stokes bandwidth factor in N_2 for different filter bandwidths.

9.4.2 Implementation and Application of Raman Thermometry

The four methods of Raman thermometry presented in the previous section can be realized through the use of a high-power pulsed laser and employment of sensitive detectors, as shown in Fig. 6.4. The implementation of the Raman scattering technique discussed in Section 6.3 equally applies to Raman thermometry. Other than the Stokes/anti-Stokes method, all other techniques mandate the use of a spectrometer as the dispersing component of the detection system because of the high spectral resolution required. Narrow band-pass filters can be used to separate Stokes and anti-Stokes bands.

The first temporally and spatially resolved single-shot temperature measurement in a combustion engine was reported by Smith (1980) using the Stokes/anti-Stokes method. The laser beam from an Nd:YAG laser was focused to a small scattering volume (0.25 × 0.5 × 1.25 mm) by a tilted 150-mm focal length lens. The scattered light was then collected through the transparent cylinder head and focused onto a spectrometer. Two photomultipliers mounted at the exit of the spectrometer detected the Stokes Raman signal at 607 ±1 nm and anti-Stokes vibrational Raman signal at 474 ± 1 nm, respectively. The temperature sensitivity of the technique was maximized by choosing very narrow band-pass filter (1 nm) so as to increase the system's temperature sensitivity (Fig. 9.5).

A gated integrator was used to reject the flame radiation. The relationship between the temperature and measured anti-Stokes/Stokes signal ratio was first calibrated with a radiation-corrected thermocouple in the postflame region of a premixed flame. The single-shot temperature uncertainty was estimated to be 125 K. When data were averaged over 350 measurements, the precision of measurements was estimated to be less than 10 K. Burned gas temperature history was obtained for four equivalence ratios. The results confirmed that slightly rich-mixture ($\phi = 0.9$) resulted in the highest combustion temperature as expected. Some burned gas temperature probability distributions were also reported.

Before the late 1980s, most Raman scattering experiments were carried out using the experimental setup shown in Fig. 6.4, where the Raman spectrum was recorded by a photomultiplier and a scanning spectrometer. This limited the application of band area and contour fit methods to laminar flows or averaged temperature measurements, because of the amount of time required to obtain a complete Raman spectrum. One such example was that by Zur Loye and Santavicca (1983), who attempted to measure in-cylinder burned gas temperature by comparing measured and theoretical nitrogen vibrational Q-branch Raman spectra. They employed a monochromator and a photomultiplier to record the Stokes Raman spectrum of N_2 from an Nd:YAG laser (50 mJ) at 532 nm. The spatial resolution along the beam was 3.3 mm, which was determined by the spectrometer slit height and the magnification of the collection optics. The measured nitrogen spectra were recorded by scanning the monochromator with a 4.4-Å slit width while the laser fired at the same crank angle for every third engine cycle. The scans took about 40 minutes, corresponding to 7500 laser pulses in each firing and nonfiring spectrum. The resulting temperature was therefore ensemble-averaged over many hundreds of cycles. Because of the long period required to perform scans, data obtained were affected by the window fouling during the test.

To determine the temperature history of the end gas in the Sandia optical engine, Smith et al. (1984) resorted to the contour fit approach. A pulse Nd:YAG laser excitation at 532.0 nm was used to generate the Raman scattering from nitrogen. The vibrational Q-branch spectrum was obtained by signal integration over many engine cycles. These spectra were then contour-fitted with a theoretical model to determine the temperature. Only the fundamental band spectra were recorded as the higher vibrational levels were sparsely populated at these end-gas temperatures. The temperatures obtained were estimated to be accurate to within ±50–70 K. Their results showed higher end-gas temperatures with knocking cycles than nonknocking cycles. However, the accuracy and ensemble-averaged nature of these temperature measurements is not ideal for knocking combustion studies because of cyclic variations in the end-gas temperature.

Within the last decade, significant advances have been made in the laser technology and multichannel detectors. The availability of powerful UV excimer lasers can significantly enhance Raman signal levels. The use of an ICCD-based imaging spectrograph system allows multiple Raman spectra to be recorded from a laser beam, from which the temperature distribution along the laser beam may be determined by one of the Raman methods shown in Fig. 9.4.

9.5 Temperature Measurement by Coherent Anti-Stokes Raman Scattering (CARS)

9.5.1 Principle of CARS Thermometry

Compared with laser Rayleigh scattering (LRS), spontaneous Raman scattering (SRS), and laser-induced fluorescence (LIF), CARS (coherent anti-Stokes Raman scattering) entails complicated theoretical treatment and complex and expensive optical setup. It is the most accurate optical technique for single-point gas temperature measurements, and its use in IC engines is limited to specialists in laser diagnostics. A detailed description of the CARS technique is beyond the scope of the book. Such information can be found in Eckbreth (1996) and a review by Goss (1993). the basic theory of CARS thermometry will be reviewed here briefly and its implementation considered. Applications of CARS thermometry to IC engines will also be discussed.

The fundamental basis of the CARS technique is outlined in Fig. 9.6. In this method, a medium is irradiated by two superimposed waves, one of a fixed frequency ω_1 and the other of tunable frequency ω_2. They interact in a nonlinear fashion through third-order susceptibility $\chi^{(3)}$ of the medium to generate a third wave at the frequency $\omega_3 = 2\omega_1 - \omega_2$, which is the CARS signal. ω_1 is termed the pump laser, and ω_2 the Stokes beam. In the CARS process, adjustment of the frequency difference $(\omega_1 - \omega_2)$ to a particular Raman transition allows CARS signals from different Raman vibrational and rotational lines to be monitored. Because of the nonlinear wave-mixing process involved, the CARS beam at frequency ω_3 is laser-like. The laser-like CARS beam is directional and allows high collection efficiency, thereby leading to a much greater signal-to-noise ratio than SRS. The adjustment of the frequency difference $(\omega_1 - \omega_2)$ can be done by scanning the frequency of the Stokes beam or using a broadband tunable laser. The use of a broadband laser generates the entire CARS spectrum by accessing simultaneously all the Raman resonance in a given spectral band.

The temperature sensitivity of the CARS technique is the result of variation in rotational-vibrational populations of the Raman transitions of the species being probed, which is reflected in the third-order susceptibility that governs the CARS process. The susceptibility can be expressed in abbreviated form

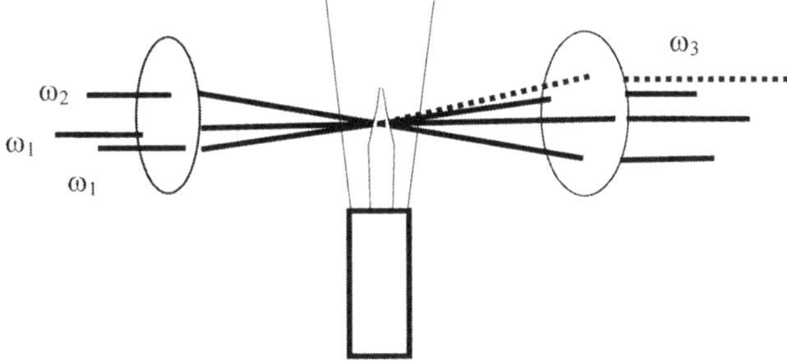

(a) Cross-beam arrangement of CARS technique

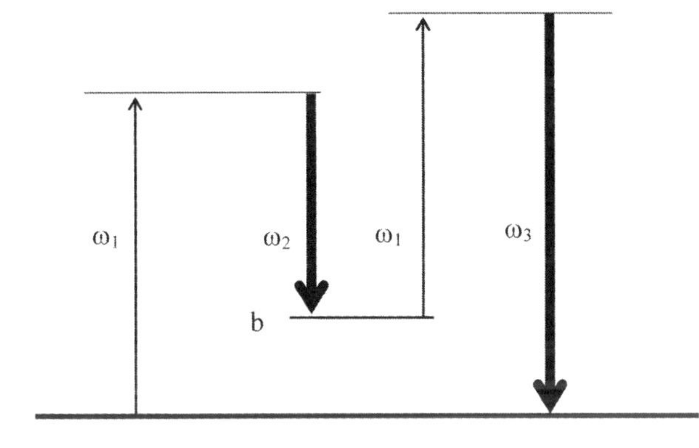

(b) Energy diagram of CARS process

Fig. 9.6 Illustration of the CARS principle.

as $\chi^{(3)} = \chi_{nr}^{(3)} + \chi_{r}^{(3)}$, where $\chi_{r}^{(3)}$ is the Raman resonant contribution and $\chi_{nr}^{(3)}$ the nonresonant contribution. The nonresonant susceptibility is proportional to the total number density of the gases in the sampling volume. The Raman resonant contribution $\chi_{r}^{(3)}$ is a complex quantity arising from Raman-allowed transitions of molecules, and proportional to population difference, $(N_a - N_b)$, between the levels involved in the wave-mixing process, as shown in Fig. 9.6. The population difference is affected by both temperature and species concentration. If the species concentration is known, the gas temperature can be found by recording the CARS signal beam spectrum and fitting it to the theoretic one.

CARS thermometry is generally performed from a major species in sufficient abundance so that nonresonant background effects are relatively unimportant. Nitrogen is the dominant constituent of the in-cylinder charge in an IC engine and is present in large concentrations. Nitrogen vibrational CARS has therefore

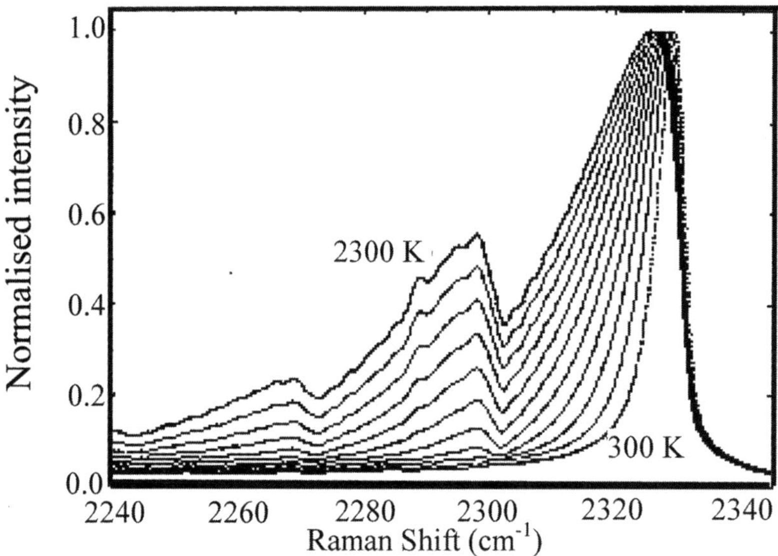

Fig. 9.7 CARS N$_2$ spectrum as a function of temperature.

been mostly used for thermometry. In addition, N$_2$ has the advantage that the spectroscopic and line-width parameters needed for accurate measurements are now relatively well known. CARS spectra for species in high concentration are qualitatively similar to Raman spectra and tend to mimic the vibrational-rotational state population distributions. Figure 9.7 shows the Q-branch spectra of N$_2$ for different temperatures. At low temperatures the N$_2$ spectrum is dominated by a strong band from $v = 1$ to $v = 0$, whose width increases with temperature because of the increased population of high rotational levels. As the temperature approaches 1000 K, the hot-band transition ($v = 2$ to $v = 1$) becomes populated. The band-shape of N$_2$ spectrum is therefore a good indicator of temperature. In practice, temperatures are derived by comparing a measured spectrum with a calculated spectrum using a curve-fitting procedure.

9.5.2 Implementation of CARS Nitrogen Thermometry

In additional to the frequency matching required between the pump and Stokes beams, the pump beams ω_1 and Stokes beam ω_2 should be arranged in such a way that the three beams are phase-matched to maximize the CARS signal generation. The phase matching means that $\bar{k}_3 = 2\bar{k}_1 - \bar{k}_2$, where k_i is the wave vector at a frequency ω_i. The wave vector is oriented along the beam propagation direction. The absolute magnitude of this vector is $| k_i | = (n_i \, \omega_i / c)$, where n_i is the refractive index of the medium at frequency ω_1, and c is the speed of light.

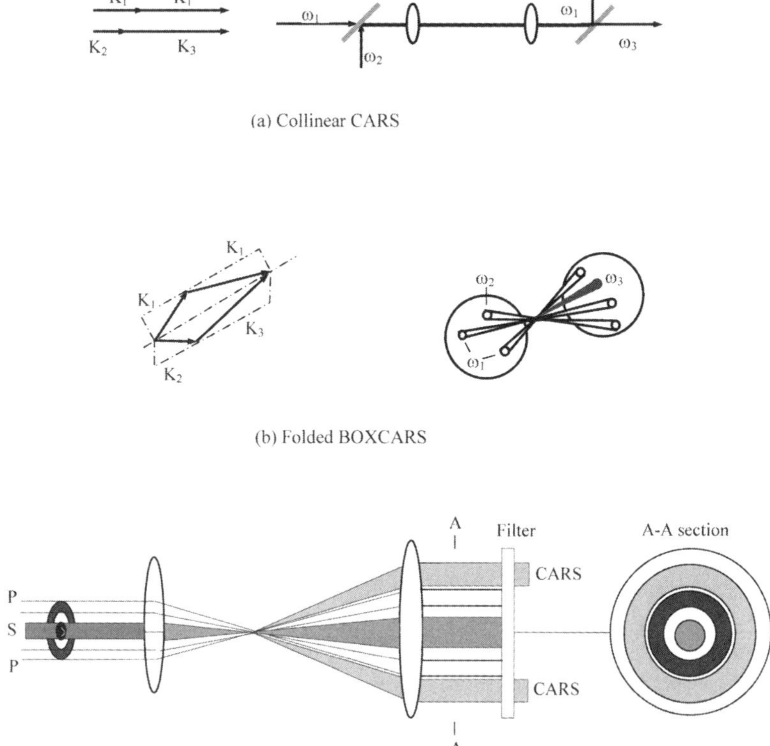

(a) Collinear CARS

(b) Folded BOXCARS

(c) USEDCARS set-up

Fig. 9.8 Examples of phase-matching arrangements for the CARS technique.

Phase matching dictates how the three beams must be oriented. Since the refractive index of gases is essentially invariant with frequency, conservation of momentum in the generation of the CARS signal at $\omega_3 = 2\omega_1 - \omega_2$ indicates that when only two input beams are used, they should be collinear, as in Fig. 9.8(a). If three input beams are used, cross-beam phase matching in gases can be used, as in Figs. 9.8(b) and 9.8(c).

Practical implementation of the CARS technique requires a trade-off between the signal intensity and desired spatial resolution. In applications where optical access is not limited, the cross-beam scheme is used because of its high spatial resolution and its ability to isolate the CARS signal from the stronger laser beams. However, this scheme is optically more complex and results in lower signal intensities because of the smaller interaction volume. The collinear CARS scheme is advantageous for applications with limited optical access. Because of its cubic relationship to the total input laser power

($P_{CARS} \propto \left|\chi^{(3)}\right| P_1^2 P_2$), the CARS signal generated at any point is strongly dependent on the flux density at that point. In CARS experiments, a region of high flux is produced by focusing the laser beams into the sample; this effectively defines an interaction volume in the form of a cylinder whose dimensions are a function of the focal length of the lens used. Therefore, it is possible to obtain good spatial resolution using a collinear beam arrangement if tight focusing is achieved with a short focus lens. In addition, collinear CARS scheme is also preferred when there is a large refractive index gradient in the probed medium, which would cause a serious beam-steering problem in cross-beam schemes.

Figure 9.9 shows the experimental setup of a broadband collinear CARS setup. The pump laser is typically a pulsed narrow-bandwidth frequency-doubled Nd:YAG laser. A broadband dye laser is used to allow the entire Q-branch of N_2 to be obtained in a single shot, thereby allowing single-shot temperature measurements to be obtained from the band-shape of the CARS signal. The 532-nm output of the Nd:YAG laser servers as the pump beam for both the CARS process and the broadband dye laser.

Part of the Nd:YAG laser output (~30%) is used as the pump beam, while the rest of it pumps the dye laser to produce a broadband Stokes beam red-shifted from the pump beam. The generated CARS signal beam is also spectrally broad, and has a blue shift with respect to the pump beam, hence termed anti-Stokes. The CARS signal beam is filtered and dispersed onto a one-dimensional diode array or CCD (charge-coupled device) to generate the CARS spectrum.

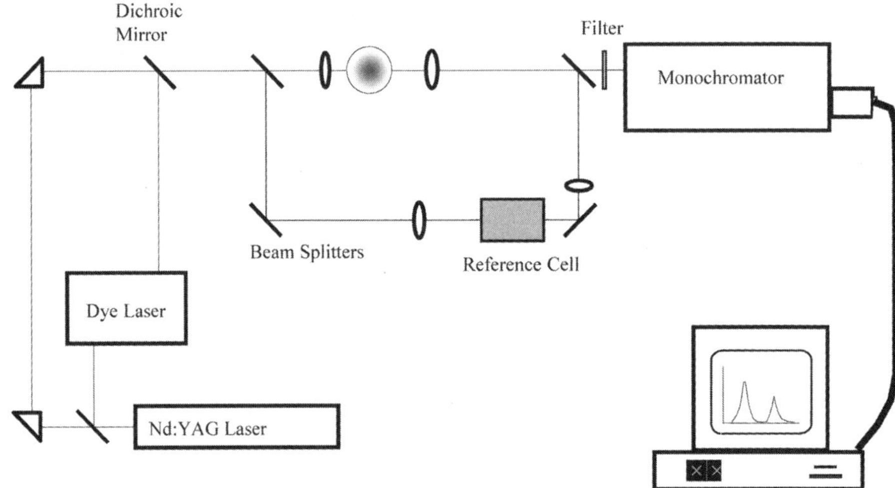

Fig. 9.9 Schematic of a collinear CARS setup.

Experience has shown that a spectrometer with a resolution of 1 cm^{-1} or better is required, which can be achieved by a 1- or 1.5-meter spectrometer with a 2400 lines/mm grating.

Raw CARS spectra should first be preprocessed to remove any offset which may arise from the detector or scattered light. In addition, the shape of the nitrogen CARS spectrum should be normalized by the CARS spectrum of the broadband dye or Stokes laser. A CARS spectrum of the latter is easily measured using a nonresonant medium such as inert gases, carbon dioxide, or propane in the reference cell. Because of the high nonresonant susceptibilities of hydrocarbons in the fuel, variations in the local air/fuel ratio can lead to changes in inferred temperature. Thus, for an accurate temperature measurement, the measured spectrum should be fitted simultaneously for temperature and nitrogen concentration.

Calibration of a CARS system is typically accomplished using a high-temperature furnace. Excellent accuracy (of order 1%) for ambient pressure CARS thermometry is possible (Goss, 1993) but less so under high-pressure conditions.

9.5.3 CARS Measurements in IC Engines

The first application of CARS to temperature measurements in an SI engine was reported by Stenhouse et al. (1979). The collinear scanned method was employed. The scanning took place over a few thousand engine cycles (several minutes). To overcome the difficulty associated with the scanning method, Klick et al. (1981) employed a broadband dye laser as the Stokes beam. Multichannel detection of the CARS signal spectrum was realized by a silicon intensified target (SIT) camera coupled to a spectrometer. The CARS signal was generated using USED (unstable-resonator spatially enhanced detection) CARS, as shown in Fig. 9.8(c). The annular cross section of the Nd:YAG pump beam, which was produced by an unstable resonator laser cavity, permitted noncollinear phase matching, thus providing spatial resolution even in the longitudinal direction. The CARS signal was emitted in an annular ring on the periphery of the pump beam annulus. A short-wavelength pass filter was used to block the Stokes and pump beams but allowed the anti-Stokes beam to pass. Both single-pulse and ensemble-averaged temperature measurements were reported in an SI engine with a compression ratio of 3:1.

Pressure has a large effect on the spectral shape of the CARS signal spectrum. Except for hydrogen, all the major species possess closely spaced and partially overlapping Raman transitions that are predominantly collision-broadened at atmospheric pressure. As the pressure increases, a phenomenon known as collisional narrowing sets in, which causes a major alteration in spectral behavior (Hall et al. 1980). To minimize the pressure effect on temperature measurement, Kajiyama et al. (1982) adapted a data analysis scheme for high-pressure spectra, which extracted temperature from the 20% peak height width of the N_2 Q-branch spectra, a parameter which was found to be affected only by temperature but not by pressure.

When the broadband CARS technique is employed, the CARS signal spectrum needs to be normalized to the broadband Stokes beam via a reference cell filled with a known gas, typically inert, at a fixed temperature and pressure, to correct the laser fluctuation and laser spectral distribution. Marie and Cottereau (1987) corrected the laser fluctuation and spectral distribution for each measurement by dividing, channel-by-channel, the CARS signal by the nonresonant CARS signal from a reference cell. The temperature was derived by fitting the complete CARS spectrum to the theoretical one, including the collisional narrowing at high pressures. The collinear CARS scheme was used in their experiment, which resulted in a resolution of 10 mm in length. This work can be considered as the first quantitative temperature measurements in IC engines by CARS thermometry. Single-shot as well as averaged gas temperature measurements were reported during the compression and expansion strokes. The end-gas temperature was also measured during knocking cycles. The uncertainty of single-shot measurements was largely due to the weak signals. The accuracy of averaged temperatures was considered within ±50°C.

Although the previous studies concentrated on the development of CARS thermometry, more recent CARS temperature measurements focused on its applications. Applications of the CARS technique have been dominated by end-gas temperature measurements. The collinear CARS technique based on Q-branch N_2 vibration spectrum was used in most studies (e.g., Kalghatgi et al., 1995; Nakada et al., 1993; Nakano et al., 1995). Bood et al., (1997) reported a dual-broadband rotational CARS process and a BOXCARS configuration for the measurement of end-gas temperature in a SI engine. In their setup, one narrowband laser beam and two broadband laser beams were focused on an intersection point that defined the probe volume (1 mm × 40 μm). One of the red dye laser beams (Stokes beams) was superimposed on the green Nd:YAG laser beam (pump beam) using a dichroic mirror. The superimposed laser beams and the single red beam were parallel to each other, separated vertically by 14 mm, before they were focused with an f = 300 mm lens. Raman transitions between rotational levels in the same vibrational state were used. The rotational CARS technique was found to give higher accuracy than vibrational CARS for nitrogen temperature measurements up to 870 K though the signal strength was less than the vibrational CARS. Using the dual-broadband rotational CARS technique, the end-gas temperatures in a spark ignition engine were measured and compared favorably with calculated temperatures.

9.6 Two-Dimensional Temperature Measurement by PLIF

9.6.1 Introduction to Fluorescence Thermometry

Because of their low-scattering cross sections, Raman and Rayleigh thermometry is limited to major species of large concentration. Clean measurement conditions are also required, particularly for Rayleigh thermometry, to minimize the interference of spurious scattered light. Compared with the Raman methods, only fluorescence offers the intensity and spectral selectivity for thermometry in moderately dirty conditions, where interference can arise owing to blackbody radiation, flame emission, or laser-induced incandescence. Because of its large-scattering cross section, LIF can be extended to achieve two-dimensional thermometry.

There are three major approaches to fluorescence thermometry, generally grouped into excitation or fluorescence scans, monochromatic, and two-line approaches. For excitation scans, the spectral bandwidth of the detector is fixed, while the excitation wavelength (laser output) is varied. For fluorescence scans, the excitation frequency is fixed, and the spectral distribution of the emissive signal is recorded by a multichannel detector. In addition to the slow scanning process, the excitation and fluorescence scans require that collisional quenching Q_{21} does not vary with excitation or fluorescence wavelength. Therefore, they are not suitable for engine applications where collisional quenching is high and often difficult to predict. In the following sections, monochromatic fluorescence thermometry will be introduced. The emphasis, however, will be on two-line fluorescence thermometry.

9.6.2 Monochromatic Fluorescence Thermometry

9.6.2.1 Principle of Monochromatic Fluorescence Thermometry

The most intuitive monochromatic fluorescence thermometry approach relies on the inherent temperature dependence of the fluorescence signal. Here we will consider linear fluorescence regime because saturated fluorescence is particularly difficult to achieve for two-dimensional applications. The expression of fluorescence signal in the linear fluorescence regime was derived in Chapter 3, shown here in Eq. 9.30:

$$P_{flu} = \eta_c \Omega V_c f_1(T) N_s B_{12} \frac{A_{21}}{A_{21} + Q_{21}} I_v \qquad (9.30)$$

where N_s represents the grouping $(\chi_m N)$, the number of fluorescing molecules. At thermal equilibrium, the Boltzmann fraction $f_1(T)$, which is the fraction of the population at the initial vibrational-rotational level of the ground electronic state, is given by

$$f_1(T) = \frac{g_J(2J+1)}{Q_{rot}\,Q_{vib}} \exp\left(-hcE(v,J)\big/kT\right)$$

(9.31)

where the rotational and vibrational partition functions are defined as the following:

$$Q_{rot} = kT/hcB$$

$$Q_{vib} = \left[1 - \exp\left(-hc\,\bar{v}_{vib}\big/kT\right)\right]$$

$E(v,J) = E_v + E_J = hc\bar{v}_{vib} + B_v J(J+1)hc$ (i.e., the energy (cm⁻¹) of the initial vibrational-rotational level)

If an excited state is chosen such that it is predissociative, the predissociation rate Q_{pre} should be included:

$$P_{flu} = \eta_c \Omega V_c\, f_1(T)\, N_s\, I_v\, B_{12} \frac{A_{21}}{A_{21} + Q_{21} + Q_{pre}}$$

(9.32)

Therefore, temperature measurements by LIF are possible by interrogating individual vibrational-rotational states through the Boltzmann fraction in Eqs. 9.31 and 9.32. They show that the influence of temperature occurs through the population of the lower vibrational-rotational level and the rate coefficients. If the fluorescence signal from a single transition is measured over a narrow bandwidth, the gas temperature can be derived from Eq. 9.32 with known values of the rate coefficients. The values of A_{21}, B_{12}, and Q_{pre} depend only on the nature of the excited molecule and can be determined accurately, whereas the quenching rate Q_{21} depends on gas composition, density, and temperature and is difficult to evaluate in most cases.

Other approaches based on monochromatic excitation include thermally assisted fluorescence (THAF) thermometry and dual fluorescence detection (Laurendeau, 1988).

9.6.2.2 Implementation and Application of Monochromatic Fluorescence Thermometry

As stated in the previous section, the major difficulty in monochromatic fluorescence thermometry is the accurate determination of the quenching rate

Q_{21}. This difficulty can be removed, however, if the loss from the excited level involved is dominated by predissociative rate Q_{pre}. Roller et al. (1995) found that the energy transfer processes of excited oxygen molecules were dominated by predissociative loss, and hence oxygen would be suitable as the fluorescence molecule for monochromatic fluorescence thermometry.

Figure 9.10 shows the experimental setup of monochromatic fluorescence thermometry used by Roller et al. (1995). The output of a tunable ArF excimer laser was tuned to the P(17) line of the (10,2) band of oxygen at 193.424 nm. The resulting fluorescence image from the (10,5) band emission (203–208 nm) was recorded by an ICCD camera. To maintain the optimum excitation efficiency throughout each experiment, 10% of the laser output was used to excite oxygen in a small laboratory Bunsen flame by a beam splitter.

To derive temperature from the fluorescence signal, the Einstein coefficients A and B were calculated, and the predissociation rate Q_{pre} and quenching rate Q_{21} were measured experimentally in separate experiments. The system response factor η_c was determined by setting the measured temperature at 77° BTDC during compression to that calculated from the measured cylinder pressure.

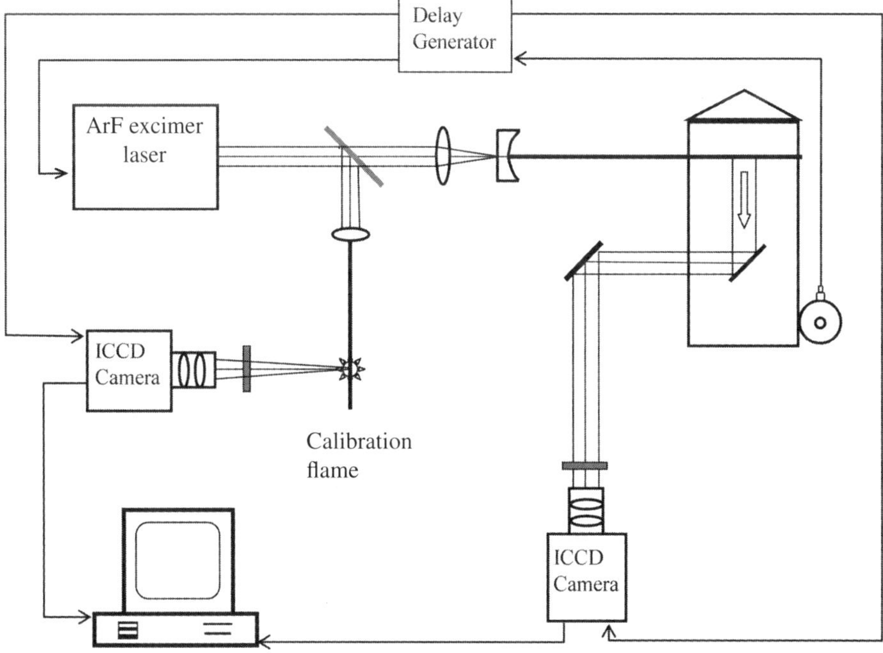

Fig. 9.10 Schematic of the experimental setup for monochromatic LIF thermometry.

Temperature measurements were carried out between 77° and 40° BTDC during the compression stroke in a diesel engine. Comparison of measured and calculated polytropic compression temperatures showed good agreement. The systematic error was estimated to be less than 10% mainly due to the uncertainty of Q_{21} and Q_{pre} at higher pressures.

The advantage of a relatively simple setup for the monochromatic is, however, compromised in several aspects. First, the application of monochromatic approach mandates the concentrations of species not vary spatially and temporally as Eq. 9.32 indicates. This requirement was fulfilled as their measurements were performed after the intake valve closed and there was no spatial variation of air in the motored diesel engine operation. The stratification of charge, resulting from the presence of burned gas and EGR (exhaust gas recirculation) or fuel, will invalidate the two-dimensional application of the monochromatic approach. Second, the accuracy of the monochromatic approach depends on the uncertainty of the quenching coefficient, which is often difficult to determine. In the work reported by Roller et al. (1995), measurements were carried out within a pressure range of 3–8 bar, where the predissociation was strong. As the pressure goes up, which is often the case at the end of the compression stroke, the quenching process will dominate and hence larger error will occur. Furthermore, like any absolute measurement techniques, the monochromatic approach suffers from errors due to the change in laser output, transmission of windows, absorption of the gas medium, and so forth. Finally, the laser beam with such a short wavelength will experience severe attenuation when fuel is present in the combustion chamber.

9.6.3 Two-Line Fluorescence Thermometry

9.6.3.1 Principle of Two-Line Molecular Fluorescence Thermometry

The principal feature of the two-line molecular fluorescence (TLMF) thermometry is shown in Fig. 9.11. The basic idea is to measure the relative populations of two states and calculate the temperature via Boltzmann expressions. Two rotational levels in the $v = 0$ level of the ground electronic state are used to infer the rotational temperature. The upper level could be a single rotational level or, more likely, a vibrational band within an excited electronic state. The excitation is performed sequentially. Level 3 is pumped from level 2, and the broadband fluorescence from level 3 to the ground electronic state is monitored; then level 3 is pumped from level 1, and the same broadband fluorescence from level 3 is measured. For turbulent combustion applications, both excitations are performed within some characteristic time scale, during which the medium is assumed to be unaltered. Dependence on quenching is avoided by exciting lower levels 1 and 2 to the same upper level. According to McKenzie and Gross (1981) and Anderson et al. (1982), if a molecule has closely spaced rotational levels, the broadband fluorescence to the ground electronic state should not depend on the particular rotational level in the excited electronic state. This is a result of the fast rotational relaxation and the weak dependence of the quenching rate on rotational levels.

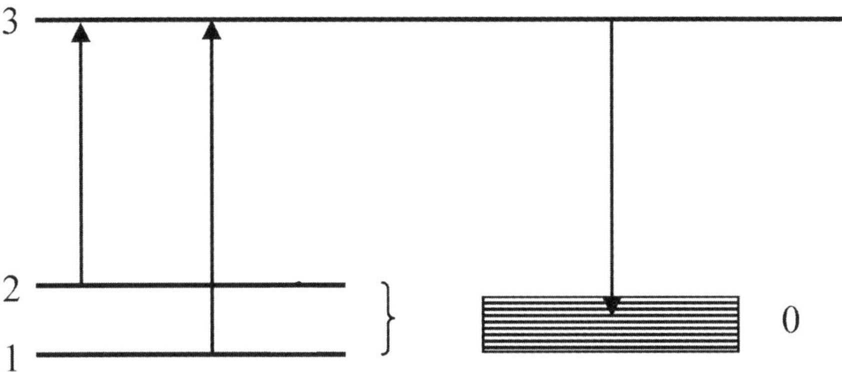

Fig. 9.11 Two-line molecular fluorescence (TLMF) thermometry.

As described in Chapter 3, the rate equation for level 3 via optical absorption from state 1 is given by

$$\frac{dN_3}{dt} = N_1 B_{13} I_{13} - N_3 \left(B_{31} I_{13} + A + Q \right) \qquad (9.33)$$

where B_{13} and B_{31} are the Einstein coefficients for stimulated absorption and emission, A is the Einstein coefficient for spontaneous emission, I_v (W cm^{-2} Hz^{-1}) is the laser spectral intensity, and Q is the nonradiative loss, mainly the quenching loss. The values of A and Q are the sum of those for individual transition lines from the excited state to the ground electronic state.

Similarly, the rate equation for pumping level 3 from level 2 is given by

$$\frac{dN_3}{dt} = N_2 B_{23} I_{23} - N_3 \left(B_{32} I_{23} + A + Q \right) \qquad (9.34)$$

The fluorescence from level 3 following the excitations from level 1 and 2 may be written respectively, assuming steady state, as

$$P_{13} = \eta_{13} N_3 A \left(\frac{h\nu}{c} \right) = \eta_{31} \frac{N_1 B_{13} I_{13} A}{\left(B_{31} I_{13} + A + Q \right)} \left(\frac{h\nu}{c} \right) \qquad (9.35)$$

$$P_{23} = \eta_{23} N_3 A \left(\frac{h\nu}{c} \right) = \eta_{32} \frac{N_2 B_{23} I_{23} A}{\left(B_{32} I_{23} + A + Q \right)} \left(\frac{h\nu}{c} \right) \qquad (9.36)$$

In the linear fluorescence regime, $B_{ij}I_v$ is much less than $(A + Q)$ and can be ignored. Using the Boltzmann fraction, Eq. 9.31, for N_1 and N_2,

$$\frac{N_2}{N_1} = \left(\frac{g_2}{g_1}\right)\left(\frac{2J_2+1}{2J_1+1}\right)\exp(-hc\Delta E / kT) \tag{9.37}$$

and radiative relations (Measures, 1984) are given by

$$A_{ij} = \left(\frac{8\pi h}{\lambda_{ij}^3}\right)B_{ij} \qquad\qquad \frac{B_{ij}}{B_{ji}} = \frac{g_j}{g_i}$$

Hence, from Eqs. 9.35 and 9.36, the fluorescence ratio is obtained:

$$\frac{P_{13}}{P_{23}} = \frac{\eta_{13}}{\eta_{23}}\frac{I_{13}}{I_{23}}\frac{B_{13}}{B_{23}}\frac{(2J_1+1)}{(2J_2+1)}\exp\left(\frac{hc\Delta E}{kT}\right) \tag{9.38}$$

Equation 9.38 provides the direct relationship between the ratio of fluorescence signals and the desired temperature. By calibrating at a known temperature, the expression also gives a measured temperature that is independent of any systematic effects on the fluorescence signals.

9.6.3.2 Implementation and Application of TLMF Thermometry

9.6.3.2.1 Selection of TLMF Molecule and Excitation Wavelengths

The selection of molecular species and energy levels involved in the excitation and fluorescence emission is critical to the successful application of TLMF thermometry. Most of original TLMF employed OH radical. Kohse-Höinghaus (1994) discussed several two-line excitation schemes that were employed for flame temperature measurements. The OH fluorescence thermometry, however, is limited to OH abundant regions and hence cannot be used for temperature measurements in any regions other than the high-temperature burned gas region.

Besides the OH radical, several other molecules have been utilized as fluorescing species for temperature measurements. Arnold et al. (1993) demonstrated that it was possible to obtain a two-dimensional temperature distribution during the compression stroke in a diesel engine by exciting $R(19)$ (10,2) and $R(29)$ (11,2) lines of oxygen using a tunable ArF excimer laser operated at 193 nm.

The most severe limitation of TLMF thermometry is its low signal-to-noise ratio, particularly at higher pressures. The shorter wavelengths required for the excitation of NO and O_2 aggravate the problem because of severe laser beam attenuation in the presence of fuels in the combustion chamber. The solution is, therefore, to find a molecule that has large fluorescence yield when excited at longer wavelengths and an excitation scheme that possesses good temperature sensitivity.

Einecke et al. (1998a, 1998b) carried out comprehensive analysis of the suitability of 3-pentanone, a fluorescence dopant widely used for mapping fuel distributions in PLIF. Figure 9.12 shows the energy states of 3-pentanone involved in the LIF process. Following the pumping from the ground electronic state S_0 to the first excited singlet state S_1, the molecule can be deactivated by

- Fluorescence emission to S_0
- Nonradiative loss to S_0
- Intersystem crossing to the triplet state T_1
- Chemical decomposition

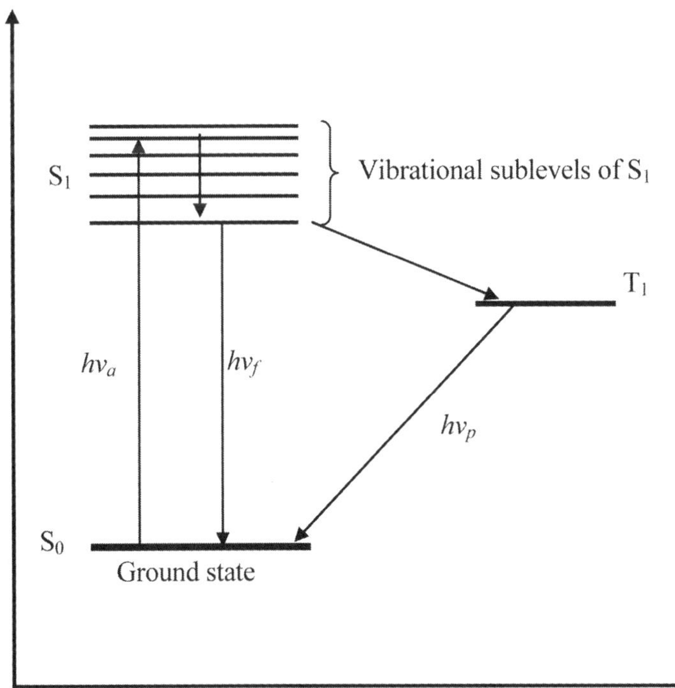

Fig. 9.12 Schematic energy diagram and mechanisms of fluorescence and phosphorescence in 3-pentanone.

The last process does not occur for temperatures below 573 K (Grossmann et al., 1996).

Figure 9.13 shows the absorption spectrum of 3-pentanone at 298 and 573 K (Grossmann et al., 1996). According to the Boltzmann distribution, the population of vibronic levels in the ground electronic state is concentrated at lower levels. As the temperature increases, higher levels become more populated. Therefore, the absorption peak at 275 nm shifts to longer wavelengths with increasing temperature (Fig. 9.13). If two laser beams are employed so that their wavelengths are positioned in the opposite-side wings of the absorption spectrum, the fluorescence signal from the shorter wavelength excitation will fall with increasing temperature, whereas the other from the longer wavelength excitation will increase. The ratio of these two signals can then be used to obtain the temperature information.

Because of overlapping vibrational-rotational bands in the electronic absorption spectra of most polyatomic molecules, broadband excitations can be used in the case of 3-pentanone. The simplification of the broadband excitation, however, is compromised by the increased overhead to obtain accurate temperature measurements. The mathematical expression of Eq. 9.38 was derived on the basis of excitation of two single transition lines to vibronic levels closely distributed in the excited electronic state. In the case of 3-pentanone, overlapping vibrational rotational bands in the absorption

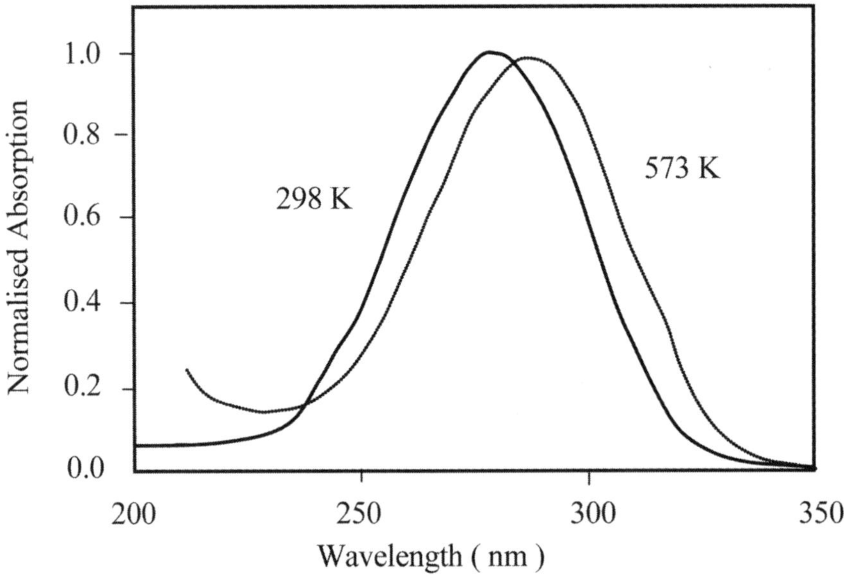

Fig. 9.13 Absorption spectrum of 3-pentanone at two different temperatures and atmospheric pressure, adapted from Grossmann et al. (1996).

spectra means that multiple vibronic levels are excited simultaneously from the ground electronic state to many different vibronic levels. The fluorescence signal in the linear regime can be rewritten as

$$P_{\lambda,flu} = \sigma_{\lambda}\left(\xi,p,T\right)\eta\,\Omega\,V_c\,N_s I_v\,\phi_{\lambda}\left(\xi,p,T\right) \tag{9.39}$$

where the first term σ represents the molecular absorption cross-section and is related to the values of Einstein coefficients B_{12} in Eq. 9.30, and the last term φ is the fluorescence quantum yield corresponding to $A/(A+Q)$ in Eq. 9.30. Both σ and φ are affected by mixture composition ξ, pressure p, and temperature T.

By taking the ratio of two fluorescence signals, most parameters in Eq. 9.39 cancels out:

$$\frac{P_{1,flu}}{P_{2,flu}} = \frac{\sigma_1\left(\xi,p,T\right) I_v^1 \phi_1\left(\xi,p,T\right)}{\sigma_2\left(\xi,p,T\right) I_v^2 \phi_2\left(\xi,p,T\right)} \tag{9.40}$$

The effect of quenching loss, which is the main factor in the fluorescence quantum yield, is largely removed. The remaining effect of pressure is related to the interlevel relaxation between different vibrational levels and is relatively weak. Grossmann et al. (1996) found that the fluorescence intensity increased substantially before it leveled off at pressures above 5 bar for excitation at 248 nm. The pressure effect became less pronounced as the excitation wavelength increased. Experiments in a high-temperature and high-pressure cell showed that a linear relationship between the ratio of two-line fluorescence intensity and gas temperature remained at pressures above 5 bar (Einecke et al., 1998b).

9.6.3.2.2 Implementation of TLMF

Figure 9.14 shows the experimental setup for TLMF thermometry measurements employed by Einecke et al. (1998b). For temporally resolved measurements, sequential excitation with a very short time delay is required. The time delay between the two lasers should be short enough to free the flow and long enough to separate the two fluorescence signals. The two laser beams can be obtained in different ways. They can be produced from two separate laser systems (Fig. 9.14), two dye lasers sharing the same pumping laser (Gross and McKenzie, 1983), or a broadband laser whose bandwidth is wide enough to simultaneously excited two transitions (Andresen et al. 1988). In the setup shown in Fig. 9.15, two gated ICCD cameras are employed to record in sequence the two fluorescence images of the same spectral characteristics. A beam splitter directs one-half of the fluorescence signal into each camera. Each camera should be synchronized with the corresponding laser excitation. For quantitative measurements, the fluorescence image recorded on each camera should be normalized to the laser output to take into account of the pulse-to-pulse variation and difference between the two lasers and ICCDs. Excitations at 248 nm and 308 nm were provided by a KrF (248-nm) laser and a XeCl (308-nm) laser with a fixed time delay of 150 ns with exposure times of 70 ns for each camera. Laser output energies were monitored on a shot-to-shot basis. Two-dimensional temperatures over 140 single shots at four different crank angles in the compression stroke were obtained from a calibration curve obtained in a calibration cell.

The output of laser sources at 248 and 308 nm are easily accessible. But the optimal wavelength combination for temperature and composition measurement over the range of engine conditions considered was shown to be 308 and 277 nm (Rothamer et al., 2008). Both of these wavelengths have low shot-to-shot laser profile variation and have higher absorption cross sections and fluorescence quantum yield at high temperatures than the more frequently used wavelengths of 248 and 308 nm. The estimated temperature for the chosen 308 and 277 nm combination was about 10 K.

Figure 9.15 shows the TLMF thermometry setup in the author's laboratory, which is similar to that used by Rothamer et al. (2008). The two excitation wavelengths at 308 and 277 nm are generated by a XeCl laser and via Raman shifting of a 248-nm KrF laser, respectively. The Raman shifter uses two lenses to focus and recollimate the laser beam. The energy conversion efficiency to the first Stokes (277 nm) is less than 15% when it is filled to 25 bar with H_2 and pumped with 400 mJ/pulse from the KrF excimer laser.

The beams' output by the two excimer lasers are rectangular (24 mm on the vertical axis and 10 mm on the horizontal axis). Large laser mirrors

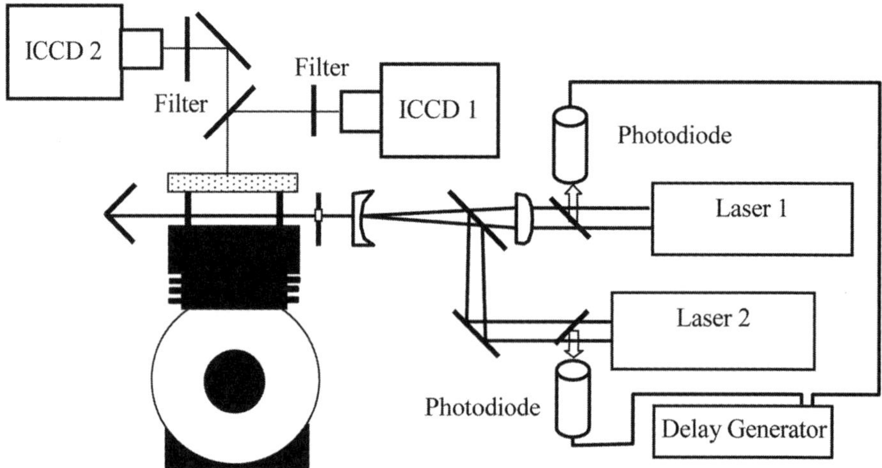

Fig. 9.14 TLMF thermometry setup with two excimer lasers and two ICCD cameras.

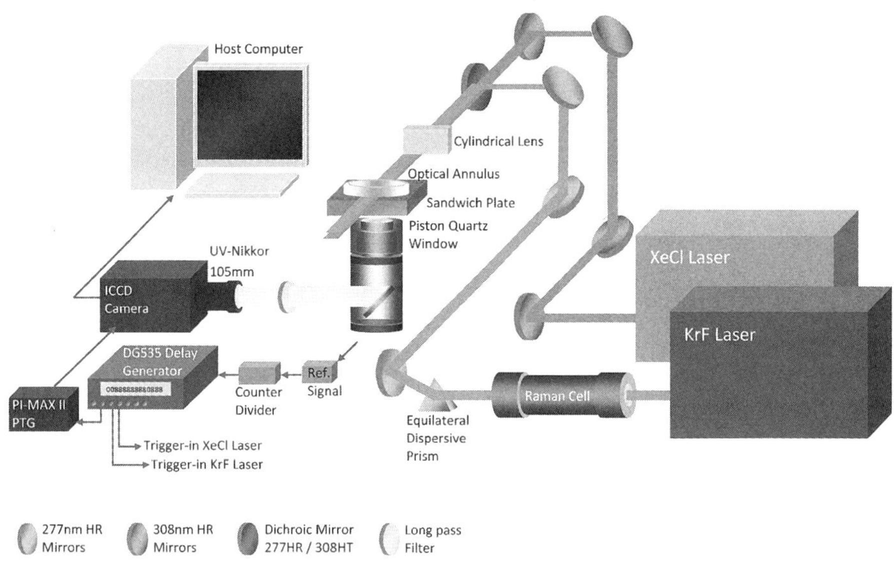

Fig. 9.15 TLMF thermometry setup with Raman cell and a single ICCD camera.

with appropriate high reflection coating (>98%) at each of two individual wavelengths are used to steer the laser beams. The output of the KrF laser is directed to the Raman shifter and then through an equilateral dispersing prism with antireflection coating at 277 nm to separate the pump and shifted wavelengths. Using three mirrors for each laser beam, the long axis of the beams are flipped, and they are raised to the height of the optical liner in the single cylinder engine.

A precise spatial overlap of the two excitation beams is crucial for the quantitative measurements, and it is accomplished with a single dichroic mirror with >97% transmission at 277 nm and >90% reflection at 308 nm. The output of the dichroic mirror passed through a single (or double) cylindrical lens that focuses the laser sheets in the vertical direction. The focused laser sheets (60-mm wide, ~0.5-mm thick) with the pulse energy of 30 and 60 mJ for the 277- and 308-nm wavelengths transmitted through the quartz liner.

The imaging system includes a camera, imaging lens, and filters. To collect the fluorescence signal and block the unwanted wavelengths, a 105-mm, 1:4.5 UV Nikkor® lens with filters such as Schott WG 320 or BG7 was used. The image intensifier is supplied with the fast decay phosphor P46 so that two images can be captured in a rapid succession with a minimum interframe time of 2 µs.

The camera and laser were synchronized to the engine by a crank angle timing signal at 4 Hz (camera full CCD readout limit). The camera's programmable timing generator (PTG) and DG535 delay generator were used to set the delays between the lasers and the camera trigger-in signals.

The TLMF thermometry with 277/308 nm combination has been applied to measure simultaneously the temperature and residual gas distribution in motored and fired HCCI engines with the negative valve overlap (NVO) strategy (Snyder et al., 2009). The same setup was also used to investigate the similarities and differences in the development of temperature stratification between motored and fired operation and the impact of direct fuel injection and cylinder wall temperature on thermal stratification (Snyder et al., 2011).

9.6.3.3 Two-Line Atomic Fluorescence (TLAF) Thermometry

For an atom, a low-lying excited electronic state (level 2) and an unpopulated electronic state (level 3) are employed in two-line atomic fluorescence (TLAF) thermometry. Detection of the fluorescence signal occurs by monitoring emission to that lower level not excited by the laser (A_{31} and A_{32}). For the atomic case, a similar analysis for narrowband detection gives

$$\frac{P_{13}}{P_{23}} = \frac{\eta_{13}}{\eta_{23}} \frac{I_{13}}{I_{23}} \frac{B_{13}}{B_{23}} \left(\frac{v_{13}}{v_{23}}\right)^4 \exp\left(hc\Delta E \middle/ kT\right) \tag{9.41}$$

TLAF thermometry can be achieved at much lower laser power than TLMF thermometry because atomic species have much higher oscillator strengths. However, measurements at lower temperatures ($T < 500$ K) are not feasible using atomic seeds because of their low vapor pressure. There is also difficulty associated with the seeding. In contrast to TLMF thermometry, the TLAF technique is not independent of fluorescence trapping (absorption of fluorescence signal) since the two detection wavelengths in TLMF are not the same. Because the populations in states 0 and 1 are normally very different at a given T, the relative amount of trapping will differ for P_{13} and P_{23} and cannot usually be accounted for in turbulent environment. Note that both Eqs. 9.38 and 9.39 are valid under the assumption that no laser absorption takes place in the probed environment and no fluorescence trapping occurs in the case of TLAF.

The most suitable atomic additive for TLAF thermometry is indium, because the gap between its two lower energy levels is such that high sensitivity is available for temperatures between 500 and 2500 K. Excitation takes place at 410 and 451 nm, which corresponds to transitions from the ground $5p^2P_{1/2}$ state to the excited $6s^2S_{1/2}$ state, and from the metastable $5p^2P_{3/2}$ state to the same excited state, respectively (Haraguchi et al., 1977). These excitation wavelengths can be readily obtained from dye lasers.

Following the previous work by Dec and Keller (1986) on TLAF thermometry in flames, Kaminski et al. (1998) demonstrated temporally resolved two-dimensional TLAF measurements in an SI engine. Indium was added to the intake charge by dissolving $InCl_3$ salt in isopropanol. The resulting solution with iso-octane was sprayed into the intake manifold. The concentration of indium was selected based on the trade-off between adequate signal-to-noise ratio and fluorescence trapping. Temporal resolution was obtained by sequentially excitation using two separate dye lasers with a time delay of 500 ns. The resulting fluorescence was split by a dichroic mirror into the 410- and 451-nm beams required for detection of indium. Each beam was directed to a gated ICCD camera, and its laser power was monitored by a PIN photodiode. The fluorescence and laser signals were analyzed on a shot-by-shot basis with online division of each fluorescence signal by the relevant laser power signal. Laser sheet profile measurements were obtained using PLIF of biacetyl, and ensemble-averaged sheet corrections were carried out to correct for variation in the beam profiles. After background subtraction, laser power, and beam profile normalization, each individual image pair was divided on a pixel-by-pixel basis so that a two-dimensional temperature distribution could be obtained. The measured temperature range covered 800–2800 K. A precision of 14% on single-shot temperature distributions was claimed, which could be significantly improved if laser beam profile was performed on a shot-by-shot basis.

9.7 Summary

Early in-cylinder gas temperature measurements were carried out entirely using line-of-sight techniques. Significant contributions were made by these temperature measurements to understanding normal and abnormal combustion processes in IC engines. The advent of laser-based techniques has made it possible to make temporally and spatially resolved temperature measurements in IC engines.

The monochromatic absorption-radiation method had been developed and employed in a number of in-cylinder gas temperature measurements before the advent of the laser diagnostics. It can provide continuous temperature readings without the emissivity of gas being known. Recently, Jeffries et al. (2010) has demonstrated the possibility to obtain in-cylinder gas temperature measurements averaged over a volume of gas near spark plug on the assumption that the absorption cross section of the absorbing species (HC fuel vapor) could be accurately modeled.

CARS thermometry is the most accurate technique for both burned and unburned gas temperature measurements. However, it is limited to single-point and single-shot measurements. For engine applications, two-dimensional temperature-imaging techniques are particularly important because of the inhomogeneous temperature distributions inside the combustion chamber. Such temporally resolved planar thermometry favors fluorescence methods because of their intrinsically high signal-to-noise ratios. Among various approaches of fluorescence thermometry, the two-line florescence thermometry techniques have demonstrated their potential for quantitative two-dimensional temperature measurements in an IC engine.

References

Agnew, W. G. (1960). "End Gas Temperature Measurement by Two-Wavelength Infrared Radiation Method." *SAE Trans*, Vol. 68, 495–513.

Anderson, W. R., Decker, L. J., and Kotlar, A. J. (1982). "Temperature Profile of a Stoichiometric CH4/N2O Flame from Laser Excited Fluorescence Measurements on OH." *Combustion and Flame*, Vol. 48, pp. 179–190.

Andresen, P., Bath, P. A., Groger, W., Lulf, H. W., Meijer, G., and ter Meulen, J. J. (1988). "Laser-Induced Fluorescence with Tunable Excimer Lasers as a Possible Method for Instantaneous Temperature Field Measurements at High Pressures: Checks with an Atmospheric Flame." *App. Opt.* Vol. 27, pp. 365–378.

Arnold, A., Brauerle, A., Buschmann, A. Decker, M., Dinkelacker, F., Heitzmann, T., Orth, A., Schafer, M., Sick, V., and Wolfrum, J. (1993). "2D-Diagnostics in Industrial Devices" *Ber. Bunsenges Phys. Chem.*, Vol. 97, No. 12, pp. 1650–1661.

Bood, J., Bengtsson, P., Mauss, F., Burgdorf, K., and Denbratt, I. (1997). "Knock in Spark Ignition Engines: End-Gas Temperature Measurements Using Rotational CARS and Detailed Kinetic Calculations of the Autoignition Process." SAE Paper No. 971669, SAE International, Warrendale, PA.

Bräumer, A., Sick, V., Wolfrum, J., Drewes, V., Zahn, M., and Maly, R. (1995). "Quantitative Two-Dimensional Measurements of Nitric Oxide and Temperature Distributions in a Transparent Square Piston SI Engine." SAE Paper No. 952462, SAE International, Warrendale, PA.

Burrows, M. C., Shimizu, S., Meyers, P. S., and Uyehara, O. A. (1961). "The Measurement of Unburnt Gas Temperatures in an Engine by an Infrared Radiation Pyrometer." *SAE Trans.*, Vol. 69, pp. 514–528.

Dec, J. E., and Keller, J. O. (1986). "High-Speed Thermometry Using Two-Line Atomic Fluorescence." *Proc. 21st Symposium on Combustion*, pp. 1737–1745. Pittsburgh: The Combustion Institute.

Drake, M. C., Lapp, M., and Penney, C. M. (1982). "Use of the Vibrational Raman Effect for Gas Temperature Measurements." In *Temperature: Its Measurement and Control in Science And Industry*, Vol. 5, edited by J. F. Schooley, pp. 631–638. New York: American Institute of Physics.

Eckbreth, A. C. (1996). *Laser Diagnostics for Combustion Temperature and Species*, 2nd ed. Amsterdam: Gordon and Breach.

Einecke, S., Schulz, C., and Sick, V. (1998a). " Dual-Wavelength Temperature Measurements in an IC Engine Using 3-Pentanone Laser-Induced

Fluorescence." In *Laser Applications to Chemical and Environmental Analysis*, Vol. 3 of OSA Technical Digest Series. Orlando: Optical Society of America.

Einecke, S., Schulz, C., Sick, V., Dreizler, A., Shiessi, R., and Maas, U. (1998b). "Two-dimensional Temperature Measurements in an SI Engine Using Two-Line Tracer LIF." SAE Paper No. 982468, SAE International, Warrendale, PA.

Goss, L. P. (1993). "CARS Instrumentation for Combustion Applications." In *Instrumentation for Flows with Combustion*, edited by A. M. K. P. Taylor, pp. 251–322. London: Academic Press.

Gross, K. P., and McKenzie, R. L. (1983). " Single-Pulse Gas Thermometry at Low Temperatures Using Two-Photon Laser-Induced Fluorescence in $NO-N_2$ Mixtures." *Opt. Lett.*, Vol. 8, pp. 368–370.

Grossmann, F., Monkhouse, P. B., Ridder, M., Sick, V., and Wolfrum, J. (1996). "Temperature and Pressure Dependence of the Laser-Induced Fluorescence of Gas-Phase Acetone and 3-Pentanone." *Appl. Phys. B,* Vol. 62, pp. 249–253.

Hall, R. J., Veerdiek, J. F., and Eckbreth, A. C. (1980). "Pressure-Induced Narrowing of the CARS Spectrum of N_2." *Opt. Commun.*, Vol. 35, 69–75.

Haraguchi, H., Smith, B., Weeks, S., Johnson, D. J., and Winefordner, J. D. (1977). "Measurement of Small Volume Flame Temperatures by the Two-Line Atomic Fluorescence Method." *Appl. Spectrosc.*, Vol. 31, 156–163.

Hofmann, D., and Leipertz, A. (1996). "Temperature Field Measurements in a Sooting Flame by Filtered Rayleigh Scattering (FRS)." *Proc. Combus. Inst.*, Vol. 26, pp. 945–950.

Hottel, H. C., and Egbert, R. B. (1942). "Radiant Heat Transmission from Water Vapor." *AIChE Trans,* Vol. 38, 531–565.

Jeffries, J., Porter, J., Pyun, S., Hanson, R., et al. (2010). "An In-cylinder Laser Absorption Sensor for Crank-Angle-Resolved Measurements of Gasoline Concentration and Temperature." *SAE Int. J. Engines,* Vol. 3, No. 2, pp. 373–382.

Kajiyama, K., Sajiki, K., Kataoka, H., Maeda, S., and Hirose, C. (1982). "N_2 CARS Thermometry in Diesel Engine." SAE Paper No. 821036, SAE International, Warrendale, PA.

Kalghatgi, G. T., Snowdon, P., and McDonald, C. R. (1995). "Studies of Knock in a Spark Ignition Engine with CARS Temperature Measurements and Using Different Fuels." SAE Paper No. 950690, SAE International, Warrendale, PA.

Kaminski, C. F., Engstrom, J., and Alden, M. (1998). "Quasi-Instantaneous Two-Dimensional Temperature Measurements in a Spark Ignition Engine Using 2-Line Atomic Fluorescence." 26th Symposium on Combustion. Pittsburgh: The Combustion Institute.

Klick, D., Marko, K. A., and Rimai, L. (1981). "Broadband Single-Pulse CARS Spectra in a Fired Internal Combustion Engine." *Applied Optics.*, Vol. 20, pp. 1178–1181.

Kohse-Höinghaus, K. (1994). "Laser Techniques for the Quantitative Detection of Reactive Intermediates in Combustion Systems." *Progress in Energy and Combustion Science*, Vol. 20, pp. 203–279.

Lapp, M. (1974). "Flame Temperatures From Vibrational Raman Scattering." In *Laser Raman Gas Diagnostics,* edited by M. Lapp and C. M. Penny, p. 107. New York: Plenum.

Laurendeau, N. M. (1988). "Temperature Measurements by Light-Scattering Methods." *Prog. Energy and Combustion Science,* Vol. 14, No. 2, pp. 147–170.

Long, D. A. (1977). *Raman Spectroscopy.* New York: McGraw-Hill.

Marie, J., and Cottereau, M. (1987). "Single-Shot Temperature Measurements by CARS in an IC Engine for Normal and Knocking Conditions." SAE Paper No. 870458, SAE International, Warrendale, PA.

McKenzie, R. L., and Gross, K. P. (1981). "Two-photon Excitation of Nitric Oxide Fluorescence as a Temperature Indicator in Unsteady Gas Dynamic Processes." *Applied Optics,* Vol. 20, pp. 2153–2165.

Measures, R. M. (1984). *Laser Remote Sensing.* New York: Wiley.

Nakada, T., Itoh, T., and Takagi, Y. (1993). "Application of CARS to Development of High Compression Ratio Spark Ignition Engine." SAE Paper No. 932644, SAE International, Warrendale, PA.

Nakano, M., et al. (1995). "Predictions of the Knock Onset and the Effects of Heat Release Pattern and Unburned Gas Temperature on Torque at Knock Limit in SI Engines." SAE Paper No. 952408, SAE International, Warrendale, PA.

Orth, A., et al. (1994). "Simultaneous 2D Single-Shot Imaging of OH Concentrations and Temperature Fields in an SI Engine Simulator." *Proc. of 25th Symposium (International) on Combustion,* pp. 143–150. Pittsburgh: The Combustion Institute.

Roller, A., Arnold, A., Decker, M., Sick, V., Wolfrum, A., Hentschel, W., and Schindler, K. P. (1995). "Nonintrusive Temperature Measurements During the Compression Phase of a DI Diesel Engine." SAE Paper No. 952461, SAE International, Warrendale, PA.

Rothamer, D. A., Snyder, J. A., Hanson, R. K., and Steeper, R. R. (2008). "Two-Wavelength PLIF Diagnostic for Temperature and Composition." SAE Paper No. 2008-01-1067, SAE International, Warrendale, PA.

Ryu, H., and Asanuma, T. (1984). " Combustion Analysis with Gas Temperature Diagrams Measured in a Prechamber Spark Ignition Engine." *Proc. 20th Symposium (International) on Combustion.* Pittsburgh: The Combustion Institute.

Sakurauchi, Y., Ryu, H., Iijima, T., and Asanuma, T. (1987). "Combustion Gas Temperature in a Prechamber Spark Ignition Engine Measured by Infrared Pyrometry." SAE Paper No. 870457, SAE International, Warrendale, PA.

Schulz, C., Sick, V., Wolfrum, J., Drewes, V., Zahn, M., and Maly, R. (1996). "Quantitative 2D Single Shot Imaging of Nitric Oxide Concentrations and Temperature in a Transparent SI Engine." *Proc. of 26th Symposium (International) on Combustion*. Pittsburgh: The Combustion Institute.

Siegel, R., and Howell, J. R. (1993). *Thermal Radiation Heat Transfer,* 3rd Ed New York: Hemisphere.

Smith, J. R. (1980). "Temperature and Density Measurements in an Engine by Pulsed Raman Spectroscopy." SAE Paper No. 800137, SAE International, Warrendale, PA.

Smith, J. R., Green, R. M., Westbrook, C. K., and Pitz, W. J. (1984). "An Experimental and Modeling Study of Engine Knock." *Proc. of 20th Symposium (International) on Combustion*. Pittsburgh: The Combustion Institute.

Snyder, J. A., Hanson, R. K., Fitzgerald, R. P., and Steeper, R. R. (2009). "Dual-Wavelength PLIF Measurements of Temperature and Composition in an Optical HCCI Engine with Negative Valve Overlap." SAE Paper No. 2009-01-0661, SAE International, Warrendale, PA.

Snyder, J., Dronniou, N., Dec, J., and Hanson, R. K. (2011). "PLIF Measurements of Thermal Stratification in an HCCI Engine under Fired Operation." SAE Paper No. 2011-01-1291, SAE International, Warrendale, PA.

Stenhouse, I. A., Williams, D. R., Cole, J. B., and Swords, M. D. (1979). "CARS Measurements in an Internal Combustion Engines." *Applied Optics,* Vol. 18, pp. 3819–3825.

Tien, C. (1968). "Thermal Radiation Properties of Gases." *Adv. Heat Transfer,* Vol. 5, pp. 254–321.

Takagi, H., Ohno, T., and Asanuma, T. (1990). "Temperature Measurements of Combustion Gas in a Spark Ignition Engine by Infrared Monochromatic Pyrometry." SAE Paper No. 900483, SAE International, Warrendale, PA.

Withrow, L., and Rassweiler, G. M. (1932). "Emission Spectra of Engine Flames." *Ind. Eng. Chem.,* Vol. 24, pp. 528–538.

Zur Loye, A., and Santavicca, D. A. (1983). "Temperature and Concentration Measurements in an Internal Combustion Engine Using Laser Raman Spectroscopy." Paper No. AIAA-83-151. AIAA 18th Thermophysics Conference, Montreal.

chapter 10

In-Cylinder Soot Concentration and Particle Size Measurement

10.1 Introduction

Diesel engines offer significant fuel economy advantages over other powerplants used for road transportation. Direct-injection (DI) gasoline engines are staging a comeback in the form of downsized engines for passenger cars because of their *improved* fuel economy. However, more stringent emission legislations are being introduced to limit not only the mass but also the number of particle emissions from both gasoline and diesel engines. The development and application of in-cylinder soot measurement techniques is therefore required to better understand the soot formation and oxidation processes in the cylinder.

One of the characteristics of combustion with soot is its high luminosity, caused by the thermal radiation of soot particles at high temperature. To take advantage of its luminous nature, the two-color method has been developed to measure diesel combustion temperature since 1940s and, more recently, DI gasoline engines. Although it is not a laser-based technique, the two-color method will be discussed here because of its simplicity and ability to measure whole field combustion temperature and soot concentration.

The light-extinction method is a simple and accurate way to measure in-cylinder soot concentration. To obtain soot particle sizing, both light-scattering and extinction measurements are necessary. The description of light-scattering theory will be kept to minimum. For complete description of light scattering from small particles, see Van de Hulst (1957), Kerker (1969), and Bohren and Huffman (1988).

Laser-induced incandescence (LII) has been developed to obtain both qualitative two-dimensional soot distribution and quantitative soot concentration and size measurements. Its principle and operational considerations will be discussed, followed by some examples of its application to in-cylinder soot measurements.

10.2 The Two-Color Method

10.2.1 Principle of the Two-Color Method

In the two-color method, thermal radiation at two different wavelengths is detected, and the soot surface temperature is then determined from their ratio by eliminating an unknown factor. However, the difference between the two temperatures is negligible (<1 K) when ambient gas and soot particles have attained thermal equilibrium.

The intensity of radiation from a blackbody varies with wavelength, and it depends on the temperature of the blackbody. This is described mathematically by Planck's law:

$$E_{b,\lambda}(T) = \frac{C_1}{\lambda^5 [e^{(C_2/\lambda T)} - 1]} \qquad (10.1)$$

where

$E_{b,\lambda}$ = monochromatic emissive power of a blackbody at temperature T (W m^{-3})

λ = wavelength (μm)

T = temperature (K)

C_1 = the first Planck's constant = 3.7418×10^{-16} W m^2

C_2 = the second Planck's constant = 1.4388×10^{-2} m K

The monochromatic emissivity of a nonblackbody is defined as

$$\varepsilon_\lambda = \frac{E_\lambda(T)}{E_{b,\lambda}(T)} \qquad (10.2)$$

where $E_\lambda(T)$ and $E_{b,\lambda}(T)$ are the monochromatic emissive power of a nonblackbody and a blackbody, respectively, at the same temperature T and wavelength λ. In other words, ε_λ is the fraction of the blackbody radiation emitted by a surface at wavelength λ.

In the two-color method, an apparent temperature T_a is introduced, defined as the temperature of a blackbody that will emit the same radiation intensity as a nonblackbody at temperature T. T_a is also known as the brightness temperature. From the definition of T_a, it follows that $E_{b,\lambda}(T_a) = E_\lambda(T)$.

Combining the definition for monochromatic emissivity with that of the apparent temperature gives

$$\varepsilon_\lambda = \frac{E_{b,\lambda}(T_a)}{E_{b,\lambda}(T)}$$

(10.3)

Combining Eqs. 10.1 and 10.3, the monochromatic emissivity is given theoretically by

$$\varepsilon_\lambda = \frac{e^{(C_2/\lambda T)} - 1}{e^{(C_2/\lambda T_a)} - 1}$$

(10.4)

In practice, ε_λ is estimated for soot particles using the empirical correlation developed by Hottel and Broughton (1932):

$$\varepsilon_\lambda = 1 - e^{-\left(KL/\lambda^\alpha\right)}$$

(10.5)

where

> K = an absorption coefficient = proportional to the number density of soot particles

> L = the geometric thickness of the flame along the optical axis of the detection system

The value of the parameter α depends on the physical and optical properties of the soot. Matsui et al. (1979) confirmed that at least in the visible range this is a correct functional relation between emissivity and wavelength.

Combining Eqs. 10.4 and 10.5 gives

$$KL = -\lambda^\alpha \ln\left[1 - \left(\frac{e^{\left(C_2/\lambda T\right)} - 1}{e^{\left(C_2/\lambda T_a\right)} - 1}\right)\right]$$

(10.6)

The unknown product KL can be eliminated by rewriting the above equation for two specific wavelengths, λ_1 and λ_2:

$$\left[1 - \left(\frac{e^{\left(C_2/\lambda_1 T\right)} - 1}{e^{\left(C_2/\lambda_1 T_{a1}\right)} - 1}\right)\right]^{\lambda_1^{\alpha_1}} = \left[1 - \left(\frac{e^{\left(C_2/\lambda_2 T\right)} - 1}{e^{\left(C_2/\lambda_2 T_{a2}\right)} - 1}\right)\right]^{\lambda_2^{\alpha_2}}$$

(10.7)

The actual flame temperature T is independent of wavelength, while the apparent temperature T_a and, possibly, the parameter α vary with wavelength.

This equation can be solved for the flame temperature T, provided the apparent temperature T_{a1} and T_{a2} for the flame are known at the two wavelengths, λ_1 and λ_2. These two apparent flame temperatures can be measured at these two wavelengths using a calibrated two-color optical pyrometer system. In fact, this is the purpose of the two-color system: to provide instantaneous measurements of the apparent temperatures T_{a1} and T_{a2} for the flame at two chosen wavelengths, λ_1 and λ_2. The choice of the two wavelengths, at which T_{a1} and T_{a2} are monitored, will be discussed later in this chapter.

Once the value of the flame temperature T has been determined using Eq. 10.7, substitution of T back into Eq. 10.6 gives an estimate for the value of the product KL, which is proportional to the soot concentration. In addition, if some assumptions are made, an estimate of the volumetric density of soot C_v (the volume of particles in a unit of space) and soot gravimetric density C_m (mass of soot per unit volume) can be obtained.

The volumetric density of particles C_v (the volume of particles per unit volume) is related to the number density of particles C_n (the number of particles in a unit of space) by

$$C_v = \frac{\pi}{6} C_n \int_0^\infty N(D)\, D^3\, dD \tag{10.8a}$$

which can be shown, according to the light scattering theory, as given by

$$C_v = \frac{1}{6\pi L \; \mathrm{Im}\left\{\dfrac{m^2-1}{m^2+2}\right\}} \frac{KL}{\lambda^{\alpha}} \tag{10.8b}$$

where m is the complex refractive index of the soot particle. Thus, using the estimate of KL, it is possible to obtain an estimate of the volumetric density of soot particles. Of course, the values of L, α, and m are needed.

The gravimetric density of soot particles C_m can also be obtained if the value of soot density ρ is known or can be estimated. The estimation of the soot number density, that is, the number of soot particles per unit volume, requires the knowledge of the size distribution of soot particles, which are normally difficult to obtain.

10.2.2 Implementation of the Two-Color Method

A typical arrangement for the implementation of the two-color method is shown in Fig. 10.1.

The radiation emitted by a cloud of incandescent soot particles passes through an optical window and aperture, and then it is guided by a fiber-optical bundle to a bifurcation where it is split into two beams. Each beam then passes through a band-pass filter (usually 5–20 nm). The instantaneous signals generated by the photodiodes are amplified, and the resulting voltages at each wavelength are recorded by a data acquisition system.

Note that the aperture inserted between the optical window and the receiving fiber defines the field of view of the system to a narrow cone that lies within the combustion chamber. The use of an aperture increases the spatial resolution, but instrument sensitivity is reduced as a result of less radiation reaching the receiving optics. Another important function of the aperture is to reduce the possibility of not filling the field of view. When only part of the field

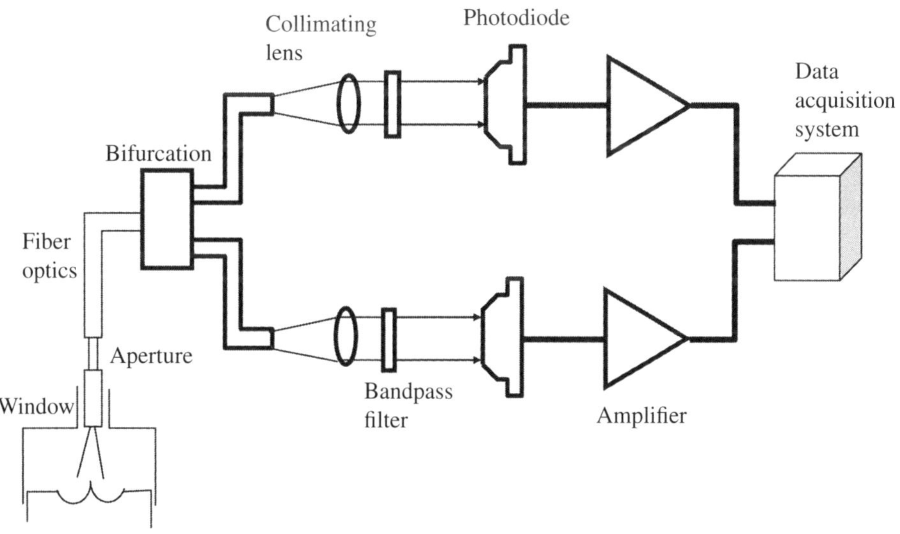

Fig. 10.1 Schematic layout of a single-point two-color system.

of view is filled with combustion gas, the brightness temperature of the flame will be underestimated.

10.2.2.1 Selection of α

Equations 10.6–10.8 show that estimation of the flame temperature, the KL factor, and the volumetric density of soot particles, depends on the values chosen for the parameter α. Thus, the choice of a suitable value for α is important.

The value of α is a function of a number of factors. It depends on the light wavelength, the soot particle size, and the refractive index of soot. Fuel type may also have an influence on the value of α (Gaydon and Wolfhard, 1970). Before selecting a value for α, consider that when the two wavelengths are

Table 10.1 Compilation of Values of the Parameter α in the Hottel and Broughton Correlation for Soot Emissivity

Value of Parameter α				
Visible Wavelengths	**Infrared Wavelengths**		**Fuel or Flame Type**	**Reference**
1.39	0.95	($\lambda > 0.8\,\mu$m)	steady luminous flame	Hottel & Broughton (1932)
1.38	0.91–0.97	($\lambda = 2$–$4\,\mu$m)	diesel engine soot	Matsui et al.(1980)
1.39	-		diesel engine soot	Yan & Borman (1988)
–	0.94–0.96	($\lambda > 0.8\,\mu$m)		Liebert & Hibbard (1970)
–	0.89, 1.04	($\lambda = 1$–$7\,\mu$m)	amyl acetate	Siddall & McGrath (1963)
–	0.77	"	avtur kerosene	
–	0.94, 0.95	"	benzene	
–	0.93	"	candle	
–	0.96, 1.14, 1.25	"	furnace samples	
–	1.06	"	petrotherm	
–	1.00	"	propane	
–	$\alpha = 0.91 + 0.28\ln\lambda$	"	various fuels	
1.43	–		acetone	Rossler and Behrens (1950)
1.39	–		amyl acetate	
1.29	–		coal gas/air	
1.23	–		benzene	
1.14	–		nitrocellulose	
0.66–0.75	–		acetylene/air	

selected to be in the visible region, the choice of the value of α is less critical to estimating the flame temperature. This is because the flame temperature given by Eq. 10.7 is little affected by the value of α in the visible range. Similarly, the estimation of KL is almost independent of α in the visible spectrum but varies significantly in the infrared (IR) region.

Table 10.1 summarizes the values of α used in the literature. From these values, it is recommended that a fixed value of 1.39 would be appropriate for most fuels when using visible wavelengths. For the IR region, the choice of the value of α is less clear-cut, but a fixed value close to unity appears to be appropriate for diesel combustion for the wavelength range of 0.8 to about 7 μm.

10.2.2.2 Selection of Wavelengths

The choice of wavelengths for the two-color method involves several considerations. In general, visible and IR wavelengths have been used in the two-color method. Visible wavelengths are preferred for many reasons. First, if the wavelengths are chosen in the visible part of the spectrum, the sensitivity of the measuring system to changes in flame temperature will be greater because the rate of change of spectral radiance with respect to temperature, dI_λ/dT, is larger in the visible than in the IR region. Also, there will be a greater difference in signal output from the system at the two visible wavelengths, because the spectral radiance with respect to wavelength, $dI_\lambda/d\lambda$, is higher in visible region for temperatures around 1000 to 2000 K. The larger separation of two wavelengths used will also improve the sensitivity of the two-color system, provided that the α parameters for the two wavelengths are known. Furthermore, estimation of temperature and KL factor is not very sensitive to the choice of the value of the parameter α, when the two wavelengths are chosen to be in the visible region.

During combustion, strong band emissions from gaseous species can be present as well as the thermal radiation from soot particles. In the reaction zones of flames, many radicals, such as OH, CH, C_2, HCO, NH, and NH_2 may be formed and give appreciable emission in the visible and near ultraviolet regions. In the IR region, wavelengths must be chosen carefully to avoid radiation or absorption from CO_2, CO, water vapor, and fuel vapor.

Photodiodes with high responsivity in the visible and near IR regions are readily available commercially. Also, there are commercially available optical fibers for use as light guides that have a high transmittance in the visible and near IR regions.

Table 10.2 shows a compilation of the wavelengths used by some investigators in the application of the two-color method.

Table 10.2 Wavelength Used by Various Investigators in the Application of the Two-Color Method

Wavelength (nm)					Reference
Visible			Infrared		
λ_1	λ_2	λ_3	λ_1	λ_2	
549	750	–	2300	3980	Matsui et al. (1980)
529	624	738	–	–	Matsui et al. (1979)
550	700	850	–	–	Yan and Borman (1988)
–	–	–	750	1000	Peterson and Wu (1986)
460	640	–	–	–	Kawamura et al. (1989)
550	650	–	–	–	Shakal and Martin (1994)
450	640	–	–	–	Arcoumanis et al. (1995)
581	631	–	–	–	Winterbone et al. (1994)
540	643	–	–	–	Miyamoto et al. (1996)
600	–	–	–	1000	Bertoli et al. (1996)
470	650	–	–	–	Shiozaki et al. (1996)

Note: Band-pass half-widths at the visible wavelengths are usually 5–25 nm.

10.2.2.3 Calibration

Recall that the two-color method measures the apparent temperatures of the flame T_{a1} and T_{a2} at wavelengths λ_1 and λ_2. These apparent temperatures can then be used in Eq. 10.7 to obtain the actual flame temperature. Since the two-color system will produce initially only voltages $V_{\lambda 1}$ and $V_{\lambda 2}$ corresponding to the radiation from the flame at the two wavelengths, the system must be calibrated so that the two voltages correspond to the apparent flame temperatures. This can be done using a source of blackbody radiation.

Using a source of blackbody radiation, a calibration curve can be generated in the form of a plot of voltage output against blackbody temperature for each wavelength. Once the calibration curves have been produced for the two wavelengths, the instantaneous voltages can be converted to the apparent temperatures T_{a1} and T_{a2}. The flame temperature T and the KL factor can then be obtained from Eqs. 10.7 and 10.6, respectively.

Often, a blackbody may not be readily available. In these circumstances, it may be more convenient to use a calibrated tungsten lamp. The lamp calibration is

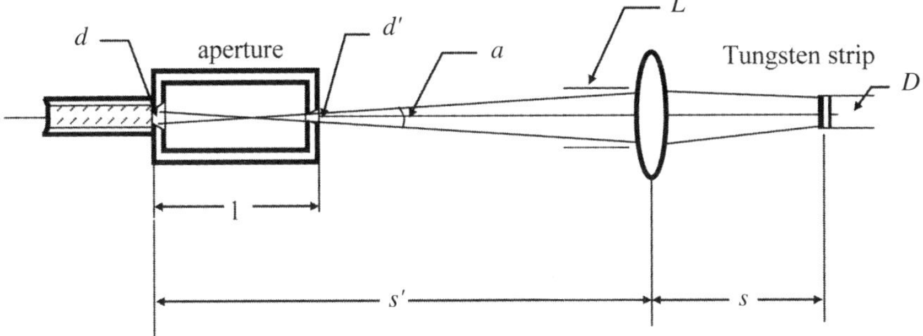

Angle α must be smaller than the acceptance angle of the optical fiber.
Angle α, dimensions d, l, D, and focal length f are specified.
Calculate d' to give required angle   from $d' = 2\,l\,tan(\alpha/2)$ - d.
Calculate distance s' from $s' = f\,[1 + (d/D)]$ where d/D is the magnification ratio.
Calculate distance s from $s = f\,[1 + (D/d)]$.
Calculate lens aperture L from $L = [\,2\,s'\,tan(\alpha/2)]$ - d
Note: if aperture is designed with both holes of equal diameter, then $d = d'$, in which
   case d is specified and l is calculated from $l = d/tan(\alpha/2)$, to give the required
   value of angle α.

Fig. 10.2 Optical setup for the calibration of a two-color system using a tungsten ribbon lamp.

often carried out at a wavelength of 650 nm only, but it is sufficiently accurate at other wavelengths close to 650 nm. For wavelengths very different from 650 nm, a new calibration curve will be required for the tungsten lamp at that wavelength (Fig. 10.2). Another difficulty when using a tungsten lamp for calibration is that the size of the tungsten ribbon filament may be too small to illuminate the whole of the field of view of the two-color system. This problem could be overcome by using a lens. The focal length and the position of the optical lens determine the size of the image of the tungsten ribbon D, which should at least be the same as the diameter of the receiving optical fiber or detector. The diameter of the lens should be chosen so that the field of view in the calibration is the same as that as defined by the aperture in the probe.

When the optical lens is used, some of the radiation from the tungsten lamp will be lost through reflection and absorption by the lens. If the transmittance of the lens is τ, the voltage output V from a photodiode will be lowered by ($1 - \tau$) than it would be without the lens for a given value of T. Thus in constructing the calibration curve for the two-color system, the quantity V/τ should be plotted against the apparent temperature, since when the soot system is operational in an engine without the lens, the voltage output will be somewhat higher.

10.2.2.4 Data Acquisition and Analysis

The output from a two-color system is normally converted into voltages by the photodiodes before subsequent calculations are carried out, to obtain the flame temperature and soot loading factor. A high-speed data acquisition system is used to collect the amplified signal outputs from the photodiodes on a crank angle basis using the output from a shaft encoder as the clock for data acquisition. Large cycle-by-cycle variation in radiation, however, is present in diesel combustion. The radiation signals could be very irregular because of the motion of soot clouds. The irregularity in radiation signals could also be caused by the random location of ignition and combustion. It has been found that the large cyclic variation of the radiation signals gives a large variation in the results of the KL factors. The flame temperature, though, was found to be less affected by the cycle-by-cycle variation. Some form of ensemble average, therefore, is required to obtain the temperature and KL measurement from the two-color method.

There are two ways to obtain an ensemble-averaged temperature and KL factor. In the first approach, which Matsui et al. (1979) used, the voltage outputs corresponding to the radiation intensity at the two wavelengths at each crank angle for 1000 cycles are ensemble-averaged. This averaged voltage outputs for the two wavelengths are then used in the calculation of the flame temperature and KL factor. An alternative approach is to use ensemble-averaged temperature and KL factor. In this method, the signal for each cycle is used to calculate the temperature variation at each degree crank angle during a particular engine cycle. Then the individual cycle temperatures are ensemble-averaged over many engine cycles. Because the flame temperature and KL factor are not linearly related to the radiation intensity, it is expected that these two approaches may not generate the same results. Yan and Borman (1988) compared these two methods and found that the difference between them was small for KL factor and temperature during most of the combustion period. In practice, the first approach is preferred because it is computationally faster and reduces the influence of noise on the calculation of temperature.

After conversion of signal outputs into apparent temperatures at the two wavelengths, the true flame temperature can be calculated using Eq. 10.6, which is highly nonlinear and can be solved by the method of Newton iteration. However, this method will not always give a convergent solution for this equation, especially when the two apparent temperatures are very close, that is, the monochromatic emissivity being near unity. As a consequence, a calculation scheme has been developed by Yan and Borman (1988). It involves the estimation of the apparent temperature at the first wavelength T_{a1} for an assumed T and measured T_{a2}, and an iteration process until the calculated T_{a1} is within 0.1°C of the measured T_{a1}, each iteration being calculated for a new estimated T.

The numerical scheme is based on the observation that the variation of T_{a1} with T at a given T_{a2} is approximately linear for a short temperature range, as shown by rearranging Eq. 10.7:

$$\frac{1}{T_{a1}} = \frac{\lambda_1}{C_2} \ln\left[1 + \left(\frac{e^{(C_2/\lambda_1 T)} - 1}{A}\right)\right] \tag{10.9}$$

where

$$A = 1 - \ln\left[1 - \left(\frac{e^{(C_2/\lambda_2 T)} - 1}{e^{(C_2/\lambda_2 T_{a2})} - 1}\right)\right]^{\alpha_2/\alpha_1}$$

It is also true that the flame temperature is greater than its apparent temperatures T_{a1} and T_{a2}. The detailed procedures to determine the flame temperature are summarized as follows:

1. Assume that the true flame temperature lies between points 1 and 2 in the figure. Point 2 should therefore be selected to be greater than the true flame temperature. Because the flame temperature in a diesel engine is around 2500 K, $T_2 = 3000$ K will be a good value. The temperature at point 1, T_1, is assigned to the measured T_{a2} at the longer wavelength, which should always be lower than the true flame temperature.

2. Evaluate the first apparent temperature at point 2, $T_{a1,2}$, with T_2 and T_{a2} using Eq. 10.9. T_{a2} should be greater than $T_{a1,m}$, which is the measured apparent temperature at the shorter wavelength.

3. Use $T_{a1,m}$ to estimate the new flame temperature by linear interpolation.

$$T_2^{new} = T_1 + \frac{(T_2 - T_1)(T_{a1,m} - T_{a21})}{(T_{a1,2} - T_{a21})} \tag{10.10}$$

4. Use the newly estimated flame temperature T_2^{new} with T_{a2} to evaluate a new $T_{a1,2}$.

5. Compare $T_{a1,2}^{new}$ with $T_{a1,m}$. If the difference is less than 0.1°C, T_2^{new} is the solution to the true flame temperature for given $T_{a1,m}$ and T_{a2}; otherwise, replace $T_{a1,2}$ with $T_{a1,2}^{new}$, and repeat the steps 3, 4, and 5 outlined above until a convergent solution is found. A few iterations are normally sufficient.

10.2.3 Accuracy

10.2.3.1 Effect of Soot Deposition on Window

Matsui et al. (1980) found that in a DI diesel engine, a reduction of 14% in transmissivity at $\lambda = 0.55$ µm resulted an error of around 1% in the flame temperature and less than 5% in the estimate of the soot concentration. As shown by Eq. 10.7, the true flame temperature is mainly determined by the ratio of flame radiation at the two wavelengths, while the KL factor is directly proportional to the absolute amount of flame radiation. The loss in transmissivity would, therefore, have a much smaller effect on the measured flame temperature than KL factors.

10.2.3.2 Wall Reflections

Note that the radiation detected by a two-color system includes not only the direct radiation from the flame (soot + gaseous species), but also the radiation reflected off the walls and the thermal radiation by the wall surfaces. Because the surface temperatures of walls are much lower than the soot particles, their effect can be ignored in the visible region. In the IR region, significant radiation from walls is present, and their effect needs to be taken into account. Using visible radiation, Matsui et al.(1980) found that the effect of reflected radiation from an opposite wall on the estimation of temperature was very small (of order 2 or 3 K), while the error in the determination of KL factor was at most 10% at very low soot concentrations, falling to 2% at high soot concentrations. When the IR wavelengths are used, the error in the estimated temperature ranged from 50 to 150 K. In addition, the error in the estimated value of the KL factor could be as high as 50%.

10.2.3.3 Effects of Nonuniform Temperature and Soot Distributions

The two-color method will only give a true flame temperature if a spatially uniform temperature exists along the line of sight. The nonuniformity of temperature and soot concentration will affect the physical meaning of the flame temperature as well as the KL measured by this technique. Detailed studies were carried out on the effects of nonuniform temperature and soot distribution on the two-color results by Matsui et al. (1979; 1980) and by Yan and Borman (1988). Their findings can be summarized as follows:

1. The effect of the soot concentration nonuniformity on the measured temperature is less severe than the effect of uneven temperature on the KL factor. If a severe nonuniform distribution of temperature exists, the KL factor will be underestimated.

2. The temperature measured by the two-color method is higher than the arithmetic mean temperature, because the rate of change of monochromatic intensity with temperature becomes greater with increasing temperature.

3. The flame temperature closer to the window will have a larger effect on the measured temperature than those of flames further away will; that is, the measured flame temperature will be biased to that near the window.

4. The measured flame temperature will be closer to the maximum flame temperature if the maximum values of temperature and soot concentration are coincident with the line of sight.

10.2.3.4 Overall Accuracy

One general conclusion that emerges from the above discussions is that the two-color method is particularly suited for accurate temperature measurements. However, the soot loading measured by the two-color method should be used with caution, because of its proneness to interference to window fouling and nonuniform soot distributions along the optical path of measurements.

Furthermore, because the thermal radiation from soot in the visible spectrum decreases rapidly with dropping temperature, the current two-color method is restricted to above 1700 K. This becomes more a constraint with highly diluted diesel combustion in diesel engines. It is, however, possible to extend the applicability of the two-color method to lower temperature regime by shifting the detection wavelengths to near IR region if the soot properties in such regions can be accurately determined. The increased contribution to the thermal radiation from surrounding environment will also need to be corrected for, which will become more pronounced at the longer wavelength region.

10.2.4 Full-Field Two-Color Measurement

10.2.4.1 Introduction

In the previous sections, the salient features associated with the two-color method have been discussed. The two-color system described so far measures the flame temperature in a small control volume as defined by the field of view of the probe. Thus, it is essentially a line-of-sight measurement. In engine combustion, however, it is desirable to obtain the two-dimensional temperature and soot distribution. This is readily achieved with the modern imaging devices, such as single-shot CCD (charge-coupled device) camera or high-speed CMOS video cameras.

10.2.4.2 Full-Field Temperature Measurement Using Solid-State Cameras

Two approaches have been adopted for full-field two-color systems with monochrome CCD cameras, as shown in Fig. 10.3. If a single CCD camera is used, the flame image is split up into two images side by side via an image doubler. Two band-pass filters are inserted in a mount in front of each half of

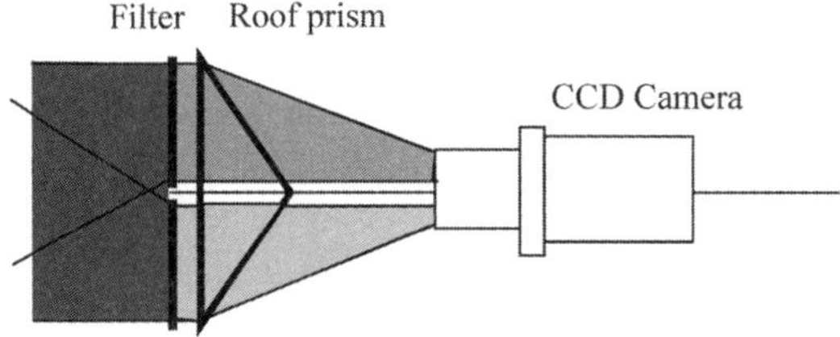

(a) Full field two-color system with one monochrome CCD camera

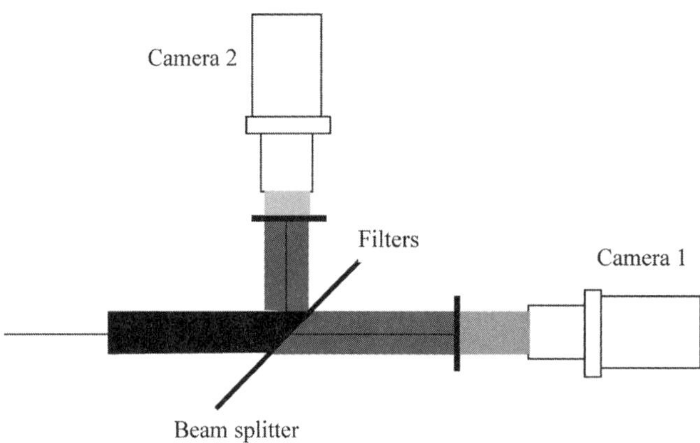

(b) Full field two-color system with two monochrome CCD cameras

Fig. 10.3 Optical layout for full-field two-color measurements using solid-state cameras.

the CCD camera, respectively, to obtain images at selected wavelengths. To obtain better spatial resolution, two CCD cameras are preferred.

The full-field two-color measurements are normally carried out through a large optical window in a single-cylinder optical engine. To overcome this difficulty, Shakal and Martin (1994) developed an endoscope-type imaging system that requires minimum engine modification. The system setup is shown in Fig. 10.4. A coherent fiber-optical bundle is used as the means for light transmission. This bundle contains many single optical fibers, whose relative positions are arranged to be the same at both ends so that image transmission is possible. The use of a fiber-optical bundle also allows a much wider field of view than viewing through a window directly. The field of view is limited to 45° by a lens mounted at the tip of the bundle on the engine side. The spatial position of the field of view is further determined by a small mirror attached to the end of the fiber-optical bundle, as shown in Fig. 10.4(a). The whole bundle is placed 1 mm away from the optical window mounted in a probe adapter, which prevents the fiber-optical bundle from being exposed to the harsh combustion environment. The dimension of the optical access port is 14 mm in diameter. The fiber-optical bundle has a diameter around 8 mm.

As shown in Fig. 10.4(b), the output from the other end of the fiber bundle is magnified by a 10X-microscope objective lens. The magnified image is than split into two side-by-side images via an image doubler. Two 10-nm bandwidth filters at 550 and 650 nm are used for the two-color measurements. Use of a gateable intensified solid-state camera is necessary because of the low radiation received by the camera, due to transmission losses in the optical system. The two images recorded on the same frame can then be used to obtain the full-field temperature and KL factor measurements using the two-color

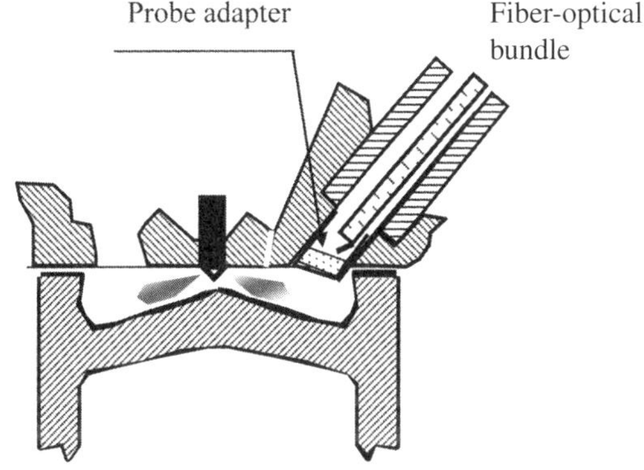

(a) Probe adapter with fiber-optical bundle installed

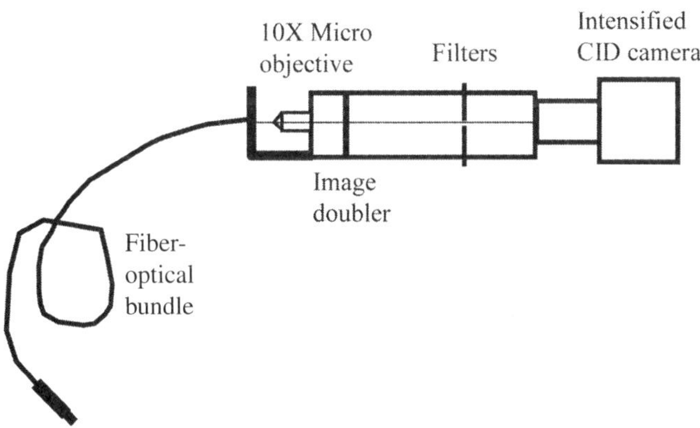

(b) Optical setup of the imaging system

Fig. 10.4 Experimental setup for the full-field temperature measurements using fiber-optics (Shakal and Martin, 1994).

method. In an alternative setup, the combustion images can be transmitted to the intensified camera by an endoscope system, which comprises an optical window, a long glass rod probe, and a four-piece cardanically linked transmissiOn arm (Bakenhus and Reitz, 1999).

To study the in-cylinder diesel combustion, a high-speed two-color imaging system has been developed in the author's laboratory, as shown in Fig. 10.5. The images are recorded by a high-speed CMOS video camera equipped with a high-repetition image intensifier through a Nikon 60-mm f2.8 lens. In front of the lens, a mounting block is installed to hold an image doubler where two half-circumference band-pass filters at 550 and 750 nm are fitted. Since the total spectral response of the system at 750 nm is significantly lower than 550 nm, two neutral-density filters with 10% transmission are used in front of the 550 nm band-pass filter to reduce its radiation intensity, so that the intensities at both 550 nm and 750 nm are of similar values. The high-speed camera records a movie over a duration of several cycles at a rate of 10,000 frames per second. Each frame is then processed to obtain the flame temperature and KL factor images, from which a movie of the flame temperature and KL factor can be produced.

The system is calibrated so that the pixel values correspond to the apparent temperatures. The calibration setup mimics the one used in the single-cylinder optical engine for the experiments. A 45° mirror is applied to simulate the mirror under the glass piston window, and a transparent piston window is placed in front of the tungsten calibration lamp, with exactly the same layout

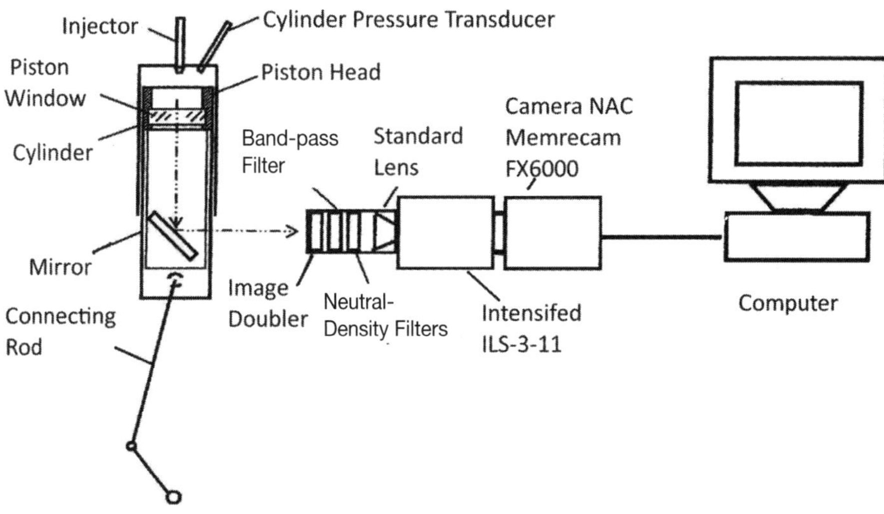

Fig. 10.5 High-speed two-color imaging system setup.

as in the single-cylinder engine. The relationship between the pixel values and the apparent temperatures is obtained for each setting of the shutter and intensifier gain combination.

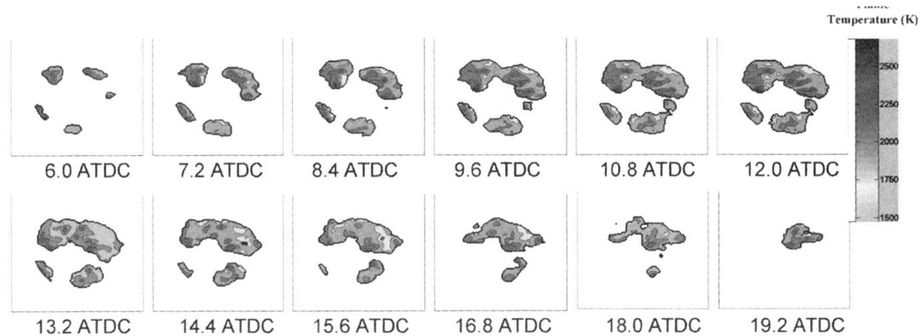

Fig. 10.6 High-speed flame temperature images of diesel combustion at 2000 rpm with 60% EGR and split injection.

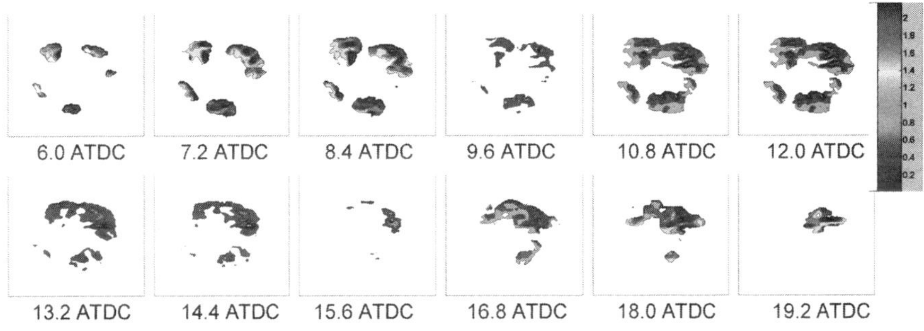

Fig. 10.7 High-speed soot loading images of diesel combustion at 2000 rpm with 60% EGR and split injection.

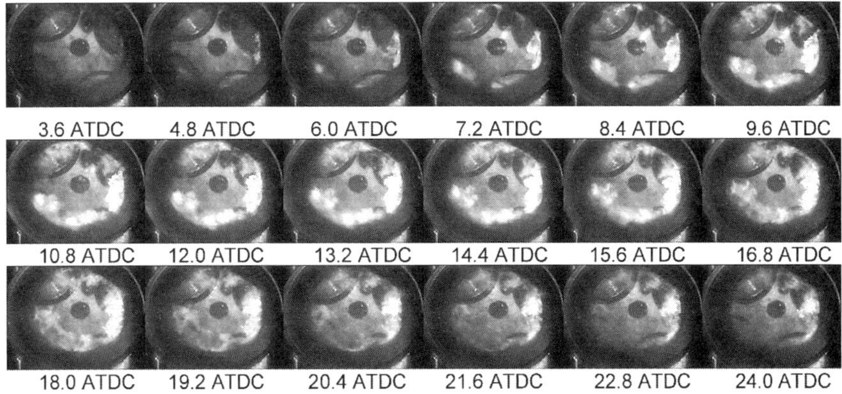

Fig. 10.8 High-speed flame images of diesel combustion at 2000 rpm with 60% EGR and split injection.

Five different injection timings for the same 10° dwell angle split injection were investigated at a part load operating condition and 2000 rpm with high EGR (exhaust gas recirculation) levels (60%).

Figures 10.6 and 10.7 present the flame temperature and *KL* factor distributions corresponding to the flame images in Fig. 10.8, taken at 2000 rpm with 60% EGR and split injection (50:50%, first injection at 11° BTDC and second at 1° BTDC). It can be seen that combustion takes place near the cylinder walls, and combustion temperature is limited to about 2250 K because of the presence of high EGR. At 12° ATDC, immediately after the peak heat release rate, the temperature of hot spots starts to decrease. From 14.4° ATDC, the flame temperature of diffusion combustion remains below 2000 K until the end of combustion, and hence the lowest NO_x emissions are measured in the exhaust. The corresponding *KL* factor images in Fig. 10.8 show that at 7.2° ATDC when premixed combustion takes place, there is a very low concentration of soot at the tip of the leading spray. The last few frames show a quite high soot concentration as a result of slower soot oxidation at a lower combustion temperature, leading to the increase of soot emissions measured in the exhaust.

10.3 Soot Concentration Measurement by the Light-Extinction Method

10.3.1 Principle of the Light-Extinction Method

When a cloud of particles are placed in a beam of light radiation, as shown in Fig. 10.9, the intensity of the light reaching the detector will be reduced. In this case, we say that the presence of the particles has resulted in extinction (attenuation) of the incident light. The attenuation of light is caused both by

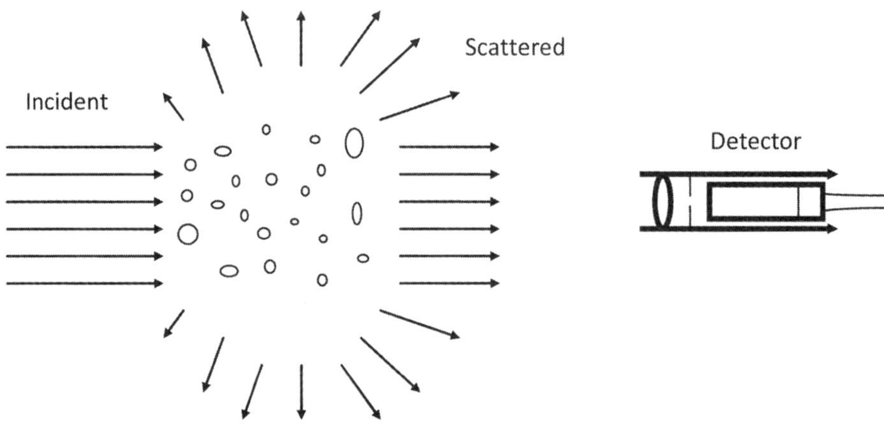

Fig. 10.9 Light extinction and scattering by a collection of particles.

scattering by the particles or by absorption in the particles, that is, extinction = scattering + absorption.

The light intensity after passing through the particle cloud is given by the Lambert-Beer's law:

$$I = I_0 \exp\left(-\int_0^L K_{ext} \, dx\right) \tag{10.11}$$

where

I_0 = the incident light intensity (W m^{-2})

K_{ext} = the extinction coefficient for a cloud of particles (m^{-1})

L = the path length of the light beam (m)

From light-scattering theory, the extinction coefficient for a cloud of particles is related to the number, size, and properties of individual particles by

$$K_{ext} = \frac{\pi}{4} C_n \int_0^\infty Q_{ext} N(D) \, D^2 \, dD \tag{10.12}$$

where

C_n = the number density (m^{-3})

Q_{ext} = the extinction efficiency of a particle

D = the particle diameter (μm)

$N(D)$ = the particle size distribution function defined such that $\int N(D) \, dD = 1$.

The value of extinction efficiency of each particle, Q_{ext}, is a complex function of its size and refractive index. The exact expressions for Q_{ext} do not exist except for some special cases. Perhaps the most important exactly soluble problem in the theory of absorption and scattering by small particles is that for a sphere of arbitrary radius and refractive index as described by the Mie theory (1908). Various approximate techniques are available, which can lead to simpler expressions for calculation, provide physical insight, and allow solutions for situations not open to rigorous analysis. One of the most common approximate methods is Rayleigh scattering. This applies to small particles with $x \ll 1$ and

$x\,|m-1|\,<<1$, where m is the complex refractive index, and $x = \pi D/\lambda$ is the particle size parameter.

Soot particles, especially those encountered in engine combustion, have a size distribution over a range from a few nanometers to 70 nm. The shape of the individual particle is roughly spherical, though they may form nonspherical agglomerates in the exhaust, which may take the form of a long chain or a cluster. When visible light with wavelength ranging between 0.4 to 0.7 μm is used, submicron soot particles lie in the regime $x\,|m|\,<<1$. For example, at wavelength of 0.5 μm and a typical refractive index of $m = 1.6 - i0.6$, we find that $x \leq 0.063$ or $x\,|m| \leq 0.107$ for $D \leq 10$ nm. For larger soot particles with $D \leq 70$ nm, then $x \leq 0.44$ or $x\,|m| \leq 0.75$. According to Kerker et al. (1978), the Rayleigh approximation is valid to within 1% for $x\,|m| < 0.2$ and to within 10% for $x\,|m| < 0.5$. Thus, we may apply the Rayleigh approximation to individual soot particles but not to any but the smallest agglomerates. The agglomerates are not spherical so that, strictly speaking, the Mie theory cannot be used. However, most researchers have tended to use the spherical assumption because of its simplicity. Their results should be considered in that light. In addition, because the composition of soot varies, the refractive index is uncertain. This leads to errors in the calculation of scattering properties and the analysis of experimental results.

Within the Rayleigh scattering regime, the extinction efficiency is given by

$$Q_{ext} = 4x\,\mathrm{Im}\left(\frac{m^2-1}{m^2+2}\right) + \frac{8}{3}x^4\,\mathrm{Re}\left(\frac{m^2-1}{m^2+2}\right) \tag{10.13}$$

The first term in Eq. 10.13 is the absorption efficiency Q_{abs}, and the second is the scattering efficiency Q_{scat}. Thus, for Rayleigh scatters the scattering efficiency is proportional to x^4, whereas the absorption efficiency is proportional to x. Since $x << 1$ in the Rayleigh regime, absorption dominates where it exists at all, that is, $Q_{abs} >> Q_{scat}$. Since soot is heavily absorbing, the extinction efficiency of a soot particle can be estimated accurately by

$$Q_{ext} = Q_{abs} = 4x\,\mathrm{Im}\left(\frac{m^2-1}{m^2+2}\right) \tag{10.14}$$

So, the Lambert-Beer's law becomes

$$I = I_0\,\exp\left(-\pi^2\,\frac{L}{\lambda}\,\mathrm{Im}\left(\frac{m^2-1}{m^2+2}\right)C_n\int_0^\infty N(D)\,D^2\,dD\right) \tag{10.15}$$

The volumetric density of particles C_v (the volume of particles per unit volume) is related to the number density of particles C_n (the number of particles in a unit of space) by

$$C_v = \frac{\pi}{6} C_n \int_0^\infty N(D)\, D^3\, dD \tag{10.16}$$

which can be obtained from the extinction measurement by

$$C_v = -\frac{2}{3L} \frac{D_{32}}{Q_m} \ln\left(\frac{I}{I_0}\right) \tag{10.17}$$

where Q_m is the mean extinction efficiency for a polydispersed particle cloud:

$$Q_m = \frac{\displaystyle\int_0^\infty N(D)\, Q(D) D^2\, dD}{\displaystyle\int_0^\infty N(D)\, D^2\, dD} \tag{10.18a}$$

D_{32} is known as either the volume/surface mean diameter or the Sauter mean diameter:

$$D_{32} = \frac{\displaystyle\int_0^\infty N(D)\, D^3\, dD}{\displaystyle\int_0^\infty N(D)\, D^2\, dD} \tag{10.18b}$$

For particles of size parameter $x < 0.5$, the difference is less than 5% between the mean extinction efficiency, Q_m, defined by Eq. 10.18a and the extinction efficiency, $Q(D_{32})$, of a particle of the Sauter mean diameter D_{32}, regardless of the size distribution function (Kontani and Gotoh, 1983, 1986). The error will be increased to 10% if the size parameter approaches 1.

Substituting Q_m with $Q(D_{32})$, the volumetric density of the particles (also known as the volume fraction), Eq. 10.17 becomes

$$C_v = -\frac{\lambda}{6\pi L\ \mathrm{Im}\left\{\dfrac{m^2 - 1}{m^2 + 2}\right\}} \ln\left(\frac{I}{I_0}\right) \tag{10.19}$$

It is obvious that the particle volume fraction can be calculated from the ratio of I/I_0 without any knowledge of particle size distributions. The only approximation will be the value of refractive index to be used. This could be the largest source of error as the refractive index varies with wavelength. Table 10.3 contains the refractive indices from Chang and Charalampopoulos (1990):

Table 10.3 Refractive indices

Wavelength, $\lambda(\mu m)$	Refractive index, $m = n - ik$
0.337	$1.48 - i0.77$
0.532	$1.75 - i0.61$
0.633	$1.80 - i0.58$
0.810	$1.85 - i0.57$
1.064	$1.91 - i0.59$

The value from Lee and Tien (1981) is given by: $m = n - ik = 1.90 - i0.55$.

A simple calculation using Eq. 10.20 shows that the soot volume fraction will be 33% lower if the refractive index is changed from $m = 1.9 - i0.55$ to $m = 1.4 - i0.55$. Thus, the value of refractive index has a significant effect on the measured value of soot concentration.

If the particle density is known or assumed, the soot mass concentration C_m can then be obtained by $C_m = C_v \rho$.

10.3.2 Implementation of the Light-Extinction Method

The setup of the light-extinction measurement is rather straightforward. All that is required is a laser and a photodetector, as shown in Fig. 10.10. A very narrowband interference filter (FWHM = 1.0 nm) is normally placed in front of the photodetector to suppress the influence of flame radiation on the measured light intensities. Using such a filter, the flame radiation reaching the detector can be reduced to the order of 0.1% of the laser beam. Light deflection resulting from hot combustion gases can be minimized to a large extent by using a large detection area detector and a focusing lens placed between the detector and measurement volume. The interference of scattering with extinction is quite a common source of error when the light-extinction measurement is carried out for large particles. It is less a problem for measurement involving soot particles, because the first lobe of scattering maximum is quite far from the optical axis. In addition, in the case of soot, absorption far outweighs scattering. For soot particles with a diameter $D < 0.1$ μm, or $x < 0.5$ in the visible region, using Eq. 10.14, the estimated ratio of the absorbed power to that scattered arriving the

Fig. 10.10 Schematic of the setup for the light-extinction measurement.

detector is greater than 10,000 if a 1 cm aperture is used at a distance of 1 m. Thus, aperture size may be neglected as a problem in the case of soot.

Since light can be attenuated due to scattering and absorption by molecules and particles, the total transmitted light through the combustion chamber would suffer from extinction by air, dust, and any liquid droplets, as well as soot particles. This is illustrated in Fig. 10.11. When the engine is motored, the light intensity is slightly reduced because of Rayleigh scattering from compressed air in the cylinder. If the engine is motored with ambient air, the second peak of the dotted line may appear during the expansion stroke because of the presence of water droplets from condensation of water vapor in the air (Hentschel and Richter, 1995). Another source of interference with light-extinction measurements is liquid fuel droplets, as shown in Fig. 10.11. It is possible to obtain the light-extinction curve (dashed line) caused by liquid fuel droplets in a nitrogen-motored engine with fuel injection. If the fuel droplet curve is subtracted from the measured total extinction curve, however, the soot concentration will be underestimated at the beginning of the measurements because the actual fuel vaporization process will be faster in a firing engine. Hentschel and Richter (1995) found that the vaporization was normally completed a few degrees CA after the end of injection under high pressure and temperature in a modern diesel engine. It can therefore be assumed that extinction soon after the end of the injection is caused by soot particles only.

Hentschel and Richter (1995) applied the light-extinction technique to a high-speed DI diesel engine. Laser beams from a HeNe laser and an argon-ion laser were introduced to the combustion chamber via a bifurcated optical fiber. The transmitted laser beams were then separated back to the two wavelengths by another bifurcated fiber. A combination of a narrow band-pass filter, short-pass filter, and neutral-density filter was employed in each path to suppress the flame radiation. Outputs from the two photodiodes were recorded by a transient recorder.

In addition to carrying out theoretical analysis on the light-extinction method, Kontani and Gotoh (1983, 1986) attempted to measure the size of soot particles using the spectral extinction or two-wavelength dispersion quotient method,

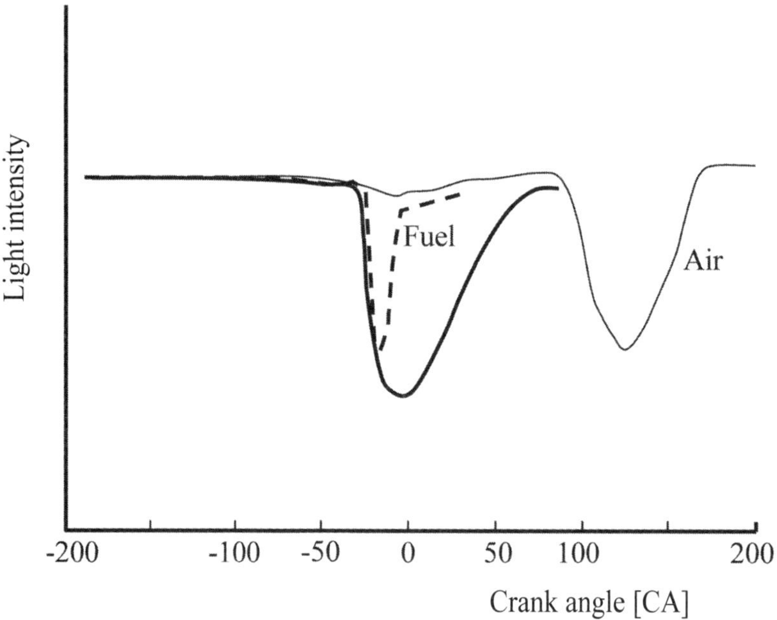

Fig. 10.11 Sources of interference in the light-extinction measurement in a diesel engine.

which assumes that the ratio of extinction efficiency at two wavelengths $(Q_{ext}(\lambda_1)\ /Q_{ext}(\lambda_2))$ is a simple function of particle size. Unfortunately, in the Rayleigh regime, $Q_{ext}(\lambda_1)/\ Q_{ext}(\lambda_2)$ is independent of particle size. Therefore, no information of particle size could be retrieved from their spectral extinction measurements.

In the previous cases, the soot volume fraction is measured along a line of points on the narrow laser beam through the soot cloud inside the combustion chamber. A matrix of measurement points are required to build up an image of soot distribution, which is tedious and time-consuming. In addition, as measurements have to be carried out at different times, only ensemble-averaged soot distributions can be measured. Therefore, it is desirable to measure the soot volume fraction at multiple points simultaneously.

It seems logical to think that infinite numbers of extinction measurements can be accomplished by means of an imaging device and an expanded light source, as shown in Fig. 10.12. An expanded parallel beam passes through the soot cloud before they reach an imaging device. The light intensity will be reduced in the region of soot cloud because of extinction by soot particles. This setup is very similar to shadowgraphy. However, the system shown in Fig. 10.12 measures the light attenuation caused by extinction only when the soot particles are suspended in a quiescent and isothermal environment. However, when there are variations of the refractive index in surrounding gases, part of the beam will be deflected. Consequently, Schlieren effect will be recorded, as well as extinction caused by soot particles.

The effect of changes in refractive index or Schlieren effect on soot visualization was studied by Nakakita et al. (1990). They compared three optical setups for soot visualization. The results are shown in Fig. 10.13. The images were taken from the swirl chamber of an IDI (indirect injected) diesel engine, which had two quartz windows at each end, by a high-speed cine camera at 4800 frames per second. The images in the top row were obtained with a Schlieren system, which was similar to the optical setup shown in Fig. 10.12 and comprises a 6-mm diameter circular stop at the focal point. These images showed fuel sprays, fuel/air mixture, flame-front, and luminous flame regions. But it was impossible to tell where the soot cloud was.

Figure 10.13(b) shows the photographs recorded without the circular stop at the focal point; that is, they are shadowgraphs of the mixture in the swirl chamber. With the removal of the circular stop, the sensitivity of those shadowgraphs is lower than Schlieren photographs. The density gradients in the mixture is still visible. Hence, dark regions in these images at 2° and 4° CA ATDC are due to the density gradients in the mixture as well as attenuation by the soot cloud. It is still difficult to separate the soot cloud from changes in density gradients.

Figure 10.13(c) shows the images taken with an uniformly diffused light through a diffuser (a frosted glass plate), using the experimental setup shown in Fig. 10.14. The flame emission was block by a narrow band-pass filter. In setting up the system in Fig. 10.14, the camera was focused to the measurement section so that the shadowgraphic effect was minimized. The photographs with this back-illuminated method do not show the variations caused by the density gradients or temperature gradients seen in Figs. 10.13(a) and 10.13(b). The Schlieren effect caused by the change in the refractive index of the mixture

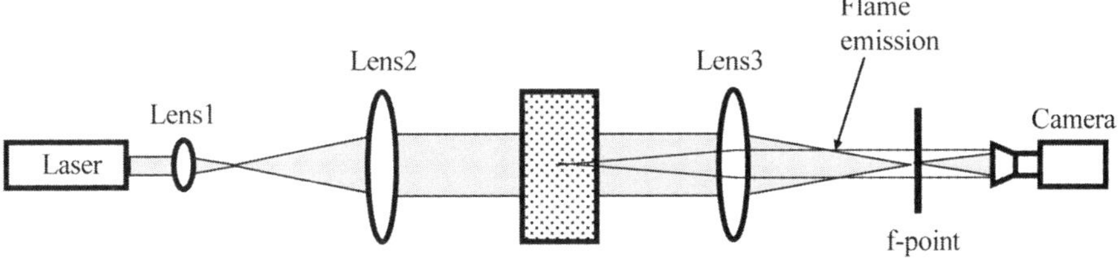

Fig. 10.12 Idealized setup for infinite numbers of light-extinction measurements.

was suppressed by the use of diffused light. The dark areas are therefore caused by extinction alone. Further comparison with photographs taken in the swirl chamber with nitrogen motored engine reveals that the top shadow was

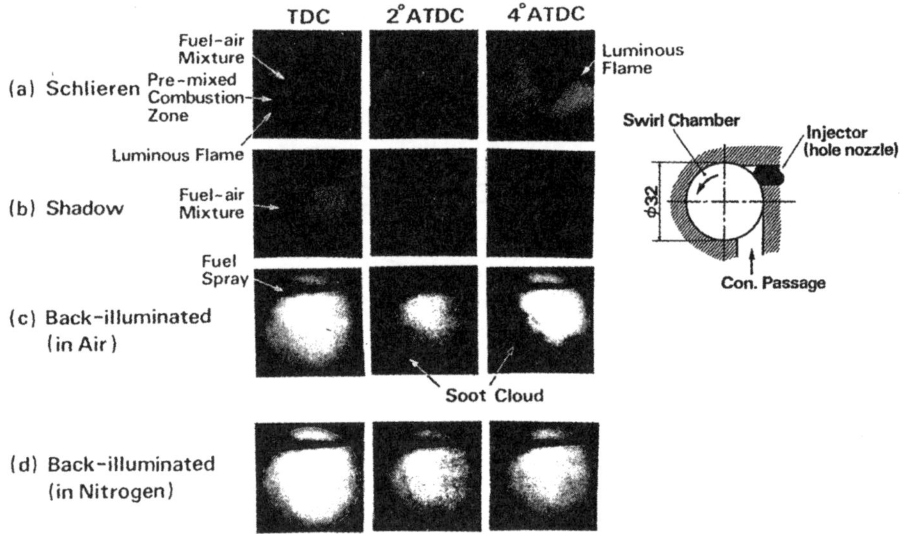

Fig. 10.13 Schlieren, shadowgraph, and light-extinction photographs of diesel combustion (Nakakita et al., 1990).

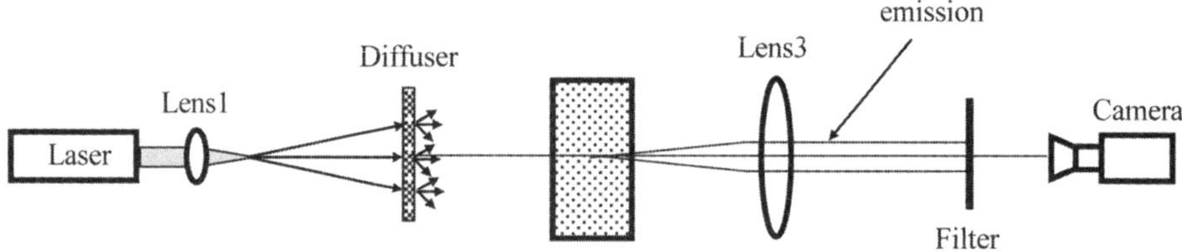

Fig. 10.14 Optical setup for soot visualization with back illumination.

caused by the extinction by liquid fuel droplets, and the other shadow areas were caused by the presence of a soot cloud.

10.4 Laser-Induced Incandescence
10.4.1 Introduction

Laser-Induced Incandescence (LII) has become the preferred method for measuring in situ soot volume fraction over the last decade because of its ability to obtain spatially and temporally resolved results in the cylinder as well as in the exhaust. In combination with the laser scattering measurement, the mean soot aggregate sizes can be determined. Furthermore, the application of time-resolved LII allows the soot primary particle sizes to be measured with high spatial and temporal resolution. Finally, LII is characterized by simple experimental setup and it can be used for measurements at a single point and determination of spatially resolved soot particle distribution in an extended area.

10.4.2 Soot Volume Fraction Measurement by LII

10.4.2.1 Principle of LII

When a cloud of soot is irradiated with an intense laser light, the soot particles can be heated to a temperature well above the surrounding gas temperature because of the absorption of laser energy. The subsequent blackbody radiation emitted by these soot particles corresponds to the elevated soot particle temperature. If laser fluence is high enough to elevate the soot particle to the vaporization temperature of carbon of 4000 K, the vaporization of small carbon fragments from the soot particle surfaces will take place. The heated soot particles will eventually cool down by exchanging heat with the surrounding environment through various heat transfer mechanisms such as sublimation, conduction, convection, and radiation, as shown in Fig. 10.15. The energy balance and the resulting temperature of the heated soot particles is determined by the rate of laser energy absorbed, the rate of conductive heat transfer to the surrounding medium, soot vaporization, and the energy loss resulting from blackbody radiation, as given by the following equation:

$$\frac{d\left(m_s C_s T_s\right)}{dt} = \dot{q}_{abs} - \dot{q}_{cond} - \dot{q}_{sub} - \dot{q}_{rad}$$

or

$$m_s \frac{d(C_s T_s)}{dt} = I_o C_{abs} - h A_s (T_s - T_g) + \frac{H_v}{M_v} \frac{dm_s}{dt} - \int_0^\infty A_s \varepsilon_{s,\lambda} E_{b,\lambda}(T_s)\, d\lambda \quad (10.20)$$

where

m_s = mass of the soot particle (kg)

C_s = specific heat of the soot particle (J/(kg·K))

T_s = temperature of the soot particle (K)

H_v = enthalpy of vaporization of the soot particle (J/mole)

M_v = molecular weight of vaporized material

I_0 = laser intensity (W/m²)

C_{abs} = absorption cross section (m²)

h = heat transfer coefficient (W/[m²·K])

A_s = surface area of the soot particle (m²)

$= \pi D^2$

T_g = temperature of surrounding medium (K)

$\varepsilon_{s,\lambda}$ = spectral emissivity of the soot particle

$E_{b,\lambda}$ = blackbody spectral irradiance (W/[m²·µm])

The continuity equation for the particle mass is given by (Hofeldt, 1993)

$$\frac{dm_s}{dt} = -\frac{N_v \pi D_s^2 M_v}{N_{Av}}$$

(10.21)

For a spherical particle, the soot particle mass changes with its diameter as given by

$$\frac{dm_s}{dt} = \frac{\pi}{2} \rho_s D_s^2 \frac{dD_s}{dt}$$

(10.22)

Combining the above equations gives

$$\frac{dD_s}{dt} = -\frac{2 N_v M_v}{N_{Av} \rho_s}$$

(10.23)

where N_v is the molecular diffusion flux, and N_{Av} the Avogadro number.

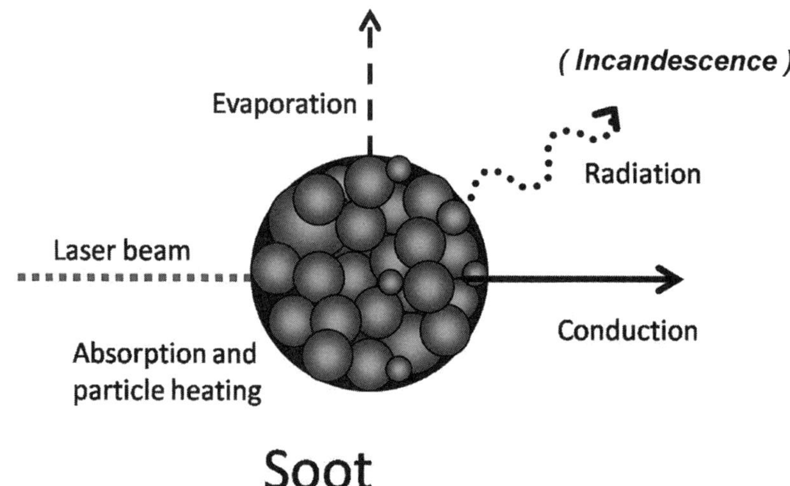

Fig. 10.15 Laser heating and cooling process of soot particles.

Therefore, Eqs. 10.20 and 10.23 are coupled nonlinear equations that must be solved simultaneously if vaporization is important, in order to obtain the time dependent particle size and temperature distributions.

The LII arises from the blackbody radiation ($\dot{q}_{rad}$) of a collection of soot particles according to the Planck's law (Eq. 10.1). Its signal S collected over a given solid angle Ω from a distribution of soot particles can be written as

$$S = V_{mv}\frac{\Omega}{4\pi}\int_0^\tau C_n(t)W(t)\int_0^\infty N(D_s)\int_0^\infty C_{abs}\,E_{b,\lambda}(T_s)\,d\lambda\;dD_s\,dt \qquad (10.24)$$

Where V_{mv} is the measurement volume in m^3, $C_n(t)$ is the soot particle number density, $W(t)$ is a signal windowing function, and $N(D_s)$ is the normalized size distribution, that is, $\int N(D_s)dD_s = 1$.

For soot particles ($D < \lambda$), *in the limit of high laser power and maximum soot particle temperature near its vaporization point*, according to Melton (1984), the LII signal can be simplified to

$$S = C_{cal}\int_0^\tau C_n(t)W(t)\int_0^\infty N(D_s)\int_0^\infty D_s^x\,d\lambda\;dD_s\,dt \qquad (10.25)$$

where C_{cal} is the calibration factor, and x is the exponential factor of LII signal's dependence on soot particle size, which can be estimated as $(3 + 0.154\lambda^{-1})$.

For λ between 0.4 and 0.7 µm, for example, the LII signal is proportional to the mean soot diameter raised to the power of 3.38 to 3.22, or approximately to the soot volume fraction, that is,

$$S_{LII} \propto C_V \approx C_n D^3 \qquad (10.26)$$

which forms the basis of soot volume fraction measurements by the LII technique.

10.4.2.2 Implementation of LII

LII can be applied either as a point, line, or two-dimensional laser sheet measurement. The principal elements of the experimental setup are similar in each case with only the differing delivering and collection optics to meet the specific measurement desired. Figure 10.16 shows the experimental setup for two-dimensional LII measurement in a diffusion flame. The output from a high-power pulsed laser (Nd:YAG) is expanded into a light sheet by a combination of spherical and cylindrical lenses. The LII signal is detected perpendicularly to the laser sheet by a gated intensified CCD camera. The low LII signal-to-noise ratio requires the use of an ICCD (intensified charge-coupled device) camera with gating down to nanoseconds. An UV (ultraviolet) enhanced camera lens is required. As will be discussed later, the best compromise between the absolute intensity of the LII signal and the ratio of the LII signal to the background luminosity is obtained at a detection wavelength of about 400 nm. A set of low-pass filters or a UV narrow band-pass filter is required to restrict the detection range of the LII signal. One or two laser mirrors are used to block off any elastic scattering from soot particles or any surfaces. Under optimum conditions, a signal-to-noise ratio of up to 1% may be achieved.

The intensity of LII signal depends on a number of factors. The dominant factor is the temperature of the laser-heated soot particles, which is determined by the rate of laser energy absorption and heat loss through conduction, radiation, and soot vaporization. The detection wavelength will also affect the LII signal intensity and its signal-to-noise ratio. The application of LII for quantitative measurement of soot volume fraction needs to consider these factors.

The excitation and detection wavelengths need to be selected to optimize the LII signal-to-noise ratio. When visible light is used, most soot particles in diesel engines scatter light in the Rayleigh region, where the particle absorption efficiency varies approximately inversely with wavelength at fixed refractive

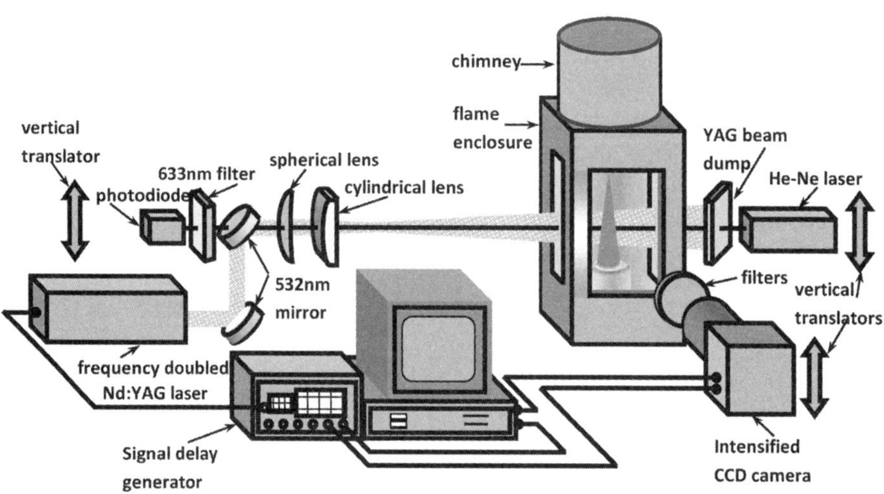

Fig. 10.16 A two-dimensional LII setup for soot measurement including a HeNe laser extinction calibration system.

index and fixed size. Hence, a shorter excitation wavelength will be more effective in heating up the soot particles. However the broadband fluorescence from polycyclic aromatic hydrocarbons (PAHs) may interfere strongly with the LII signal if an UV excitation is to be used. This imposes severe interference with the LII signal. The use of the second harmonic of an Nd:YAG laser at 532 nm is a good compromise between the excitation efficiency and signal-to-noise ratio. With excitation at 532 nm, any fluorescence from PAHs or other molecules will be at longer visible wavelengths and near IR region if there is any. Because the LII signal is normally observed at around 400 nm (for reasons to be discussed later), the interference from the fluorescence can be minimized by using a short-pass filter. Because of the energy loss in frequency doubling process, the fundamental output of an Nd:YAG laser at 1064 nm is often used so that a less powerful laser can be used. However, use of an IR laser output does introduce some inconvenience, as the laser beam is not visible and alignment is more tedious.

In the LII technique, the temperature difference between the laser-heated and not-heated soot particles is used to reject the background luminosity from the soot not in the probe volume. In a diffusion flame or diesel combustion process, the temperature of nonheated soot particles is about 2200 K or less, whereas the temperature of laser-heated soot particles could reach 4000 K. Assuming that blackbody radiation takes place on soot surfaces, one can compare the blackbody radiation at 4000 K to that at 2200 K. It can be shown that the ratio of radiation from laser-heated soot particles to that of nonheated soot particles increases as the detection wavelength becomes shorter. Therefore, the detection of LII signal at shorter wavelengths is preferred in order to reject unwanted luminosity. On the other hand, the overall intensity of the collected light will also decrease as the cutoff wavelength is reduced. The LII signal is normally detected around 400 nm to achieve a good signal-to-noise ratio. For measurements in the engine exhaust stream, the lack of flame radiation permits the detection of LII signal at longer wavelength, allowing more sensitive and low cost detectors to be used.

A short-pass filter or a narrow band-pass filter is normally used. Since the LII signal-to-noise ratio is very low (<1%), a small amount of elastically scattered laser light from soot and dust passing through the filter may cause serious interference with an LII signal. A laser mirror, which reflects light at the laser wavelength but allows the others to pass, will be needed to block the elastically scattered laser light. A holographic notch filter can be used as a very effective laser mirror because of its extremely high blockage at the laser wavelength.

The laser output affects the intensity of a LII signal in two ways: laser fluence and laser beam shape. Below a certain laser fluence threshold, no observable LII signal is detected. Just above the laser fluence threshold, the LII signal exhibits a rapid increase with increased laser fluence. This is because the soot particle temperature increases as a function of the laser fluence, which causes a corresponding increase in the LII signal.

For a rectangular beam, the signal decreases at very large laser fluences as a result of the soot particles losing significant mass. When a Gaussian profile laser beam is used, the LII signal is nearly independent of laser power above the threshold fluence (about $10 \text{ MW}/\text{cm}^2$). This can be explained by the fact that for a Gaussian profile beam, the laser intensity is high in the middle and gradually falls toward the edge of the beam. Increasing laser energy results in an increase in the effective laser beam diameter or the LII probe volume, in which the laser fluence exceeds the threshold of the LII signal. A similar result was also reported by Quay et al. (1994).

Since the LII signal overlaps with flame radiation, synchronization between the laser output and detection of the LII signal is required. Unlike elastic scattering (Rayleigh or Mie scattering), which lasts as long as the laser pulse, LII can last a few hundred nanoseconds. Figure 10.17 shows the temporal profiles of LII signals averaged from a small region in the center of a laser sheet in a diffusion diesel flame (Bryce et al., 2000). The intensifier gate was set to a width of 5 ns, and the delay was incremented to provide images spanning the duration of the signal. After the initial rise of signal during the laser pulse, the signal shows a decay that is characterized in accordance with the energy balance Eq. 10.20 and stems from the two mechanisms of vaporization and conduction. If, as in this case, the laser intensity is sufficient to heat the soot above the sublimation point, the initial, rapid decay is caused by the dominant mechanism of vaporization. The particle decreases in size during this time. Once the particle cools to temperatures below the sublimation point, the signal trails off as a result of the second, slower decay that is due to conduction. Increasing the laser intensity will lead to a spike in the LII signal, which occurs nearer the beginning of the laser pulse. Decreasing the laser intensity will increase the proportion of signal that occurs after the laser pulse, especially when the radiating volume is no longer decreased by vaporization.

One might be tempted to collect signals after the laser pulse, thus removing interferences from other laser-induced signals, such as fluorescence and scattering. However, as well as decreasing signal-to-noise ratio, the signal would now be subject to increased error because of fluctuations in temperature and composition that affect the signal decay as a result of conduction. There is also an increased risk of bias toward larger particles, which take longer to heat and cool less rapidly. Finally, and perhaps most importantly, with delayed detection, a residual structure can be observed on images because of spurious signal induced by weak intensity flare about the laser sheet. The signal should therefore be collected synchronously to the laser pulse over a comparable time scale (e.g., 30 ns).

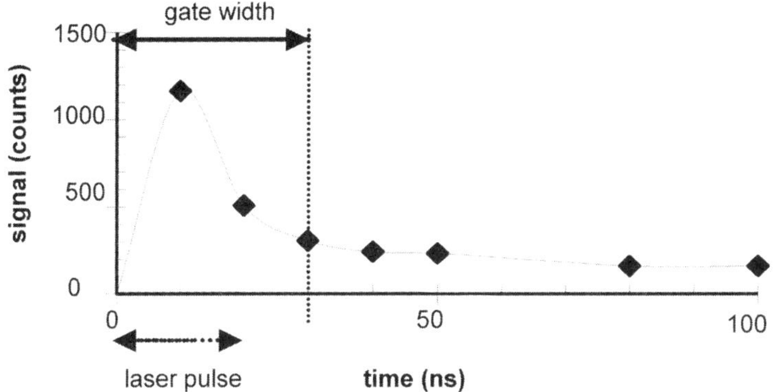

Fig. 10.17 Temporal profile of an LII signal obtained in a diffusion diesel flame.

Additional considerations arise from the laser beam profile and reflected light off walls. Since a laser beam is typically characterized by a Gaussian intensity distribution, the LII signal resulting from the edge of the laser beam is characterized by a longer rise time. Delayed detection will include more contributions from low laser fluence region. Because the low laser fluence does not heat the soot particles to near vaporization temperature, the dependence of LII signal on particle size to $D^{3.+0.1154/\lambda}$ is no longer valid and soot volume fraction cannot be accurately determined from LII signals. The same consideration applies to the laser beam reflected from surfaces, which results in erroneous soot volume fraction measurements because of its reduced laser fluence and longer LII rise time.

10.4.2.3 Calibration of LII for Quantitative Soot Volume Fraction Measurement

Since the LII technique involves the absolute measurement of incandescent light, calibration of LII signal is required to obtain the soot volume fraction. The calibration of LII measurement is performed using the light-extinction method and it is typically carried out in a sooting laminar flame (Santoro et al., 1983 and 1987, Bryce et al., 2000).

Since the LII signal is proportional to the soot volume fraction, scaled by a constant of proportionality K_{cal}, the integrated soot volume fraction along a line of-sight in the sooting flame is given by

$$\int_0^L C_V \, dx = K_{cal} \int_0^L S_{LII}(x) \, dx = K_{cal} \sum_{pixel=1}^{n} S_{LII}(i) \, \Delta x \qquad (10.27)$$

where L is the optical path length, and Δx is the size of a pixel/photosensor. The soot distribution profile $S_{LII}(i)$ is the signal value of pixel i in the LII image.

Using Eqs. 10.11 and 10.27 and writing out the imaginary part in Eq. 10.14 explicitly, it can be shown that

$$K_{cal}\Delta x \sum_{pixel=1}^{n} S_{LII}(i) \frac{6\pi}{\lambda} \left[\frac{6nk}{\left(n^2+k^2+2\right)^2 + 4n^2k^2} \right] = \ln\left(\frac{I_0}{I}\right) \quad (10.28)$$

Therefore, in the experimental setup shown in Fig. 10.16 the calibration factor can be determined from the light-extinction measured along a line in the flame corresponding to the row of pixels (pixel 1 to pixel n) in a CCD camera. A tomographic correction procedure can then be used in the determination of soot volume fraction profiles from multiple-point extinction measurements as to be discussed in the next section.

In practice, laser-extinction measurement can be done using a small powered CW (continuous wave) laser. In Fig. 10.16, a HeNe laser beam is set to be horizontal, bisecting the flame axis, and it is positioned at a known height above the burner. A photodiode is placed behind a 532-nm mirror to collect the light from the HeNe laser passing through the flame. By this method, the extinction of the HeNe laser through the flame is determined so the LII results can be scaled to soot volume fraction. The pixels on the camera are scaled in millimeters by setting the HeNe beam to a series of known heights and viewing the narrow beam on the camera as it passes through the flame. Thus the pixels can be calibrated accurately to 1 pixel = 0.0679 mm.

10.4.2.4 Signal Trapping Correction and Calibration Procedure

Another effect that requires consideration, before the soot volume fraction distribution can truly be inferred, is the *trapping* of LII signal due to its extinction by the soot field between the laser sheet and the detection system. To quantify this effect, a procedure has been developed in the author's lab so that LII results could be corrected for signal trapping as well as calibrated to obtain quantitative soot volume fractions, using one single laser extinction measurement in the same plane as the laser sheet, with details found in Bryce et al. (2000).

10.4.3 Soot Particle Sizing Using LII and Laser Scattering (LS)

It has been shown that the LII signal is proportional to the soot volume fraction $C_v \approx C_n \times D_s^3$, if the laser power is high enough for the soot to reach its sublimation temperature. On the other hand, the laser scattering (LS) signal from soot particles is proportional to $C_n \times D_s^6$ according to Rayleigh approximation for soot particles. Therefore, from the measured LII and LS intensities, one can calculate a relative value of the particle size, D_{63}, expressed as the ratio of the D^6 and D^3, $D_{63} = (D_s^6/D_s^3)^{1/3} \approx (I_{scat}/I_{LII})^{1/3}$ and the number density of soot particles, $C_n \approx I_{LII}^2/I_{scat}$.

In the experimental setup shown in Fig. 10.16, this was realized by carrying out both LII and LS measurements at the same location in the flame. First the 500-pulse-averaged LII signal was collected; then the scattering signal in a separate averaged measurement at the same location was obtained by replacing the filters in front of the intensified CCD camera with a 532-nm band-pass filter and a series of neutral-density filters.

Figures 10.18 and 10.19 show the measured soot volume fractions in gray scales and the size and number density profiles along a diesel diffusion flame. Each flame was fueled with different oxygenated diesel fuels, and LII results were

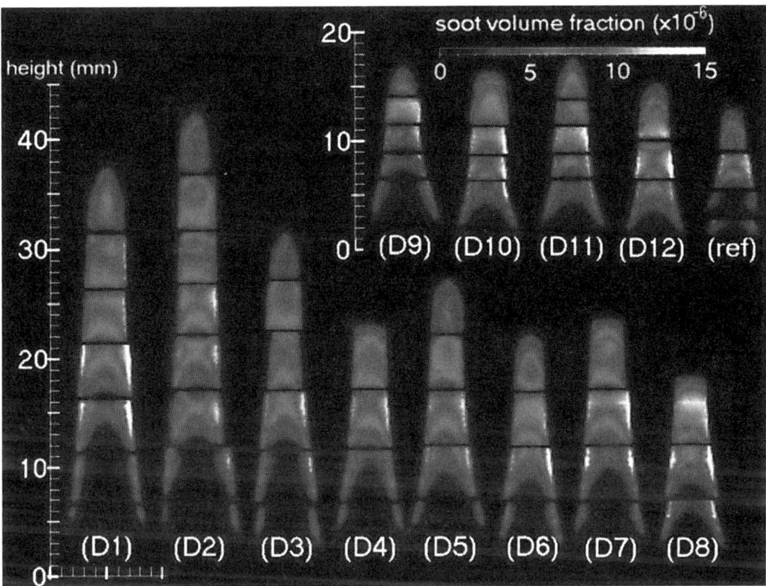

Fig. 10.18 Soot volume fraction distribution in diffusion flames burning different diesel fuels.

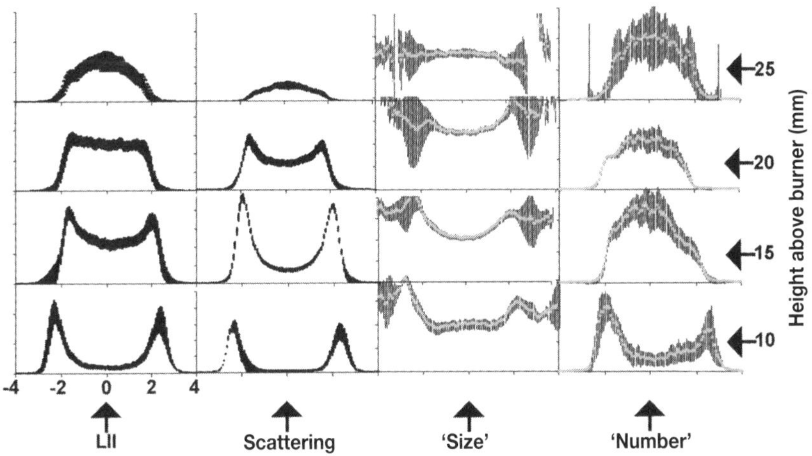

Fig. 10.19 Soot particle size and number density profiles along a diesel diffusion flame determined from the LII and laser scattering measurements.

made up images of 5 mm in height. Such images show quantitatively the soot distribution in such flames and their formation and oxidation path.

10.4.4 Particle Sizing by Time-Resolved LII (TiRe-LII)

10.4.4.1 Principle of TiRe-LII

As discussed previously, the basic principle of time-resolved LII (TiRe-LII) is to heat soot particles closer to their evaporation temperature by using a short-pulsed high-energy laser beam, causing the heated soot particles emit radiation according to Planck's law. The heated soot particles will eventually cool down by exchanging heat with the surrounding environment through various heat transfer mechanisms, such as sublimation, conduction, convection, and radiation. Immediately after the laser pulse, the soot particles are at a high temperature, and their cooling is dominated by evaporation. As the temperature falls, heat conduction becomes the most important heat loss mechanism. As shown by the second term on the right-hand side of Eq. 10.20, the heat conduction from soot particles to the surrounding media is proportional to the surface area of the soot particle. The smaller the soot particles, the greater the heat loss becomes and hence the faster the LII signal decays. Thus, by measuring the temporally resolve LII signal with a fast response photomultiplier, the particle sizing can be obtained through the decay rate of the LII signal.

The energy and the mass balance equations of laser-heated soot particles can be solved numerically to obtain a time-dependent variation of temperature and size of the laser-heated soot particle. Substituting the calculated particle temperature and size in the Planck's law will provide a relation corresponding to the spectral radiance that is emitted by a laser-heated soot particles at a given wavelength:

$$I_\lambda = A\varepsilon \frac{2hc^2}{\lambda^5} \left[\exp\left(\frac{hc}{\lambda kT} \right) - 1 \right]^{-1} \tag{10.29}$$

Where A is the surface area of the particle ($4\pi r^2$). The intensity of spectral radiation I_λ is proportional to absorption efficiency Q_{abs} of laser-heated soot particles, given by Eq. 10.14. The energy absorbed by soot particles is dependent upon the intensity of the laser beam q_l.

Since the temperature and diameter of laser-heated soot particle vary as a function of time, the LII signal from a single soot particle can be expressed as a spectral function $S(D_s, t, q, \lambda_{laser}, \lambda_{em}, \Delta\lambda)$, where λ_{em} is the central detection wavelength, and $\Delta\lambda$ the spectral bandwidth of the detection system:

$$S(D_s,t,q,\lambda_{laser},\lambda_{em},\Delta\lambda) = \frac{\pi^2 c^2 h}{2\lambda_{em}^5} D_s^2(t) C_{abs} \left\{ \left[\exp\left(\frac{hc}{\lambda_{em} kT_s(t)} \right) - 1 \right]^{-1} - \left[\exp\left(\frac{hc}{\lambda_{em} kT_g(t)} \right) - 1 \right]^{-1} \right\}$$

(10.30)

The last term in the square brackets is included to take into account of the thermal radiation from hot gases. The total radiation of multiple soot particles in the probing volume can be obtained by directly integrating the equation with an assumed size distribution $P(D_s)$ over a defined spectral wavelength. Thus the theoretical time-resolved LII signal ($J\left(D_s,t,\lambda\right)$) can be obtained numerically as given by

$$J\left(D_s,t,\lambda\right) = \int_0^\infty N_p P(r) S(D_s,t,\lambda,T) dD_s$$

(10.31)

where N_p is the particle number density, and $P(D_s)$ is the size distribution, which could be defined by monodispersive or Gaussian or a log-normal distribution function. The log-normal distribution has been shown to provide more realistic behavior compared to monodispersive distribution (Bladh and Bengtsson, 2004; Liu et al., 2006):

$$P(D_s) = \frac{1}{\sqrt{2\pi} D_s \ln\sigma_g} \exp\left\{ -\frac{\ln D_s - \ln CMD^2}{2\left(\ln\sigma_g\right)^2} \right\} D_s$$

(10.32)

Where CMD is the counter median diameter, and σ_g geometric standard deviation. By fitting the experimental decay of the TiRe-LII signal with the numerically calculated TiRe-LII signal $J(D_s, t, \lambda)$, one will be able to obtain the soot particle size through the curve fitting techniques (e.g., the least square fitting, Dreier et al., 2006; multidimensional nonlinear regression, Lehre et al., 2003).

To carry out the fitting of the experimental data with numerically calculated LII, it is required of the initial temperature that is attained by the soot particle shortly after laser irradiance. This peak soot temperature (T_s^o) is required as an initial value to solve the energy and mass balance equation numerically using a Runge-Kutta scheme and could be experimentally determined using a two-color method described in Section 10.2.

10.4.4.2 Implementation and Application of TiRe-LII for Soot Particle Sizing

Figure 10.20 shows the TiRe-LII setup used for in-cylinder soot measurements in a single-cylinder optical common rail DI diesel engine in the author's lab. The fundamental output of an Nd:YAG laser is steered to pass through one side window in the cylinder head and stopped by a beam dump at the opposite window. The LII signal is collected from another side window at the right angle and split into two optical paths with appropriate band-pass filters. Two fast photomultipliers are used to detect the temporal LII signals for the subsequent postprocessing. Both the initial temperature of heated soot and the unheated soot (gas) temperature are obtained by the two-color method using the ratio of PMT (photomultiplier tube) outputs recorded at 415- and 665-nm wavelengths. However, in many applications of the TiRe-LII technique, a single PMT has been used to obtain the LII signal decay time and an averaged soot particle size (Dankers and Leipertz, 2004; Schraml et al., 2000; Snelling et al., 2000; Witze et al., 2005).

In addition, TiRe-LII can be extended for two-dimensional semiquantitative visualization of soot particle size distribution by the ratio of two LII images recorded with different time delays from the laser pulse (Will et al., 1998). Because of the lack of size dependence in the initial LII signal decay curve, it is better to set the time for the first LII image after the first tens of nanoseconds. As the actual LII signal level is strongly influenced the exact value of

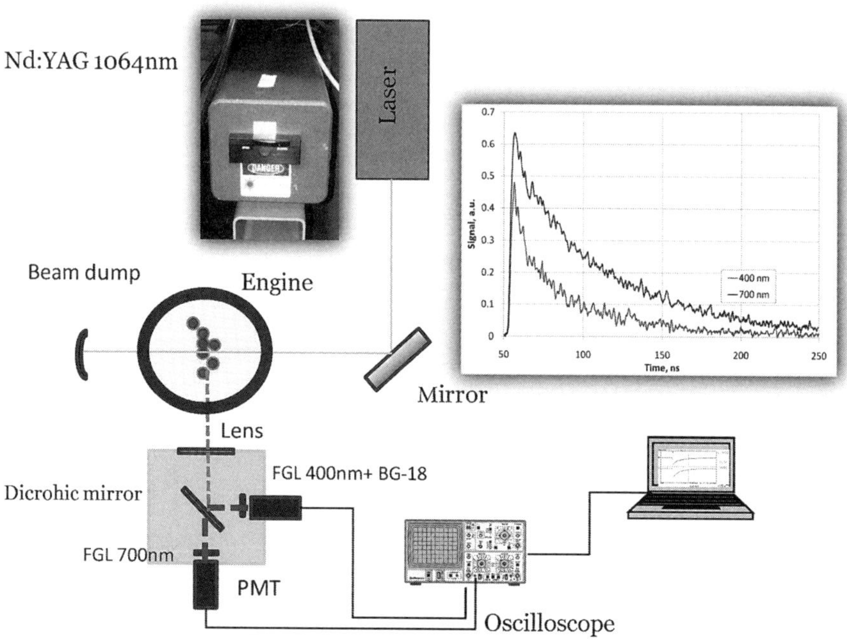

Fig. 10.20 Experimental setup of the TiRe-LII system for in-cylinder soot particle sizing.

vaporization, surrounding medium, and particle size, the time delay of the second image needs to be adjusted according to the measurement conditions.

Under atmospheric conditions, particularly at the exhaust of the engines, the mean free path is large, and the cooling occurs in the free-molecular regime and the decay rates of the LII signal are of the order of 700 to 800 ns (Snelling et al., 2000; Vander Wal and Choi, 1999).

In diesel engines, the cylinder pressures are very high during combustion and the mean free path of gas molecules is significantly shorter. This leads to a higher degree of interaction between the laser-heated soot particulates and the surrounding gases. The mode of heat losses between the particle surface and the surrounding are different from ambient environment, as a result of which the decay rates of the LII signals are faster and are of the order of 50 to 100 ns (Bougie et al., 2006a, 2007; Kock et al., 2006). So under high-pressure conditions, the time response of the detection system has to be carefully considered in relation to the decay rates of the LII signal. To avoid the influence of the response of the detector system on the measured LII signal, the measured signal should be deconvolved under high-pressure operation. It has been shown by Bougie et al. (2006b) that by deconvolving the raw high-pressure LII signal, a 10% reduction in the fitted particle size was seen.

10.4.5 Further Considerations of LII Techniques

Over the last decade, the LII technique has been demonstrated to be capable of measuring both the soot volume fraction and soot particle size with high spatial and temporal resolution. There are still many uncertainties in the quantitative interpretation of LII results.

Most of the LII models are based on the assumption that soot particles in the probing zone are isolated primary particles, but in reality they are polydispersive aggregates of polydispersive primary soot particles. More detailed investigations are still required to understand aggregate effects on the sublimation process (Yoder et al., 2005), heat transfer (Liu et al., 2006), and the heat absorption process.

The performance of LII models at high pressures have not been completely understood (Hofmann et al., 2008). The uncertainty of physical parameters such as soot absorption coefficient, molecular weight, density, and the specific heat of soot and its vapor significantly alters the calculated energy losses and thereby influences the estimated particle size (Kuhlmann et al., 2006; Smallwood et al., 2001; Snelling et al., 2004).

Influence of laser heating on the changes in the soot morphology and their effects on heat loss during conduction and evaporation has not been completely explored (Vander Wal and Choi, 1999). Insights are needed on the fluence at which laser heating affects soot vaporization (Yoder et al., 2005) and fragmentation. Soot evaporation may lead to the formation of secondary nanoparticles (Michelsen et al., 2007), and their effects on particle size measurements using LII needs to be explored.

In light of the great interest in LII, a website called LIIscience.org (http:// liiscience.org/) has been set up to foster best practices and make LII quantitative. LIIscience.org contains information about modeling the nanoscale heat- and mass-transfer processes, experimental design, signal evaluation, and interlaboratory comparisons, as well as many papers on LII published by the members of the LII workshop series.

References

Arcoumanis, C., Bae, C., Nagwaney, A., and Whitelaw, J. (1995). "Effect of EGR on Combustion Development in a 1.9L DI Diesel Optical Engine." SAE Paper No. 950850, SAE International, Warrendale, PA.

Bakenhus, M., and Reitz, R. (1999). "Two-color Combustion Visualization of Single and Split Injections in a Single-Cylinder Heavy-Duty D.I. Diesel Engine Using an Endoscope-Based Imaging System." SAE Paper No. 1999-01-1112, SAE International, Warrendale, PA.

Bertoli, C., Beatrice, C., di Stasio, S., and Del Giacomo, N. (1996). "In-Cylinder Soot and NOx concentration Measurements in DI Diesel Engine Fed by Fuels of Varying Quality." SAE Paper No. 960832, SAE International, Warrendale, PA.

Bladh, H., and Bengtsson, P. (2004). "Characteristics of Laser-Induced Incandescence From Soot in Studies of a Time-Dependent Heat- and Mass-Transfer Model." *Applied Physics B*, Vol. 78, No. 2, pp. 241–248.

Bohren, C. E., and Huffman, D. R. (1988). *Absorption and Scattering of Light by Small Particles*. New York: Wiley-Interscience.

Bougie, B., Ganippa, L. C., Dam, N. J., and ter Meulen, J. J. (2006a). "On particle characterization in a heavy duty diesel engine by time-resolved laser-induced incandescence." *Applied Physics B*, Vol. 83, pp. 477–485.

Bougie, B., Ganippa, L. C., van Vliet, A. P., Meerts, W. L., Dam, N. J., and ter Meulen, J. J. (2006b). "Laser-Induced Incandescence Particle Size Measurements in a Heavy Duty Diesel Engine." *Combustion and Flame* Vol. 145, pp. 635–637.

Bougie, B., Ganippa, L.C., van Vliet, A. P., Meerts, W.L., Dam, N.J., and ter Meulen, J.J. (2007). "Soot Particulate Characterisation in a Heavy Duty Diesel Engine for Different Engine Loads by Laser Induced Incandescence.", *Proceedings of the Combustion Institute, Vol.*31, pp. 685 – 691.

Bryce, D. J., Ladommatos, N., and Zhao, H. (2000). "Quantitative Investigation of Soot Distribution by Laser-Induced Incandescence." *Applied Optics* Vol. 39, pp. 5012–5022.

Chang, H., and Charalampopoulos, T. (1990). "Determination of the Wavelength Dependence of Refractive Indices of Flame Soot." *Proc. R. Soc. Lond. A* Vol. 430, pp. 577–591.

Dankers, S. and Leipertz, A. (2004). "Determination of Primary Particle Size Distributions From Time-Resolved Laser-Induced Incandescence Measurements." *Applied Optics*, Vol. 43, pp. 3726–3731.

Dreier, T., Bougie, B., Dam, N., and Gerber, T. (2006). "Modeling of Time-Resolved Laser-Induced Incandescence Transients for Particle Sizing in High-Pressure Spray Combustion Environments: A Comparative Study." *Applied Physics B*, Vol. 83, pp. 403–411.

Gaydon, A., and Wolfhard, H. (1970). *Flames*. London: Chapman and Hall.

Hentschel, W., and Richter, J. (1995). "Time-Resolved Analysis of Soot Formation and Oxidation in a Direct-Injection Diesel Engine for Different EGR-Rates by an Extinction Method." SAE Paper No. 952517, SAE International, Warrendale, PA.

Hofeldt, D. (1993). "Real-Time Soot concentration Measurement Technique for Engine Exhaust Streams." SAE Paper No. 930079, SAE International, Warrendale, PA.

Hofmann, M., Kock, B. F., Dreier, T., Jander, H., and Schulz, C. (2008). "Laser-Induced Incandescence for Soot Particle Sizing at Elevated Pressure." *Applied Physics B*, Vol. 90, pp. 629–639.

Hottel, H., and Broughton, F. (1932). "Determination of True Temperature and Total Radiation from Luminous Gas Flames." *Ind. and Eng. Chem.*, Vol. 4–2, pp. 166–174.

Kawamura, K., Saito, A., Yaegashi, T., and Iwashita, Y. (1989). "Measurement of Flame Temperature Distribution in Engines by Using a Two-Color High-Speed Shutter TV Camera System." SAE Paper No. 890320, SAE International, Warrendale, PA.

Kerker, M. (1969). *The Scattering of Light*. New York: Academic Press.

Kerker, M., Scheiner, P., and Cooke, D. D., J. (1978). "The Range of Validity the Rayleigh and Thomson Limits for Lorentz-Mie Scattering." *Opt. Soc. Am.*, Vol.68, pp.135-237.

Kock, B. F., Tribalet, B., Schulz, C., and Roth, P. (2006). "Two-Color Time-Resolved LII Applied to Soot Particle Sizing in the Cylinder of a Diesel Engine." *Combustion and Flame*, Vol. 147, 79–92.

Kontani, K., and Gotoh, S. (1983). "Measurement of Soot in a Diesel Combustion Chamber by Light Extinction Method and In-Cylinder Observation by High-Speed Shadowgraphy." SAE Paper No. 831291, SAE International, Warrendale, PA.

Kontani, K., and Gotoh, S. (1986). "Effects of Particle Size Distribution on Soot Particle Measurement by Transmissive Light-Extinction Method." SAE Paper No. 861234, SAE International, Warrendale, PA.

Kuhlmann, S.-A., Reimann, J., and Will, S. (2006). "On Heat Conduction Between Laser-Heated Nanoparticles and a Surrounding Gas." *Aerosol Science,* Vol. 37, pp. 1696–1716.

Lee, S., and Tien, C. (1981). "Optical Constants of Soot in Hydrocarbon Flames." *Proc. 18th Symposium (International) on Combustion,* pp. 1159–1166. Pittsburgh: The Combustion Institute.

Liebert, C. H., and Hibbard, R. R. (1970). NASA TN D-5647.

Liu, F., Daun, K. J., Snelling, D. R., and Smallwood, G. J. (2006). "Heat Conduction Form a Spherical Nano Particle: Status and Modeling Heat Conduction in Laser-Induced Incandescence." *Applied Physics B,* Vol. 83, pp. 355–382.

Matsui, Y., Kamimoto, K., and Matsuoka, S.(1979). "A Study on the Time and Space Resolved Measurement of Flame Temperature and Soot Concentration in SI Diesel Engine by the Two-Color method." SAE Paper No. 790491, SAE International, Warrendale, PA.

Matsui, Y., Kamimoto, K., and Matsuoka, S. (1980). "A Study on the Application of the Two-Color Method to the Measurement of Flame Temperature and Soot Concentration in Diesel Engines." SAE Paper No. 800970, SAE International, Warrendale, PA.

Melton, L. A. (1984). "Soot Diagnostics Based on Laser Heating." *Applied Optics,* Vol. 23, pp. 2201–2208.

Michelsen, H. A., Liu, F., Kock, B. F., Bladh, H., Boiarciuc, A., Charwath, M., Dreier, T., Hadef, R., Hofmann, M., Reimann, J., Will, S., Bengtsson, P.-E., Bockhorm, H., Foucher, F., Geigle, K. P., Mounaim-Rousselle, C., Schulz, C., Stirin, R., Tribalet, B. and Suntz, R. (2007). "Modeling Laser-Induced Incandescence of Soot: A Summary And Comparison of LII Models." *Applied Physics B,* Vol. 87, pp. 503–521.

Mie, G. (1908). "Beitrage zur Optik truber Medien speziell kolloidaler Metallosungen." *Ann. Phys.,* Vol. 25, pp. 377–445.

Miyamoto, N., Ogawa, H., Arima, T., and Miyakawa, K. (1996). "Improvement of Diesel Combustion and Emissions with Addition of Various Oxygenated Agents to Diesel Fuels." SAE Paper No. 962115, SAE International, Warrendale, PA.

Nakakita, K., Nagaoka, M., Fujikawa, T., Ohsawa, K., and Yamaguchi, S. (1990). "Photographic and Three Dimensional Numerical Studies of Diesel Soot Formation Process." SAE Paper No. 902081, SAE International, Warrendale, PA.

Peterson, R., and Wu, K. (1986). "The Effect of Operating Conditions on Flame Temperature in a Diesel Engine." SAE Paper No. 861565, SAE International, Warrendale, PA.

Quay, B., Lee, T., Ni, T., and Santoro, R. J. (1994). "Spatially resolved Measurements of Soot Volume Fraction Using Laser-Induced Incandescence." *Combustion and Flame*, Vol. 97, pp. 384–392.

Rossler, F., and Behrens, H. (1950). "Bestimung des Absorptionskoeffizieten von russteilchen verschiedenser flammen." *Optik*, Vol. 6, p. 145.

Santoro, R. J., Semerjian, H. G., and Dobbins, R. A. (1983). "Soot Particle Measurements in Diffusion Flames." *Combustion and Flame*, Vol. 51, pp. 203–218.

Santoro, R. J., Yeh, T. T., Horvath, J. J., and Semerjian, H. G. (1987). "The Transport and Growth of Soot Particles in Laminar Diffusion Flames." *Combus. Sci. Technol.*, Vol. 53, pp. 89–115.

Shakal, J., and Martin, J. (1994). "Imaging and Spatially Resolved Two-Color Temperature Measurements Through a Coherent Fiberoptic: Observation of Auxiliary Fuel Injection Effects on Combustion in a Two-Stroke DI Diesel." SAE Paper No. 940903, SAE International, Warrendale, PA.

Schraml, S., Heimgärtner, C., Will, S., Leipertz, A., and Hemm, A. (2000). "Application of a New Soot Sensor for Exhaust Emission Control Based on Time-Resolved Laser-Induced Incandescence (TIRE-LII)." SAE No. Paper 2000–01–2864, SAE International, Warrendale, PA.

Shiozaki, T., et al. (1996). "The Analysis of Combustion Flame Under EGR Conditions in a DI Diesel Engine." SAE Paper No. 960323, SAE International, Warrendale, PA.

Siddall, R. G., and McGrath, I. A. (1963). "The Emission of Luminous Flames." *Proc. of 9th Symposium on Combustion*, pp. 102–110.

Smallwood, G. J., Snelling, D. R., Liu, F., and Gulder, O. L. (2001). "Clouds Over Soot Evaporation: Errors in Modeling Laser-Induced Incandescence of Soot." *Journal of Heat Transfer*, Vol. 123, pp. 814–818.

Snelling, D. R., Liu, F. S., Smallwood, G. J., and Gulder, O. L. (2004). "Determination of the Soot Absorption Function and Thermal Accommodation Coefficient Using Low Fluence LII in a Laminar Coflow Ethylene Diffusion Flame." *Combustion and Flame*, Vol. 136, pp. 180–190.

Snelling, D. R., Smallwood, G. J., Sawchuk, R. A., Neill, W. S., Gareau, D., Clavel, D. J., Chippior, W. L., Liu, F., and Gulder, O. L. (2000). "In-Situ Real Time Characterization of Particulate Emissions From a Diesel Engine Exhaust by Laser-Induced Incandescence." SAE Paper No. 2000–01–1994, SAE International, Warrendale, PA.

Van de Hulst, H. C. (1957). *Light Scattering by Small Particles*. London: Chapman and Hall.

Vander Wal, R. L., and Choi, M. Y. (1999). "Pulsed Laser Heating of Soot: Morphological Changes." *Carbon*, Vol. 37, pp. 231–239.

Winterbone, D., Yates, D., Clough, E., Rao, K., Gomes, P., and Sun, J. (1994). "Combustion in High-Speed Direct-Injection Diesel Engines–a Comprehensive Study." *Proc. Instn. Mech. Engrs.*, Vol. 208, p. 223.

Witze, P., Shimpi, S., Durrett, R., and Farrell, L. (2005). "Time-Resolved Laser-Induced Incandescence Measurements for the EPS Heavy-Duty Federal Test Procedure." *JSME International Journal Series B*, Vol. 48, pp. 632–638.

Yan, J., and Borman, G. (1988). "Analysis and In-Cylinder Measurement of Particulate Radiant Emissions and Temperature in a Direct-Injection Diesel Engine." SAE Paper No. 881315, SAE International, Warrendale, PA.

Yoder, G. D., Diwakar, P. K., and Hahn, D. W. (2005). "Assessment of Soot Particle Vaporization Effects During Laser-Induced Incandescence with Time-Resolved Light Scattering." *Applied Optics*, Vol. 44, pp. 4211–4219.

Will, S., Schraml, S., Bader, K., and Leipertz, A. (1998). "Performance Characteristics of Soot Primary Particle Size Measurements by Time-Resolved Laser-Induced Incandescence." *Applied Optics*, Vol. 37, pp. 5647–5658.

Index

355

About the Author

Professor Hua Zhao has been involved in combustion engine research and the development of laser diagnostics for more than 25 years. As the Director of the Centre for Advanced Powertrain and Fuels Research at Brunel University in London, he leads a team of 20 academic staff and researchers on the fundamental studies and applied research of cleaner and more efficient combustion engines. His research includes CAI/HCCI (controlled autoignition / homogeneous charge compression ignition) gasoline and diesel engines, two- and four-stroke combustion systems, highly downsized direct-injection (DI) gasoline engines, cost-effective air hybrid powertrain, and the development and application of advanced laser measurement techniques to internal combustion (IC) engines. Professor Zhao has led a great many research projects in the UK and EU and directed key national projects in China on CAI/HCCI combustion engines. He has written or edited five books on IC engines and published more than 200 papers at international journals and conferences. He has received publication awards from the Institution of Mechanical Engineers, Institute of Energy, and Professional Engineering Publishing. He has collaborated widely with the automotive industry and provided consulting services to several vehicle, engine, and oil companies in Europe and China.

Professor Zhao graduated from Tianjin University in China and obtained his PhD from the University of Leeds in the UK. He worked as a research fellow at Cambridge University and Imperial College in London before joining Brunel University in 1994. In 2009, he was awarded the higher doctorate degree of Doctor of Science (DSc) for his outstanding contribution to combustion engine research. He is a Fellow of SAE International and the Institution of Mechanical Engineers.